Data Mining and Computational Intelligence for E-learning and Education

Data Mining and Computational Intelligence for E-learning and Education

Guest Editors

Antonio Sarasa Cabezuelo
Ramón González del Campo Rodríguez Barbero

MDPI

Basel • Beijing • Wuhan • Barcelona • Belgrade • Novi Sad • Cluj • Manchester

Guest Editors
Antonio Sarasa Cabezuelo
Department of Computer Systems and Computing
Complutense University of Madrid
Madrid
Spain

Ramón González del Campo Rodríguez Barbero
Department of Computer Systems and Computing
Complutense University of Madrid
Madrid
Spain

Editorial Office
MDPI AG
Grosspeteranlage 5
4052 Basel, Switzerland

This is a reprint of the Special Issue, published open access by the journal *Data* (ISSN 2306-5729), freely accessible at: https://www.mdpi.com/journal/data/special_issues/eLearning_education.

For citation purposes, cite each article independently as indicated on the article page online and as indicated below:

Lastname, A.A.; Lastname, B.B. Article Title. *Journal Name* **Year**, *Volume Number*, Page Range.

ISBN 978-3-7258-4030-4 (Hbk)
ISBN 978-3-7258-4029-8 (PDF)
https://doi.org/10.3390/books978-3-7258-4029-8

Contents

About the Editors

Antonio Sarasa Cabezuelo

Antonio Sarasa Cabezuelo is an Associate Professor in the Department of Computer Systems and Computing at the Complutense University of Madrid. His research career has been shaped by a strong interdisciplinary focus, integrating computer science with digital humanities, educational technology, and accessibility. His main scientific interests include semantic web technologies, linked data, digital preservation, inclusive learning environments, and the application of artificial intelligence to improve access to information and education.

Over the past two decades, he has contributed to and led numerous research and development projects at the national and European levels. These projects have addressed challenges in areas such as the digitization and open dissemination of cultural heritage, the design of accessible educational platforms, and the development of ontologies and metadata schemas to enhance interoperability in digital archives. His work frequently bridges the gap between technological innovation and social impact, with a focus on inclusivity and accessibility.

He is currently involved in projects that explore the use of semantic technologies and artificial intelligence to support adaptive and accessible digital learning systems, as well as initiatives for the integration of cultural data sources using linked open data. His academic output includes a substantial number of scientific publications and technical reports and participation in expert committees related to digital transformation in education and cultural institutions.

His research contributions have been recognized with two national research evaluation awards ("sexenios") granted by the Spanish Ministry of Science and Innovation, acknowledging the quality and impact of his scientific output.

Ramón González del Campo Rodríguez Barbero

Ramón González del Campo Rodríguez Barbero holds a degree in Physics, specializing in Automatic Computation, from the Complutense University of Madrid (1996) and a PhD in Computer Science from the same university (2012). The title of his doctoral thesis was "Generalizations of Specificity Measures and T-Transitivity for Interval-Valued Fuzzy Sets".

His research focuses on the foundations of fuzzy logic. He is currently participating in the following research project: Techniques for Obtaining, Processing, and Representing Fuzzy Information for Decision-Making.

His main lines of research are as follows: specificity measures for fuzzy sets and interval-valued fuzzy sets; k-specificity measures for fuzzy sets; concepts of T-transitivity and transitive closure for interval-valued fuzzy sets; and the study of relations for dubious fuzzy sets, as well as concepts of T-transitivity and transitive closure for dubious fuzzy sets and computable aggregations.

In the teaching field, he has taught courses on algorithms, databases, and web applications in the Faculties of Computer Science, Statistical Studies, Mathematics, and Tourism.

Preface

Artificial intelligence has become a cornerstone of innovation in the 21st century, and its influence continues to expand into diverse fields—education being one of the most dynamic among them. As digital transformation accelerates, new possibilities emerge to better understand, support, and enhance the learning process through intelligent systems.

The use of data mining and computational intelligence in education enables the extraction of meaningful insights from complex educational data. These insights can reveal patterns in student learning, identify potential causes of academic success or failure, and provide predictive capabilities that support decision-making in both teaching and institutional management. Such methods are particularly powerful in online learning environments, where the volume of data generated is vast and diverse.

Moreover, artificial intelligence is also being used to automate aspects of the educational process itself. From adaptive learning platforms to intelligent tutoring systems and chatbots capable of guiding or mentoring students, these technologies are reshaping the way learners interact with content and instructors. As these systems become more sophisticated, they not only deliver personalized learning experiences but also raise important ethical questions about transparency, data privacy, and the role of human educators in AI-supported environments.

This Reprint, Data Mining and Computational Intelligence for E-learning and Education, brings together a curated selection of works that highlight the latest developments and practical applications in the field. The contributions come from diverse educational contexts and present both theoretical advancements and real-world case studies. Together, they reflect the growing maturity and multidisciplinary nature of research at the intersection of AI and education.

Our aim with this collection is to offer a platform for ongoing discussion and discovery. It is intended for researchers, practitioners, educators, and policy-makers who are exploring how computational intelligence can contribute to more adaptive, efficient, and inclusive educational systems. As the challenges and opportunities of AI in education continue to evolve, we hope this volume will serve as a valuable reference and source of inspiration for future work in this important field.

Antonio Sarasa Cabezuelo and Ramón González del Campo Rodríguez Barbero
Guest Editors

data

MDPI

Data Descriptor

Knowledge Discovery and Dataset for the Improvement of Digital Literacy Skills in Undergraduate Students

Pongpon Nilaphruek and Pattama Charoenporn *

Department of Computer Science, School of Science, King Mongkut's Institute of Technology Ladkrabang, Bangkok 10520, Thailand; 63605014@kmitl.ac.th
* Correspondence: pattama.ch@kmitl.ac.th

Abstract: For over two decades, scholars and practitioners have emphasized the importance of digital literacy, yet the existing datasets are insufficient for establishing learning analytics in Thailand. Learning analytics focuses on gathering and analyzing student data to optimize learning tools and activities to improve students' learning experiences. The main problem is that the ICT skill levels of the youth are rather low in Thailand. To facilitate research in this field, this study has compiled a dataset containing information from the IC3 digital literacy certification delivered at the Rajamangala University of Technology Thanyaburi (RMUTT) in Thailand between 2016 and 2023. This dataset is unique since it includes demographic and academic records about undergraduate students. The dataset was collected and underwent a preparation process, including data cleansing, anonymization, and release. This data enables the examination of student learning outcomes, represented by a dataset containing information about 45,603 records with students' certification assessment scores. This compiled dataset provides a rich resource for researchers studying digital literacy and learning analytics. It offers researchers the opportunity to gain valuable insights, inform evidence-based educational practices, and contribute to the ongoing efforts to improve digital literacy education in Thailand and beyond.

Dataset: https://dx.doi.org/10.21227/370s-1s37

Dataset License: CC-BY 4.0

Keywords: digital literacy dataset, IC3 certification; improvement; learning analytics; RMUTT

Citation: Nilaphruek, P.; Charoenporn, P. Knowledge Discovery and Dataset for the Improvement of Digital Literacy Skills in Undergraduate Students. *Data* **2023**, *8*, 121. https://doi.org/10.3390/data8070121

Academic Editors: Antonio Sarasa Cabezuelo and Ramón González del Campo Rodríguez Barbero

Received: 25 June 2023
Revised: 15 July 2023
Accepted: 17 July 2023
Published: 20 July 2023

1. Summary

Digital literacy is a personal skill regarding one's ability to use a present digital technology for daily use, which includes operating, understanding, accessing, communicating, searching, and processing information technology [1]. In the 21st century, this set of skills and competencies is very important for professional life, Industry 4.0, and work in academic fields [2]. Nowadays, digital technology consists of hardware, software, and information. The technology can include personal computers, mobile phones, tablets, computer programs, and online media. Digital literacy is the primary factor affecting quality of life in the digital age. If a country fails to adopt and utilize information and communication technologies (ICTs), it will encounter digital exclusion as it cannot access conventional mainstream information sources [3].

The policy of Thailand 4.0 considers the country's economic development, providing a model for the development of the national economy by relying on the production structure and the occupational basis of people in Thai society [4]. Also, according to such policy, youth groups and students play an important role in the development of the country, as the youth population is three times greater than the working-age population. However, the main problem is that the ICT skill levels of the youth are rather low; this is a factor that greatly influences the upgrading of the Thailand 4.0 policy. Similarly, the digital transformation process still encounters problems in many areas, and it is necessary for the population to develop fundamental digital skills to make the digital transformation process more efficient [5].

The Rajamangala University of Technology Thanyaburi (RMUTT) aligns itself with the vision of Thailand 4.0 and places great importance on students acquiring digital literacy skills. In line with this policy, RMUTT strives for a high success rate, aiming for nearly one hundred percent proficiency in digital literacy skills among the student population. RMUTT students have high expectations of the university, envisioning an educational environment that fosters excellence in the 21st century, with a particular emphasis on the development of digital literacy skills [6]. The IC3 Digital Literacy Certification (IC3) is a globally recognized standard utilized to certify individuals at entry and employee levels with sufficient ICT skills. In Thailand, the IC3 certification is widely adopted as a measure of digital literacy proficiency [7]. During the pilot phase, the IC3 certification was implemented at RMUTT as a testing standard as part of a short-term program aimed at enhancing students' digital skills. Although the percentage of passing examinations was at an acceptable level, the number of students participating in the program was still very small compared to the total number of students at the university.

To address this issue and expand student involvement, the university introduced a new general education subject relating to digital literacy, titled "Computer and Information Technology Skill" (RMUTT CITS course), during the first phase of the program in 2019. This initiative aims to increase the number of students engaging with and acquiring essential digital skills. Students from all faculties can register for this subject freely, and they also use IC3 as a testing standard in mid-year and final examinations. Moreover, the RMUTT Learning Management System (LMS) was employed as the primary platform used for learning this course to develop the ICT skills of students and lecturers with regard to using a digital platform. This LMS is not provided for full self-learning. It is used for learning activities such as online assignment submission and the provision of online resources. However, despite these efforts, the students' IC3 pass rate remained disappointingly low. This study investigated what factors influence students' digital skills and how we can elucidate the relationships among these factors.

From primary to higher education, the LMS has been utilized for years to facilitate the establishment of a good learning environment. With the rapid advancement of information technology, large-scale data collection on student populations is feasible. Several scientific researchers have studied the influence of student data analysis in recent years. This demonstrates the significance of open datasets, which provide a consistent method for comparing and visualizing results. There are several publicly accessible data sets discussed in previous studies. Table 1 showcases the dataset's contents, encompassing demographic information, academic records, and results from ICT skill tests. In contrast, the RMUTT Digital Literacy Dataset (RMUTT-DLD) offers a broader scope by including data on RMUTT students from 2016 to 2023. This extended dataset encompasses demographic information, academic learning records, and certification outcomes, providing a more comprehensive view of students' digital literacy progression over time.

Education has a substantial impact on economic growth and employment prospects. With the aim of providing students with the best learning resources, an abundance of predictive analytical educational research articles has been released in recent years. Over the past several years, effective statistical and machine learning approaches have been widely applied to educational datasets. For example, high school and college dropout rate datasets have been proposed by several researchers [8–10]. These datasets can be used to develop a model for predicting the dropout rate, which in turn may allow for the dropout rate to be lowered if the needs of students are better met. There was also a study that investigated the student dropout rate at the University Faculty of Electrical and Computer Engineering (FECE) from 2001 to 2015 [11]. This is why decreasing the number of students who drop out before graduating is so crucial. Using data mining techniques, [12] suggested a novel recommendation system based on student data aimed at enhancing the number of university graduates by offering suitable subject selections.

In addition, higher-education students are continuously expected to improve their ICT competencies amongst the rapid development of the digital technology era. In 2017, [13] proposed a dataset that includes data from 22 courses presented by 32,593 Open University students (OU). The dataset contains demographic information as well as clickstream data gathered from student interactions in a virtual learning environment (VLE). In order to assess the impact of a VLE on learning outcomes, the VLE dataset was proposed. Some studies [11,12,14] have suggested datasets containing observations of students' ICT skill usage and evaluations of students' new technology learning skills. Digital Kids Asia Pacific (DKAP) published a new dataset encompassing 1061 observations of students' information and communication technology competence rates from several high schools across five Vietnamese regions and cities. The dataset includes responses from thousands of students who were asked to rate their digital literacy. Consequently, in order to address and analyze the university's digital literacy and provide the best quality education, our dataset, spanning from 2016 to 2023, consists of three main sections concerning the students' demographics, academic records, and IC3 digital literacy exam results.

Table 1. Comparison of recent datasets in the academic area.

Dataset	Year	High School	Undergraduate	Number of Observations	Purpose	Location
Open University Learning Analytics Dataset [13]	2017	-	✓	22 courses, 32,593 students	Students' interactions in the virtual learning environment (VLE)	Open University (OU)
Digital Competency Observation Dataset [15]	2019	✓	-	1061 students	Digital competency	Vietnam
Academic Performance Evaluation Dataset [11]	2020	✓	✓	12,411 students	Observe the influence of social variables and the evolution of students' learning skills	Colombia
Video Conferencing Tools Acceptance Dataset [14]	2020	-	✓	277 records	Video conferencing tools (VTCs)	Vietnam
High-School Dropout Rate Dataset [10]	2022	✓	-	1613 records	Student Dropout rates	United States
C# Programming Examination Dataset [12]	2022	-	✓	Unspecified	Academic results in C# programming language	Iraq, Sudan, Nigeria, South Africa, and India
Undergraduate and High-School Dropout Rate Dataset [9]	2022	✓	✓	50 records, 143,326 records	Student dropout rate	Mexico
* RMUTT-DLD	2023	-	✓	45,603 records	IC3 Digital Literacy Certification	Thailand

Note: * The dataset in this study is called the RMUTT Digital Literacy Dataset (RMUTT-DLD).

2. Data Description

To fully comprehend the proposed dataset, a description of the RMUTT digital literacy learning process must be provided. RMUTT is one of Thailand's public universities, with approximately 25,000 students enrolled in various programs. The RMUTT LMS is used to deliver digital literacy-related learning resources. The database stores instructor and student interactions with course materials and assignments. It allows for the frequency of online assignment submissions in related modules to be lowest, low, medium, or high.

Students are aware of the policies regarding data protection and the ethics code in the use of student data recorded in databases for learning and research analysis. They are provided with crucial details on how their data is used and the possibility of data sharing with other academics for research purposes that can be disclosed to students. Additionally, this RMUTT-DLD dataset has been anonymized and cannot be used to identify specific pupils and lecturers.

This dataset comprises two distinct learning process application periods. The first term ran from 2016 to 2018, and the second from 2019 to 2023. Figure 1 depicts the learning process approach for the first period, in which all students studying the RMUTT CITS course had to register for a schedule of IC3 certification exams to evaluate their digital literacy skills. After receiving the schedule, students took the certification exam and received a score which equates to either a "Fail" or "Pass" grade. Also, the scores from such examinations were partially used for grading the course.

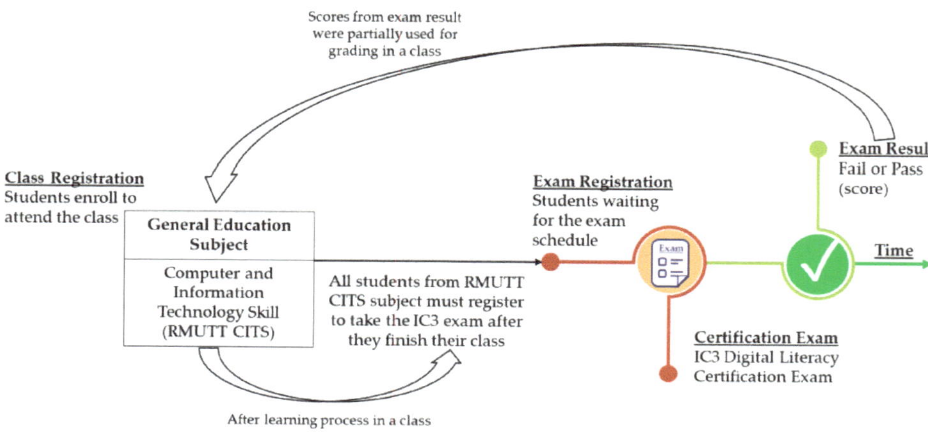

Figure 1. Digital literacy learning procedure—first period (2016–2018).

Figure 2 demonstrates the learning procedure for the second phase. As mentioned in the previous section, the initial phase of implementation did not achieve the desired outcomes. Therefore, RMUTT created a digital literacy improvement platform, including two modules. First, the self-e-Learning module was designed based on the standard gamification concept, and learners can study using that module completely on their own. Second, an intensive tutoring module was provided for a certain period. Typically, any student can register for the self-e-Learning module without registering for the RMUTT CITS course. For students who meet the qualification criteria, there is the option to participate in the intensive tutoring module. Additionally, students who are deemed qualified by the board of lecturers from the RMUTT CITS course can also directly access the intensive tutoring module. Then, students can take the IC3 certification exam in the first period.

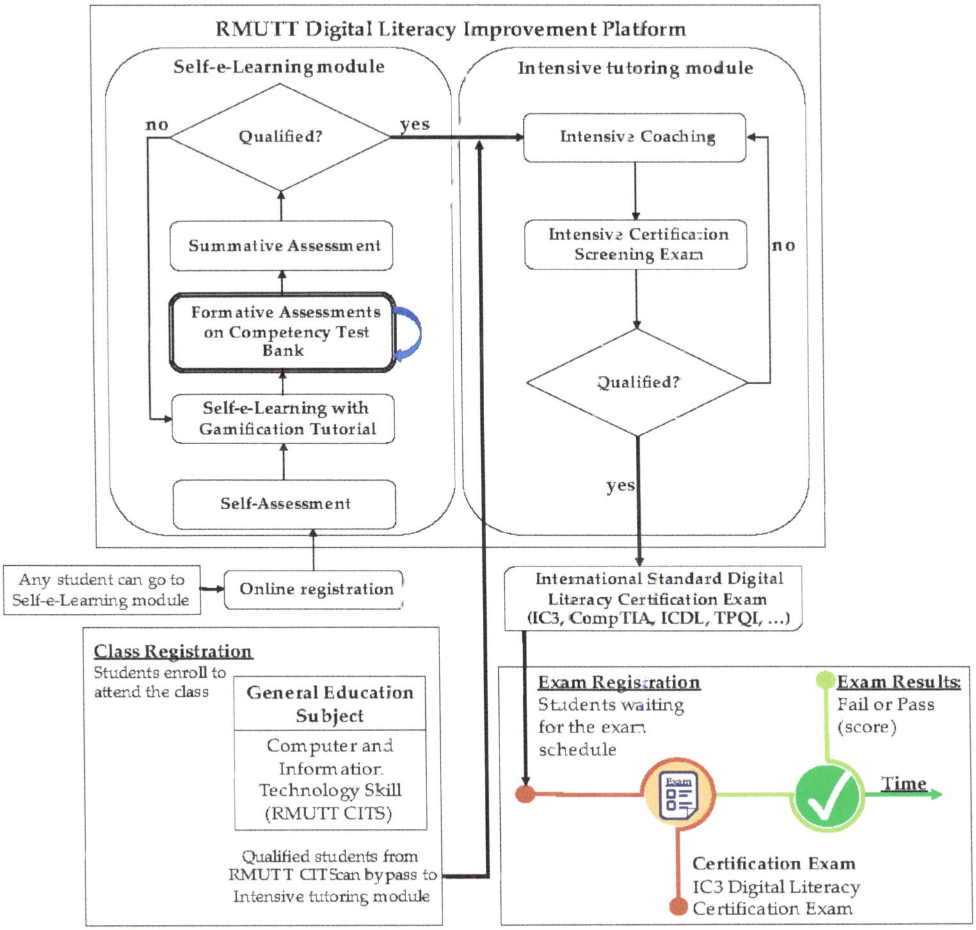

Figure 2. Digital literation learning procedure—second period (2019–2023).

Table 2 shows the detailed structure of the RMUTT-DLD dataset, consisting of the field name, data type, description, and data scope. The dataset is a collection of anonymous students' profiles, academic records, and IC3 digital literacy exam results, spanning from 2016 through 2023, as shown in Figure 3. The dataset focuses on students; hence, students are the focal point. Each record within the data corresponds to a student who registered for the IC3 certification exam in a specific module. The dataset includes a variety of demographic information, consisting of the student's encoded identifier, first-entry GPA, current GPA, admission year, faculty name in Thai and English, department name in Thai and English, home province name in Thai, home district name in Thai, and contact zip code in Thailand. The prefix "STD" was added to the field names of these data. The records of the IC3 exam results were combined with the students' profiles, which can refer to other fields, and the prefix "IC3" was added. The IC3 exam has three main modules, including 'IC3 GS5—Computing Fundamentals', 'IC3 GS5—Key Applications', and 'IC3 GS5—Living Online'. Information regarding the language, score, result, used time, station, and year of the IC3 examinations were also recorded. Furthermore, there are six fields of academic records, including the class identifier, teacher's encoded identifier in each class, number of students who enrolled in a class, year of class opening, semester period, and frequency of online assignment submissions. The prefixes "CLASS" and "ONLINE" were added to the academic records.

Table 2. The detailed structure of the RMUTT-DLD dataset.

No.	Field Name	Data Type	Description	Data Scope
1	STD_ENCODE_ID	Text	Record of student's encoded identifier.	There are 45,603 IC3 examination records that were recorded.
2	IC3_MODULE_NAME	Text	IC3 certificate module name. This field has only three modules.	IC3 GS5—Computing Fundamentals IC3 GS5—Key Applications IC3 GS5—Living Online
3	IC3_EXAM_LANGUAGE	Text	Language for examination.	English/ Thai.
4	IC3_SCORE	Integer	IC3 certificate score for each module.	0 to 1000 points.
5	IC3_RESULT	Text	IC3 certificate result. Scores $\geq$ 700 pass; otherwise, fail.	Fail/ Pass.
6	IC3_EXAM_TIMEUSED	Integer	The time that was used during the examination.	0 to 3000 s.
7	IC3_EXAM_STATION	Text	Station of the test taker, mostly including building and computer name. For example, IWORK-201-01 is IWORK building, room number 201, and computer number 01.	There are 997 stations. Some are not in the standard format because they may use an extra building or computer.
8	IC3_EXAM_YEAR	DateTime (Year)	Year of IC3 examination in yyyy format, such as 2023.	2016 to 2023 A.D.
9	STD_ENTRY_GPA	Float	Student's first-entry GPA	1.0 to 4.0 on a 4.0 scale.
10	STD_CURRENT_GPA	Float	Student's current GPA during the IC3 examination.	0.0 to 4.0 on a 4.0 scale.
11	STD_ADMIT_YEAR	DateTime (Year)	Student's admission year in yyyy format, such as 2022.	2012 to 2022 A.D.
12	STD_FACULTYNAME_THAI	Text	Student's faculty name in Thai.	There are 13 faculties.
13	STD_FACULTYNAME_ENG	Text	Student's faculty name in English.	There are 13 faculties.
14	STD_DEPARTMENTNAME_THAI	Text	Student's department name in Thai.	There are 43 departments.
15	STD_DEPARTMENTNAME_ENG	Text	Student's department name in English.	There are 43 departments.
16	STD_HOME_PROVINCENAME	Text (GEO)	Student's home province name in Thai.	There are 77 provinces in Thailand.
17	STD_HOME_DISTRICT	Text (GEO)	Student's home district name in Thai.	There are 988 districts.
18	STD_CONTACT_ZIPCODE	Text (GEO)	Student's contact zip code in Thailand. In general, some districts have the same contact zip code.	There are 855 contact zip codes. Some values are NA, which is undefined.
19	CLASS_ID	Text	Class identifier is used for classifying a class/section for RMUTT CITS.	There are 788 sections for the RMUTT CITS class.
20	CLASS_TEACHER_ENCODE_ID	Text	Record of teacher's encode identifier. This field can distinguish a lecturer from each other.	There are 76 teachers who taught many classes and have different name IDs.
21	CLASS_ENROLLSEAT	Integer	Number of students who enrolled in a class.	Between 3 and 78 students in a class.
22	CLASS_ACADEMIC_YEAR	DateTime (Year)	Year of class opening in yyyy format, such as 2022.	2015 to 2022 A.D.
23	CLASS_SEMESTER	Integer	Semester period in which the class opens.	Semester 1, 2, or 3.
24	ONLINE_ASSIGNMENT _SUBMISSION_FREQUENCY	Text	Frequency of online assignment submissions in related modules. This field was transformed to include four levels.	Lowest/Low/Medium/High

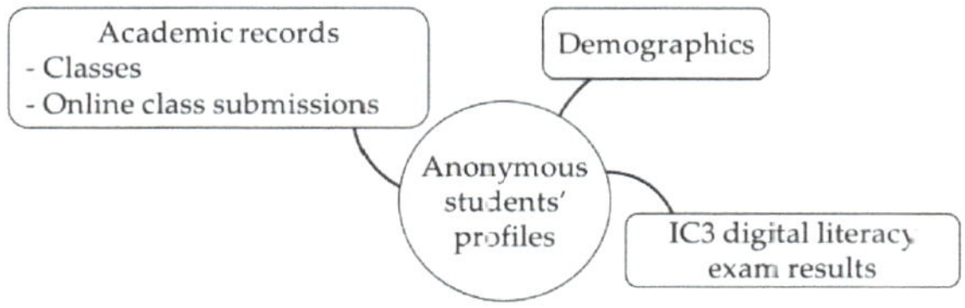

Figure 3. Overall dataset structure diagram.

The dataset is available in the .xlsx format and comprises three modules with 45,603 enrolled students. It can be freely downloaded by visiting the provided link via the file named "RMUTT-DLD-dataset-master.xlsx". This dataset can be imported into any application for further analysis or use, making it applicable to various scenarios. It facilitates the evaluation of predictive models to anticipate students' certification exam results and allows for model comparisons with those created by other researchers.

One interpretation of the dataset can be seen in Figures 4 and 5, and there are significant differences between the results of the certification exams before (2016–2018) and after (2019–2023) the digital literacy platform improvement. This is due to differences in the digital literacy learning procedure, which was explained in the previous paragraph. The differences are clear; the pass rate in 2019–2023 was better than the previous period.

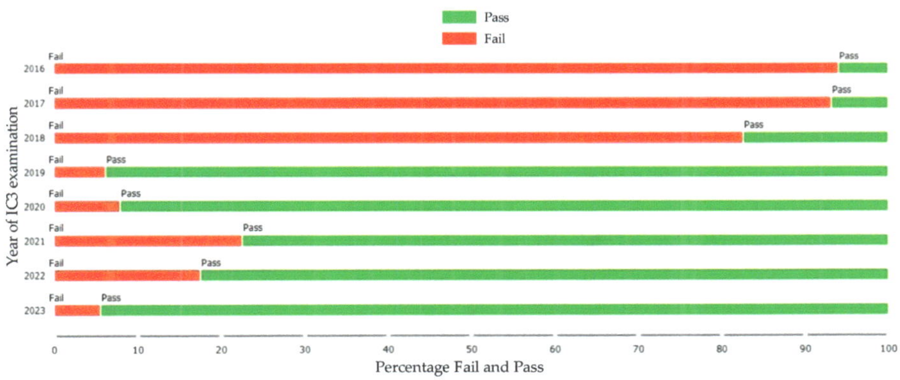

Figure 4. Certification exam results from 2016 to 2023 (percentage).

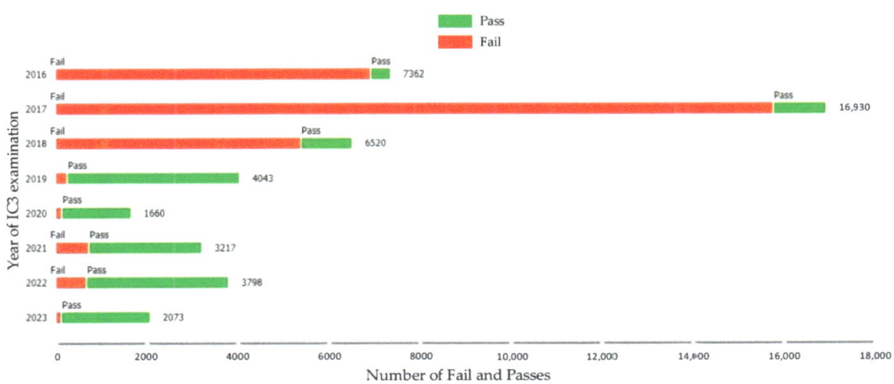

Figure 5. Certification exam results from 2016 to 2023 (number).

In Figure 6, an interpretation of the data is presented, showing the relationship between the number of assignments and the pass rate of students who took the IC3 exam. The data visualization divides the data into two categories: before and after the platform improvement. Furthermore, Figure 7 illustrates the distribution of the student population across Thailand and IC3 exam pass rates based on their province of residence. It is evident that students residing in the central Thailand area exhibited superior digital literacy skills compared to other provinces, on average.

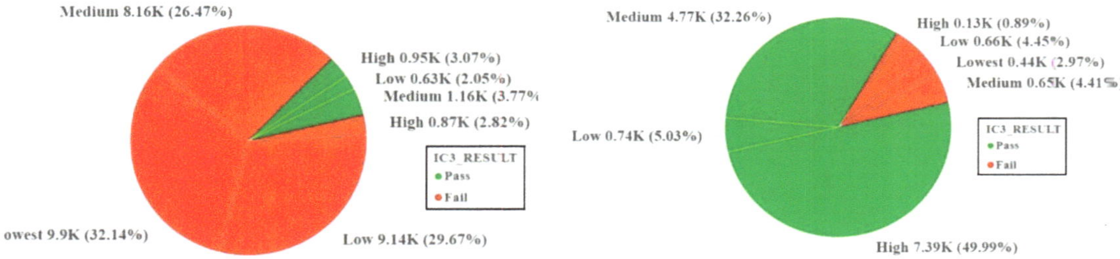

(**a**) (**b**)

Figure 6. Online assignment submission frequency and IC3 result rate. (**a**) Before improvement (2016–2018); (**b**) after improvement (2019–2023).

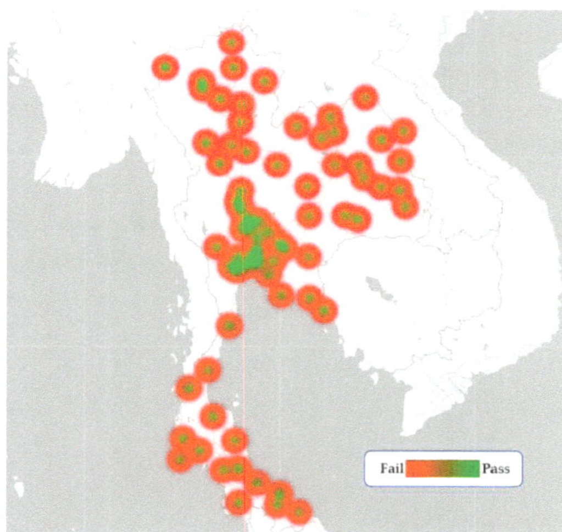

Figure 7. The demographic makeup of RMUTT students in the Thailand region.

3. Methods

3.1. Raw Data

The data preparation process involved three key stages: raw data handling, data cleansing, data anonymization, and release, as shown in Figure 8. The first stage was raw data handling, which encompassed the collection, extraction, and initial storage of data from various sources. Students at RMUTT have access to a variety of information system technologies that can be used to support their academic activities. As mentioned in the previous section, RMUTT has a data center for collecting all information due to the significant variation between information systems. In the dataset utilized for this article, three distinct types of data are distinguished:

- Demographic data—represent basic information on the students, such as name, age (date of birth), home province, home district, first-entry GPA, current GPA, faculty name, etc.
- Academic data—show the records of enrollment information of a student's education at RMUTT, including information on teachers, classes, and activities in RMUTT LMS.
- IC3 digital literacy exam data—are a record of student exam results according to digital literacy abilities.

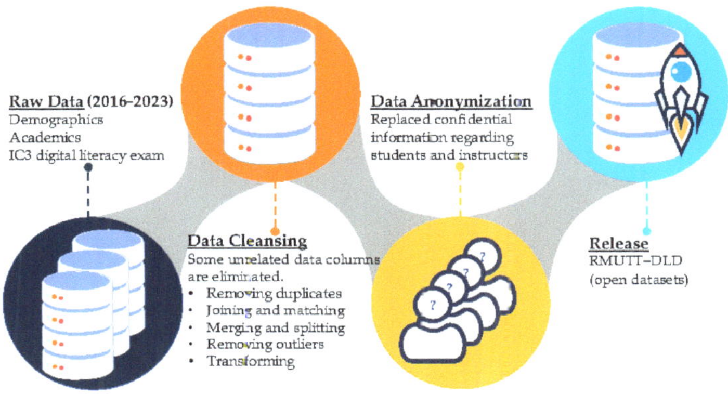

Figure 8. Dataset preparation process overview.

3.2. Data Cleansing

Data cleansing describes the activity of detecting and correcting mistaken records in a dataset. The data center has collected demographic, academic, and digital literacy exam data on students since 2016. We compiled information on digital literacy exams given at RMUTT between 2016 and 2023. Due to the records coming from various sources, they were combined with student ID, which can represent a specific source. Insignificant fields were also removed because there are some repeated values, such as payment type, voucher, and exam level. Some examples of data cleansing processes used in the study include:

- Removing the duplicated data and unused columns from the raw dataset.
- Joining, merging, and splitting the data among sources using student ID as a key.
- Removing outliers from data sources. For example, the minus values of GPA on a 4.0 scale were removed because the data were sometimes entered incorrectly from the beginning.
- Transforming some local data to international data units, such as year in B.E. into A.D. format, and the number of assignments submitted into the four simplified levels.

3.3. Data Anonymization and Release

The dataset anonymization procedure was built in accordance with RMUTT's ethical and privacy guidelines. The entire process of creating and releasing datasets is overseen by the RMUTT administration and approved by the Academic Resources and Information Technology (ARIT) departments. Self-anonymization is accomplished through a series of stages. The first step is to replace student and instructor personal information. This includes the student's ID number, instructor's name, and RMUTT-specific identification.

4. Data Evaluation

As a preliminary evaluation, a correlation matrix analysis was conducted on the dataset. This analysis is performed to identify relationships, explore data, select variables, and make data-driven decisions. The correlation matrix heatmap, as depicted in Figure 9, is a visual representation of the correlation values between different variables in a dataset. Each cell in the heatmap corresponds to the correlation coefficient between two variables.

The correlation coefficient ranges from negative one to one, indicating the strength and direction of the relationship between the variables. The correlation analysis (Figure 9) reveals several noteworthy findings concerning the relationships between different variables:

(1) The variables IC3_Score, IC3_Result, and IC3_Exam_Timeused exhibit a high correlation with each other, indicating that a negative correlation is observed between IC3_Exam_Timeused and performance, suggesting that students who take more time to complete the exam tend to have lower scores.

(2) Variables such as IC3_Exam_Year, Std_Admit_Year, Class_Id, and Class_Academic_Year demonstrate a positive correlation with IC3_Score and IC3_Result. This implies that students who enrolled after the implementation of the digital literation learning procedure achieved better scores and higher pass rates.

(3) Std_Entry_GPA and Std_Current_GPA also show a positive correlation with IC3_Score and IC3_Result. This suggests that students with strong entry and current GPAs tend to obtain higher IC3 scores and pass the exam.

(4) The variable Class_Teacher_Encoded_Id plays a role in determining IC3_Score and IC3_Result. This indicates that the selection of a teacher can influence a student's grades and overall success, as different teachers may vary in their delivery of course materials.

(5) The frequency of Online_Assignment_Submission is also correlated with IC3_Score and IC3_Result. A lower frequency of assignments given in a class is associated with lower scores and pass rates for students.

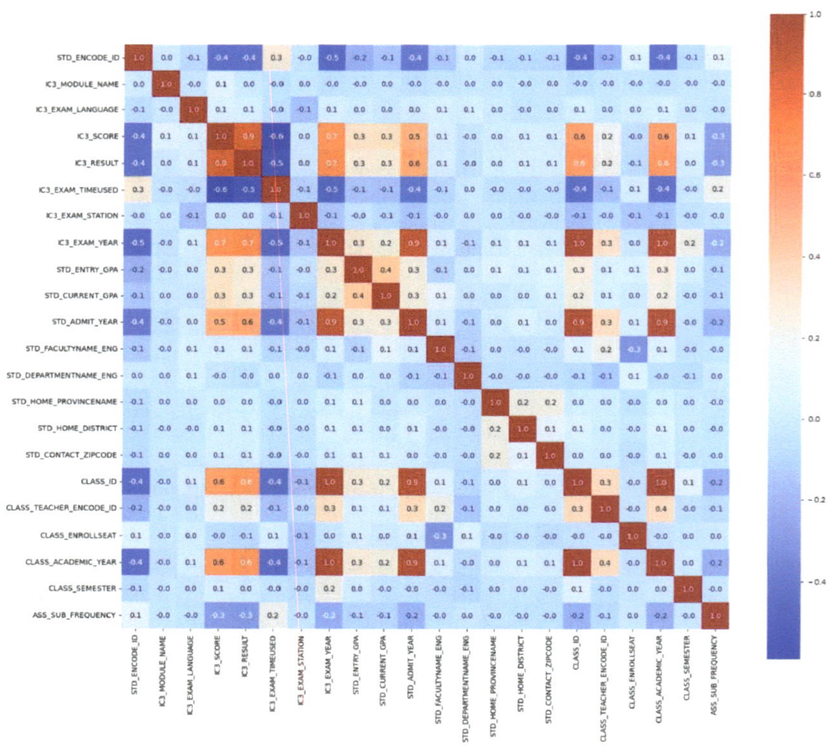

Figure 9. Dataset correlation matrix heatmap.

As a further method of evaluation, an open-source Orange [16] application was used to evaluate this dataset. Data may now be dynamically analyzed and more aesthetically visualized using Orange. Additionally, supported by this program are a number of machine learning methods that may be quickly and easily set up using a visual workflow. Figure 10 depicts the workflow used in this study. Six algorithms, Naïve Bayes [17], Logistic Regression [18], kNN [19], Random Forest [20], Support Vector Machine (SVM) [21], and Neural Network [22], were used to assess the accuracy of student certification results as predictors. Using a stratified tenfold cross-validation sample type with the average across classes as the target class, the data population was randomly chosen to be a sample dataset. Figure 11 shows the features and target used, based on the correlation analysis.

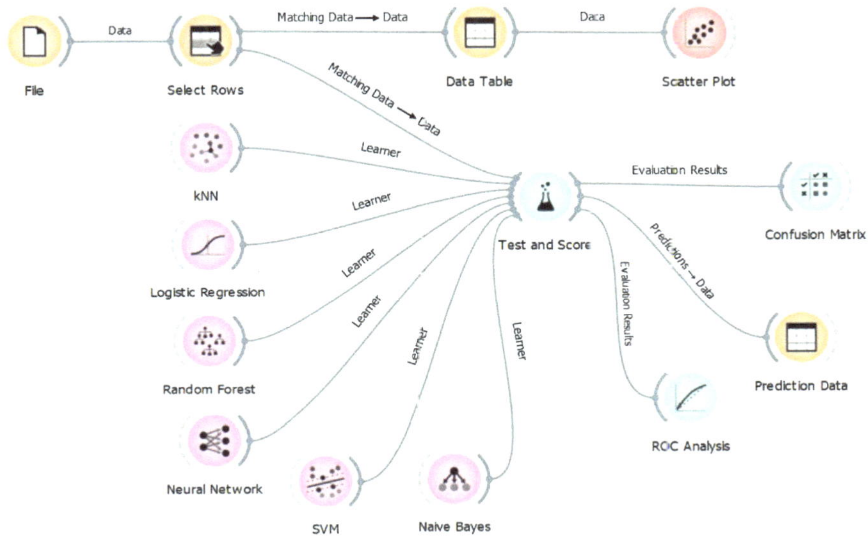

Figure 10. Classification model workflow using Orange.

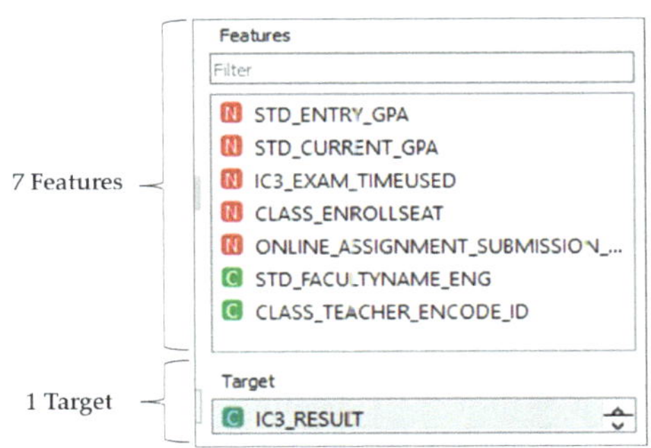

Figure 11. Dataset features used as evaluation.

Algorithm performance comparison can be seen through the Receiver Operating Characteristic (ROC) curve [23]. The evaluation results are then presented in the form of a confusion matrix based on Equations (1)–(6) regarding accuracy, true positive (*TP*) rate, false positive (*FP*) rate, recall, precision, and F1 measure [24]. Figure 12a–f shows each of the confusion matrices of the six algorithms used. A comparative evaluation of the six algorithms is presented in Table 3. Classification accuracy (CA), precision rate, area under the ROC curve (AUC), F1 score, and recall were the metrics used to evaluate the data mining classifiers.

$$Accuracy = \frac{TP + TN}{TP + FP + TN + FN} \quad (1)$$

$$True\ positive\ rate = \frac{TP}{TP + FN} \quad (2)$$

$$False\ positive\ rate = \frac{FP}{FP + TN} \quad (3)$$

$$Recall = \frac{TP}{TP + FN} \quad (4)$$

$$Precision = \frac{TP}{TP + FP} \quad (5)$$

$$F_1 Measure = \frac{2 \times Precision \times Recall}{Precision + Recall} \quad (6)$$

Figure 12. Confusion matrix of evaluation results. (a) Naïve Bayes. (b) Logistic Regression. (c) kNN. (d) Random Forest. (e) SVM. (f) Neural Network.

Table 3. Evaluation results.

Model	AUC	CA	F1	Precision	Recall
Logistic Regression	0.976	0.925	0.926	0.930	0.925
kNN	0.976	0.921	0.921	0.922	0.921
Random Forest	0.974	0.914	0.914	0.915	0.914
Neural Network	0.974	0.902	0.902	0.902	0.902
Naïve Bayes	0.952	0.896	0.896	0.899	0.896
SVM	0.934	0.889	0.889	0.892	0.889

The ROC curve can be used to graphically assess the accuracy of predictions. Plotting the anticipated true positive (TP) rate against the predicted false positive (FP) rate as a gauge of the effectiveness of the classification algorithm led to the creation of the ROC curve. Figure 13a,b presents the ROC curve for the prediction analysis of pass and fail student certification scores, illustrating the differences in the predictive performance of the six methods. As the final stage of evaluation, visualization of the data was carried out into a scatter plot so that the data could be read more easily. Figure 14 shows the relationship between the IC3 score and IC3 exam time used, illustrating that students who spend more time tend to have lower scores, as mentioned in the correlation matrix heatmap. Meanwhile, Figure 15 shows the relationship between the faculties and the results of the IC3 certification result, where almost all students from the faculty of Fine and Applied Arts experience failure. This is because the majority of the education provided by this faculty is not primarily related to basics of ICT skills. Moreover, Figure 16 shows the relationship between the teachers who teach the course and the IC3 certification result. This means that teachers also affect the students' experience of failure or success.

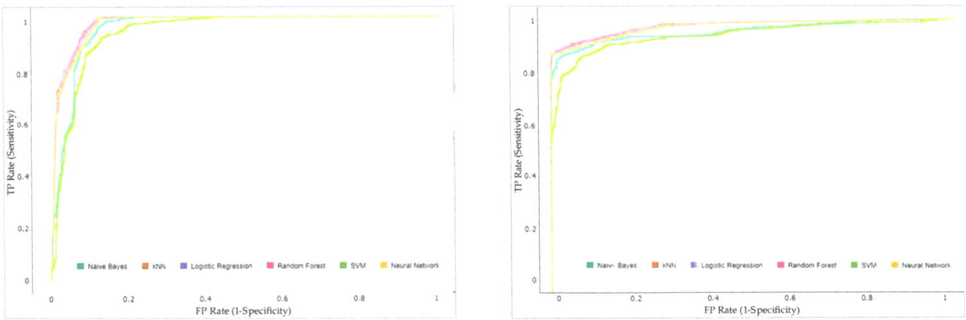

(a) (b)

Figure 13. ROC curve. (**a**) Fail. (**b**) Pass.

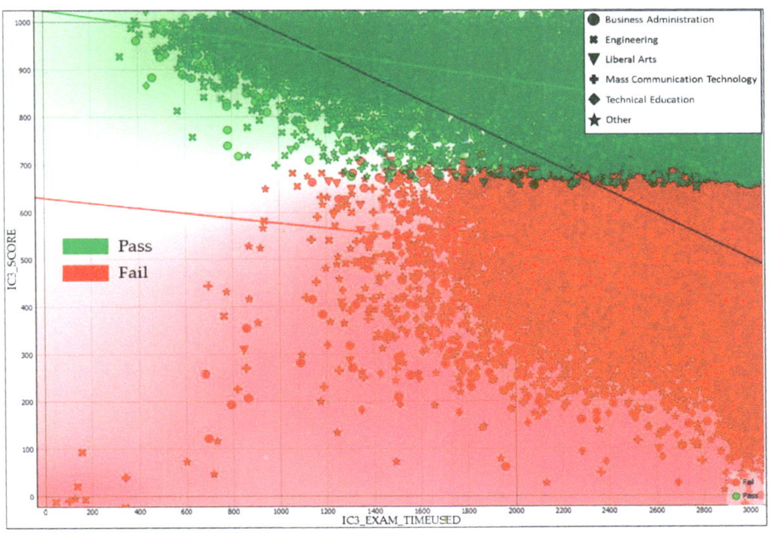

Figure 14. Scatter plot of IC3 score to IC3 exam time used.

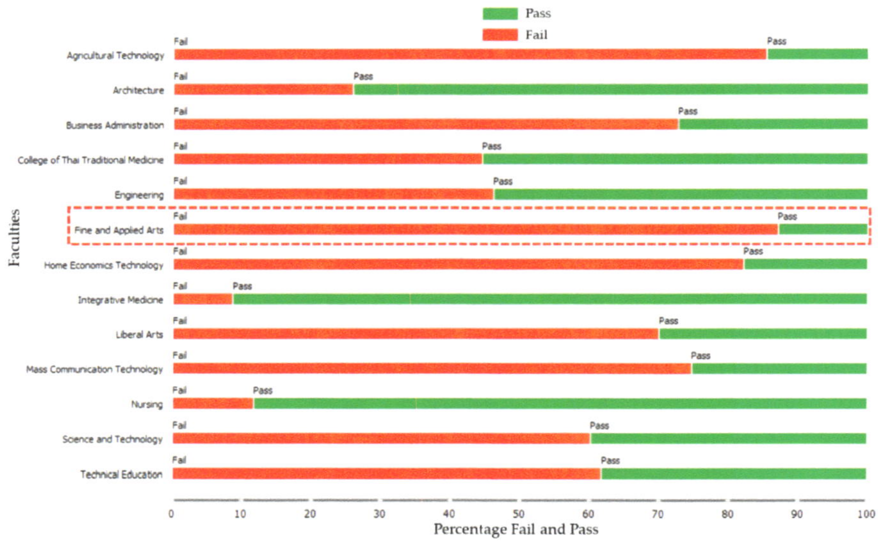

Figure 15. Certification exam results based on faculties (percentage).

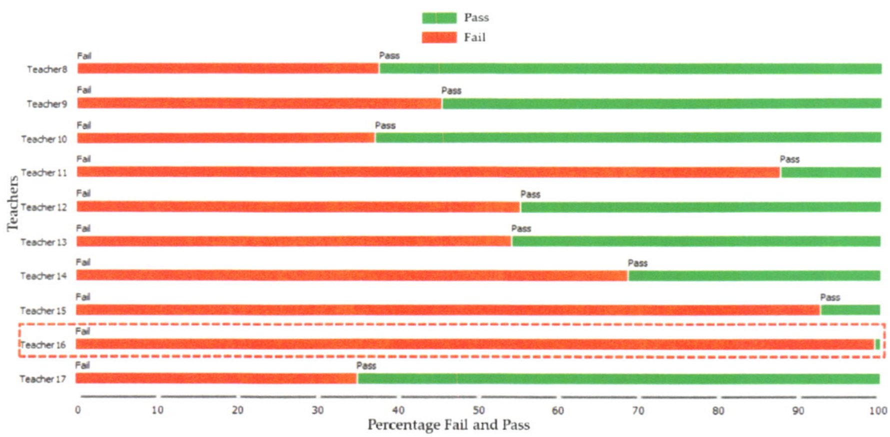

Figure 16. Certification exam results based on teachers (percentage).

5. Conclusions

This data descriptor presents a dataset created based on data obtained from the Rajamangala University of Technology Thanyaburi (RMUTT) called the RMUTT-DLD dataset, including the collection methodology for data preparation. This dataset is an amalgamation of several separate databases related to IC3 digital literacy certification results for students enrolled in the RMUTT CITS course. This dataset contains 45,603 records with 24 main variables and was collected between 2016 and 2023, including students' profiles and demographics, academic records, and IC3 digital literacy exam results. Also, the digital literacy learning procedure used between 2016 and 2018 was changed to the new implementation for improvement used between 2019 and 2023. Evaluation of the dataset was carried out by applying six machine learning algorithms. Making the right model based on this dataset will benefit students by implementing the right strategy to support student certification pass rates, especially in the field of digital literacy. To predict student/instructor performance and recognize pupils at risk of failing, new or improved models are required. In summary, the availability of the RMUTT-DLD dataset, along with the detailed methodology and evaluation results, presents numerous opportunities for teachers, universities, and researchers. It enables them to leverage the dataset for research, replicate the methodology for data collection in their own contexts, and gain insights to improve digital literacy programs and support student success. Furthermore, this dataset is useful for researchers who wish to conduct comparative studies on the performance of student digital literacy competencies and for training in the field of machine learning.

Author Contributions: Conceptualization, P.N. and P.C.; methodology, P.N. and P.C.; software, P.N. and P.C.; validation, P.N. and P.C.; formal analysis, P.N. and P.C.; investigation, P.N. and P.C.; resources, P.N. and P.C.; data curation, P.N. and P.C.; writing—original draft preparation, P.N.; writing—review and editing, P.N. and P.C.; visualization, P.N.; supervision, P.C.; project administration, P.C. All authors have read and agreed to the published version of the manuscript.

Funding: This research received no external funding.

Institutional Review Board Statement: Privacy issues related to the collection, curation, and publication of student data were validated with RMUTT Data Owners and the Academic Resources and Information Technology (ARIT) departments.

Informed Consent Statement: Not applicable.

Data Availability Statement: The data presented in this study are openly available at https://dx.doi.org/10.21227/370s-1s37 (accessed on 25 June 2023).

Acknowledgments: We would like to express our deepest gratitude to King Mongkut's Institute of Technology Ladkrabang (KMITL), Rajamangala University of Technology Thanyaburi (RMUTT), and Academic Resources and Information Technology RMUTT for the support and facilities that were provided for this research.

Conflicts of Interest: The authors declare no conflict of interest.

References

1. Tinmaz, H.; Lee, Y.T.; Fanea-Ivanovici, M.; Baber, H. A Systematic Review on Digital Literacy. *Smart Learn. Environ.* **2022**, *9*, 21. [CrossRef]
2. Ramaila, S.; Molwele, A.J. The Role of Technology Integration in the Development of 21st Century Skills and Competencies in Life Sciences Teaching and Learning. *Int. J. High. Educ.* **2022**, *11*, 9. [CrossRef]
3. Alhassan, M.D.; Adam, I.O. The Effects of Digital Inclusion and ICT Access on the Quality of Life: A Global Perspective. *Technol. Soc.* **2021**, *64*, 101511. [CrossRef]
4. Wittayasin, S. Education Challenges to Thailand 4.0. *Int. J. Integr. Educ. Dev.* **2017**, *2*, 29–35.
5. Tripopsakul, W. Preparing for Industry 4.0-Will Youths Have Enough Essential Skills? An Evidence from Thailand. *Int. J. Instr.* **2020**, *13*, 89–104.
6. Metee, P. Expectations of Hands-on Instructional Quality in the 21st Century Amongst Undergraduate Student: A Case Study at RMUTT. *Adv. Sci. Lett.* **2018**, *24*, 4507–4510.
7. Daungtod, S. A Study of Digital Literacy of 1st Year Computer Education Students Faculty of Education Nakhon Phanom University. In Proceedings of the ACM International Conference Proceeding Series; Association for Computing Machinery: New York, NY, USA, 2019; pp. 241–244.
8. Fernández-García, A.J.; Rodríguez-Echeverría, R.; Preciado, J.C.; Conejero Manzano, J.M.; Sánchez-Figueroa, F. Creating a Recommender System to Support Higher Education Students in the Subject Enrollment Decision. *IEEE Access* **2020**, *8*, 189069–189088. [CrossRef]
9. Alvarado-Uribe, J.; Mejía-Almada, P.; Masetto Herrera, A.L.; Molontay, R.; Hilliger, I.; Hegde, V.; Montemayor Gallegos, J.E.; Ramírez Díaz, R.A.; Ceballos, H.G. Student Dataset from Tecnologico de Monterrey in Mexico to Predict Dropout in Higher Education. *Data* **2022**, *7*, 119. [CrossRef]
10. Stein, M.; Leitner, M.; Trepanier, J.C.; Konsoer, K. A Dataset of Dropout Rates and Other School-Level Variables in Louisiana Public High Schools. *Data* **2022**, *7*, 48. [CrossRef]
11. Delahoz-Dominguez, E.; Zuluaga, R.; Fontalvo-Herrera, T. Dataset of Academic Performance Evolution for Engineering Students. *Data Brief* **2020**, *30*, 105537. [CrossRef]
12. Ibrahim, W.; Abdullaev, S.; Alkattan, H.; Adelaja, O.A.; Subhi, A.A. Development of a Model Using Data Mining Technique to Test, Predict and Obtain Knowledge from the Academics Results of Information Technology Students. *Data* **2022**, *7*, 67. [CrossRef]

13. Kuzilek, J.; Hlosta, M.; Zdrahal, Z. Data Descriptor: Open University Learning Analytics Dataset. *Sci. Data* **2017**, *4*, 170171. [CrossRef] [PubMed]
14. Pho, D.-H.; Nguyen, X.-A.; Luong, D.-H.; Nguyen, H.-T.; Vu, T.-P.-T.; Nguyen, T.-T.-T. Data on Vietnamese Students' Acceptance of Using VCTs for Distance Learning during the COVID-19 Pandemic. *Data* **2020**, *5*, 83. [CrossRef]
15. Le, A.V.; Do, D.L.; Pham, D.Q.; Hoang, P.H.; Duong, T.H.; Nguyen, H.N.; Vuong, T.T.; Nguyen, H.K.T.; Ho, M.T.; La, V.P.; et al. Exploration of Youth's Digital Competencies: A Dataset in the Educational Context of Vietnam. *Data* **2019**, *4*, 69. [CrossRef]
16. Wahbeh, A.H.; Al-Radaideh, Q.A.; Al-Kabi, M.N.; Al-Shawakfa, E.M. A Comparison Study between Data Mining Tools over Some Classification Methods. *IJACSA Int. J. Adv. Comput. Sci. Appl. Spec. Issue Artif. Intell.* **2020**, *18*, 72–76.
17. Chen, S.; Webb, G.I.; Liu, L.; Ma, X. A Novel Selective Naïve Bayes Algorithm. *Knowl.-Based Syst.* **2020**, *192*, 105361. [CrossRef]
18. Christodoulou, E.; Ma, J.; Collins, G.S.; Steyerberg, E.W.; Verbakel, J.Y.; van Calster, B. A Systematic Review Shows No Performance Benefit of Machine Learning over Logistic Regression for Clinical Prediction Models. *J. Clin. Epidemiol.* **2019**, *110*, 12–22. [CrossRef]
19. Chen, Y.; Hu, X.; Fan, W.; Shen, L.; Zhang, Z.; Liu, X.; Du, J.; Li, H.; Chen, Y.; Li, H. Fast Density Peak Clustering for Large Scale Data Based on KNN. *Knowl.-Based Syst.* **2020**, *187*, 104824. [CrossRef]
20. Tyralis, H.; Papacharalampous, G.; Langousis, A. A Brief Review of Random Forests for Water Scientists and Practitioners and Their Recent History in Water Resources. *Water* **2019**, *11*, 910. [CrossRef]
21. Cervantes, J.; Garcia-Lamont, F.; Rodríguez-Mazahua, L.; Lopez, A. A Comprehensive Survey on Support Vector Machine Classification: Applications, Challenges and Trends. *Neurocomputing* **2020**, *408*, 189–215. [CrossRef]
22. Yamazaki, K.; Vo-Ho, V.K.; Bulsara, D.; Le, N. Spiking Neural Networks and Their Applications: A Review. *Brain Sci.* **2022**, *12*, 863. [CrossRef] [PubMed]
23. Sarlis, N.v.; Skordas, E.S.; Christopoulos, S.R.G.; Varotsos, P.A. Natural Time Analysis: The Area under the Receiver Operating Characteristic Curve of the Order Parameter Fluctuations Minima Preceding Major Earthquakes. *Entropy* **2020**, *22*, 583. [CrossRef] [PubMed]
24. Maxwell, A.E.; Warner, T.A.; Guillén, L.A. Accuracy Assessment in Convolutional Neural Network-Based Deep Learning Remote Sensing Studies—Part 1: Literature Review. *Remote Sens.* **2021**, *13*, 2450. [CrossRef]

data

MDPI

Article

Data Balancing Techniques for Predicting Student Dropout Using Machine Learning

Neema Mduma

Department of Information and Communication Sciences and Engineering, The Nelson Mandela African Institution of Science and Technology, Arusha P.O. Box 447, Tanzania; neema.mduma@nm-aist.ac.tz

Abstract: Predicting student dropout is a challenging problem in the education sector. This is due to an imbalance in student dropout data, mainly because the number of registered students is always higher than the number of dropout students. Developing a model without taking the data imbalance issue into account may lead to an ungeneralized model. In this study, different data balancing techniques were applied to improve prediction accuracy in the minority class while maintaining a satisfactory overall classification performance. Random Over Sampling, Random Under Sampling, Synthetic Minority Over Sampling, SMOTE with Edited Nearest Neighbor and SMOTE with Tomek links were tested, along with three popular classification models: Logistic Regression, Random Forest, and Multi-Layer Perceptron. Publicly accessible datasets from Tanzania and India were used to evaluate the effectiveness of balancing techniques and prediction models. The results indicate that SMOTE with Edited Nearest Neighbor achieved the best classification performance on the 10-fold holdout sample. Furthermore, Logistic Regression correctly classified the largest number of dropout students (57348 for the Uwezo dataset and 13430 for the India dataset) using the confusion matrix as the evaluation matrix. The applications of these models allow for the precise prediction of at-risk students and the reduction of dropout rates.

Keywords: student dropout; prediction; machine learning; classification; data sampling; imbalanced datasets

Citation: Mduma, N. Data Balancing Techniques for Predicting Student Dropout Using Machine Learning. *Data* **2023**, *8*, 49. https://doi.org/10.3390/data8030049

Academic Editors: Antonio Sarasa Cabezuelo and Ramón González del Campo Rodríguez Barbero

Received: 28 January 2023
Revised: 19 February 2023
Accepted: 21 February 2023
Published: 27 February 2023

1. Introduction

This paper presents a novel approach for predicting student dropout using machine learning (ML) methods and data balancing techniques. The proposed method has been tested on real-world datasets collected from Tanzania and India. Additionally, this paper provides a unique contribution by suggesting the use of data balancing techniques to improve the accuracy of machine learning models for student dropout prediction. This research can contribute to environmental sustainability by providing better education planning and policymaking. It can also help in understanding the impact of climate change on student dropout by providing better predictions of the risk factors associated with it, by taking into consideration supervised learning applications. The majority of supervised learning applications face the problem of classifying unbalanced datasets, where one class is underrepresented relative to another [1–8]. This problem is common in the real-world applications of telecommunications, the web, finance, ecology, biology, medicine, etc., with a negative impact on the classification performance of machine learning models [2,9,10]. In the context of education, the classification of an imbalance problem occurs in the field of student dropout because the number of students enrolled is higher than the number of dropouts [11,12]. Student dropout is one of the challenges facing several schools in developing countries [13,14]. It is more common in girls than boys, and in lower secondary schools as compared to higher levels [15]. According to [16], the imbalance ratio is around 1:10, and, in most cases, the minority class usually represents the target group [2]. Regarding improving the predictive accuracy of the minority class as one of the greatest learning interests, many researchers have focused on developing solutions for the problem of class imbalance. The data sampling technique among the developed solutions aims to balance data before model development [17]. It consists of undersampling techniques i.e., Random Under Sampling (RUS), oversampling techniques i.e., Random Over Sampling (ROS) together with Synthetic Minority Over Sampling Technique (SMOTE) and it also includes hybrid techniques i.e., Synthetic Minority Over Sampling Technique with Edited NearestNeighbor (SMOTE ENN) and Synthetic Minority Over Sampling Technique with Tomek links (SMOTE TOMEK). RUS is a non-heuristic technique that selects a subset of the majority class to create a balanced class distribution [18]. In this technique, examples are randomly selected from the majority class for exclusion, with no replacement until the outstanding number of examples is thoroughly combined with that of the minority class. The main advantage of this technique, especially in Big Data, is the reduction in execution cost due to the decrease in data size caused by removing some examples. However, by excluding certain examples from the majority class, potential information may be lost that could have an impact on the learning process. On the contrary, the ROS technique is more commonly used than the RUS technique since undersampling tends to eliminate important information from the data. ROS tends to randomly balance the distribution of data up until the number of chosen examples, plus the original examples of the minority class is roughly equal to that of the majority class [19]. Despite its ability to balance class distribution, ROS tends to cause overfitting problems. On the other hand, SMOTE emphasizes the creation of examples of synthetic minorities for inclusion in the original dataset [12]. This technique forms new examples of minority classes by combining several examples of minority classes [20]. SMOTE has become the most frequently used technique, but the limitation of this technique, similar to ROS, is to assume equal importance for all minority instances. SMOTE TOMEK hybrid technique combines both SMOTE and Tomek links. Tomek links were proposed to be applied in an oversampled training set as a data cleaning technique in order to come up with a better defined class cluster [21]. This technique tends to delete examples that form Tomek links between the two classes. In the meantime, SMOTE ENN combines SMOTE and Edited Nearest Neighbor (ENN) [22]. The motive behind this technique is similar to that of SMOTE TOMEK; however, ENN is used to expel examples from both classes, so any example that has been misclassified by its three nearest neighbors is removed from the training set. This technique should help to further clean up the data, as ENN tends to eliminate more examples than Tomek links. Apart from data sampling techniques, data imbalance can also be handled by using algorithmic modification techniques that focus on changing the learning algorithm to adapt the imbalance data settings [18] and cost-sensitive learning techniques that focus on minimizing costs associated with the learning process [23]. While there are several approaches to dealing with imbalanced datasets, data sampling techniques are simple to use to deal with the problem of class imbalance [12].

In addressing the problem of student dropout, different machine learning models such as Multi-Layer Perceptron (MLP), Random Forest (RF), and Logistic Regression (LR) have been used [24–42]. MLP is an Artificial Neural Networks (ANN) that consists of an input layer, one or more hidden layers, and an output layer [43–45]. This model is commonly used for classification problems because of its low complexity and ability to produce an appropriate outcome for nonlinear relationships [46,47]. The model is a feed-forward artificial neural network classifier with forward connections, and every perceptron is connected to all the perceptrons in the next layer except the output layer that directly gives the result [48].

On the other hand, RF is an ensemble classification model that is made up of several randomized decision trees [49–53]. It is a widely used overall model because of its efficient implementation and its ability to reduce overfitting [54–58]. The performance of the RF model is determined by the tuning of its parameters and the feature selection [59]. This model is a non-parametric tree model, which is somewhat required when dealing with high-dimensional datasets [60]. Since RF is based on the definition of several independent trees, it is straightforward to obtain a parallel and faster application of the RF method, in which many trees are built in parallel on different cores [61]. As well, LR is among the classification approaches used to model the probability of discrete (binary or multinomial) outcomes [62]. This model works very similarly to linear regression by analyzing the relationship between multiple independent variables and a categorical dependent variable and calculating the probability of the existence of an event by fitting data to a logistic curve [63,64].There are two kinds of logistic regression: binary logistic regression (as in the present study) and multinomial logistic regression [64,65]. Despite the ability of these models (MLP, RF and LR) to predict student dropout problems, data imbalance was ignored in many studies and needs to be addressed in order to improve the predictive results of machine learning models.

For the evaluation of the performance of machine models, one of the key factors guiding the algorithmic modeling is the evaluation criteria. Accuracy as a statistical measure to quantify the level of accuracy has been used as a common metric by many researchers [66,67]. However, in the imbalanced data domain, this metric is no longer an appropriate measure, for it has less effect on the minority class than the majority class, and combined with the fact that it cannot distinguish between the magnitude of errors. In the context of imbalanced datasets, standard measures using particular measures are used to account for class distribution. The confusion matrix saves the results for examples correctly and incorrectly recognized by each class in a binary class problem [68]. This matrix is an important tool for assessing prediction results in a way that is very easy to understand [69]. In addition, the Geometric Mean (G_m) of actual rates measures the capacity of the model to balance sensitivity (TPrate) and specificity (TNrate) [1]. G_m is at a maximum when TPrate and TNrate are equal. F-measure (F_m) is a harmonic mean of precision and recall [66]. This metric ensures the TPrate changes more in the positive predictive value (precision) than in the True Positive rate (TPrate). A high value of F_m shows that both precision and recall are sensibly high. On acquiring the highest TPrate without excessively minimizing the TNrate, the Adjusted Geometric Mean (AG_m) was introduced [2]. Despite the ability of these metrics to evaluate the performance of the machine learning models, other studies have reported their limitations in terms of the effects on the minority classes in the imbalance datasets [2,70,71]; hence, the application of several metrics is highly recommended when evaluating the performance of the machine learning models.

Therefore, this paper presents several data balancing techniques for predicting student dropout using datasets from developing countries. The research problem is to identify how to effectively use machine learning models for predicting student dropout when the dataset is imbalanced. The objective of the paper is to explore the use of various data balancing techniques to improve the accuracy of machine learning models for predicting student dropout. The novelty of the paper lies in its comparison of the performance of different data balancing techniques to address the issue of imbalanced datasets.

The next section presents related works that applied data balancing techniques to addressing the problem of student dropout. Section 3 introduces the materials and methods used to conduct this study. The results are presented and discussed in Section 4. Finally, the article presents the conclusion and prospective future directions in Section 5.

2. Literature Review

The use of data balancing techniques to predict student dropout using machine learning has been applied in several studies, as summarized in Figure 1. A study by [11] used machine learning to predict student dropout and academic success. The study used a dataset to build machine learning models for predicting academic performance and dropout. Imbalanced data were identified, and different techniques for handling this problem were proposed, such as data-level techniques including Synthetic Minority Over Sampling Technique (SMOTE) and Adaptive Synthetic Sampling Approach (ADASYN), or algorithm-level techniques including Balanced Random Forest and SMOTE-Bagging. Another study by [72] used data balancing techniques to predict student dropout at a university in Turkey. A dataset of 1510 student records was used, and different classifiers such as decision trees and support vector machines were applied. Data balancing techniques such as oversampling and undersampling were used to improve the accuracy of the models. The results showed that the use of data balancing techniques improved the accuracy of the models and reduced the bias in the data.

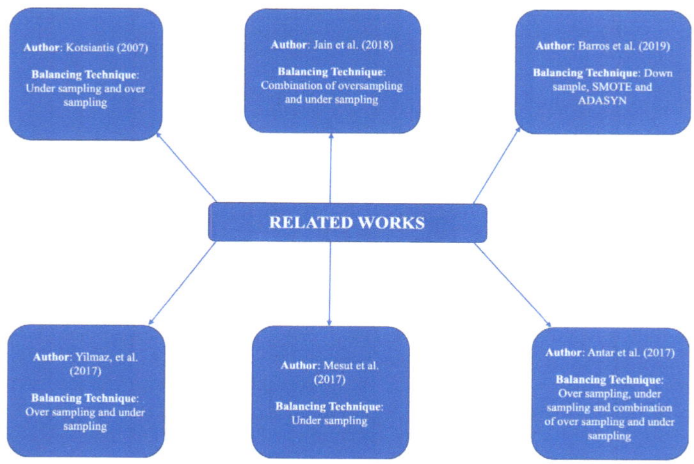

Figure 1. Summary of the related works. Note: Yilmaz, et al. (2017) [72]; Mesut et al. (2017) [73]; Antar et al. (2017) [74]; Jain et al. (2018) [75]; Barros et al. (2019) [76]; Kotsiantis (2007) [77].

A study by [73] used machine learning and applied data balancing techniques to predict student dropout. The study used an unbalanced dataset from a real university and applied an undersampling technique to balance it. The study used a decision tree algorithm to predict student dropout and obtain an accuracy of 83.2%.

Another study by [74] used machine learning and applied data balancing techniques to predict student dropout. The study used a dataset of student records collected from a university and applied oversampling, undersampling, and a combination of both techniques to balance it. The study applied a Random Forest algorithm to predict student dropout and obtain an accuracy of 81.2%.

A study by [75] used machine learning and applied data balancing techniques to predict student dropout. The study used an imbalanced dataset from a university and applied a combination of oversampling and undersampling techniques to balance it. The study used a decision tree algorithm to predict student dropout and obtain an accuracy of 85.3%.

One study by [76] developed predictive models for imbalanced data. The study applied data mining techniques to forecast dropout rates. The study used a decision tree, neural networks, and balanced bagging. Classifiers were tested with and without the use of data balancing techniques, including downsample, SMOTE, and ADASYN data balancing. The results showed that the geometric mean and UAR provide reliable results when predicting dropout rates using balanced bagging classification techniques. Finally, a study by [77] applied data balancing techniques to predict student dropout using machine learning. The study used a dataset of 3420 student records from a university in Greece. A variety of classification algorithms were tested, including Naïve Bayes, C4.5, and Support Vector Machines. Furthermore, data balancing techniques such as undersampling and oversampling were applied to remove the bias and improve the accuracy of the models. The results showed that the use of data balancing techniques improved the accuracy of the models for predicting student dropout.

Despite the fact that many studies applied data balancing techniques to addressing the problem of student dropout, many of them were carried out in developed countries using developed countries' datasets.

3. Materials and Methods

3.1. Dataset

To address student dropout, this study used two publicly available datasets from developing countries. The first dataset was Uwezo data [1] on learning at the country level in Tanzania, which was collected in 2015 with the objective of assessing children's learning levels across hundreds of thousands of households. The second dataset was collected in 2016 with the aim of assessing student dropout in India [2]. The Uwezo dataset consisted of 61,340 samples, of which 98.4% were retained and 1.6% were dropouts, and the India dataset consisted of 11,257 samples, of which 95.1% were retained and 4.9% were dropouts. Therefore, these two datasets were highly imbalanced, as presented in Figure 2a,b, respectively.

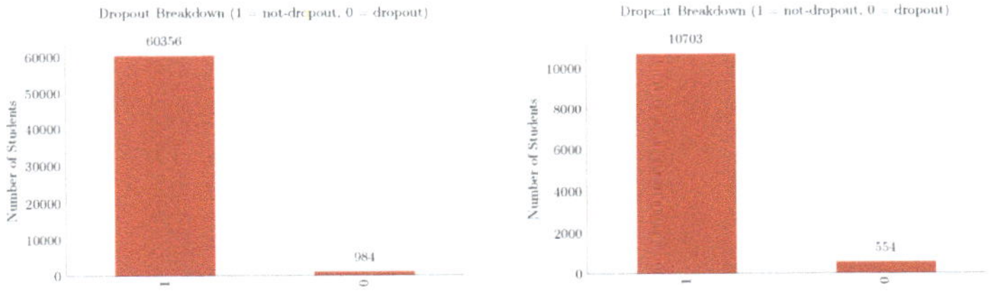

Figure 2. Dropout distributions for the Uwezo and India datasets; (**a**) Dropout distribution: Uwezo dataset; (**b**) Dropout distribution: India dataset.

The Uwezo dataset consisted of 18 variables: Main source of household income (Income), Boy's Pupil Latrines Ratio (BPLR), School has a privacy room for girls (SGR), Region, District, Village, Student gender (Sex), Parent check child's exercise book once a week (PCCB), Household meals per day (MLPD), Student read a book with his/her parent last week (SPB), Parent discuss child's progress last term with the teacher (PTD), Student age (Age), Enumeration Area type (EA area), Household size (HH size), Girl's Pupil Latrines Ratio (GPLR), Parent Teacher Meeting Ratio (PTMR), Pupil Classroom Ratio (PCR), Pupil Teacher Ratio (PTR) and Dropout. India dataset consisted of variables: Continue drop, Student id, Gender, Caste, Mathematics marks, English marks, Science marks, Science teacher, Languages teacher, Guardian, Internet, School id, Total students, Total toilets, and Establishment year.

3.2. Data Pre-Processing

Data from the two datasets were pre-processed prior to obtaining a final training set. This process was carried out as a precautionary measure to ensure that datasets are properly cleaned and accurate prior to model development. The data clean-up was carried out by removing information that could reveal the identity of individuals to the end-user. Missing values were replaced with medians and zeroes. The following variables were identified with missed values: Pupil Teacher Ratio (PTR), Pupil Classroom Ratio (PCR), Girl's Pupil Latrines Ratio (GPLR), Boy's Pupil Latrines Ratio (BPLR), Parent Teacher Meeting Ratio (PTMR), Main source of household income (Income), and Enumeration Area type (EA area).

Parent who checks his/her child's exercise book once a week (PCCB), Parent who discusses his/her child's progress last term with the teacher (PTD), Student who read a book with his/her parent last week (SPB), School has a privacy room for girls (SGR), Household meals per day (MLPD). On handling missing values, PTR, PCR, GPLR, and BPLR were imputed with medians, and PTMR, Income, EA area, PCCB, PTD, SPB, SGR, and MLPD were imputed with zeros. In addition, data samples with nominal variables were converted to numerical values to comply with Scikit-learn.

3.3. Data Sampling Techniques

Five data balancing techniques were employed to address the issue of data imbalance in the datasets. These techniques were employed before model development due to their ability to provide in-depth data cleaning, produce straight-forward and satisfactory results when handling data imbalance, address the overfit problem, and reduce running time and cost. RUS, ROS, SMOTE, SMOTE ENN, and SMOTE TOMEK have been implemented. RUS was performed by randomly selecting examples from the majority class for exclusion with no replacement until the outstanding number of examples were thoroughly combined with those of the minority class. This approach was chosen due to its ability to reduce the cost of execution by decreasing the size of the data through the removal of a few examples. ROS was performed by randomly balancing the distribution of data over the application of minority data duplication up to when the number of chosen examples plus the original examples of the minority class was roughly equal to that of the majority class. This approach was chosen based on its ability to not eliminate important information from the data. SMOTE was selected to form new minority class examples by incorporating several minority class examples. Furthermore, SMOTE TOMEK was selected to remove examples that form Tomek links from both classes, and SMOTE ENN was selected to expel examples from both classes; therefore, any example that has been misclassified by its three nearest neighbors was removed from the training set. This technique was anticipated to give more in-depth data cleaning, as ENN tends to eliminate more examples than Tomek links.

3.4. Classification Models

Three popular classification models: Logistic Regression (LR), Random Forest (RF), and Multi-Layer Perceptron (MLP) were assessed on a set of supervised classification datasets in order to see which model would perform better with consideration of the data imbalance problem. The selection of the three models took into consideration the supervised learning approach, particularly with respect to the classification problem. These models were selected because they were able to give satisfactory results on the prediction of student dropout. LR was selected to represent the linear model and was used to model the probability of binary outcomes (dropout/not dropout). In addition, RF represented an ensemble model and was chosen to reduce the overfitting problem and handle high-dimensional data. The MLP, on the other hand, represented an artificial neural network and was selected to reduce complexity.

3.5. Evaluation Metrics

To assess the performance of classification models, three popular metrics were used: Geometric Mean (G_m), F-measure (F_m), and Adjacent Geometric Mean (AG_m). Furthermore, a confusion matrix was used to determine the best model based on the actual number of samples correctly and improperly classified. These metrics were chosen with an emphasis on the imbalance domain and as a standard measure in class distribution. G_m was selected to measure the ability of the model to balance TPrate and TNrate. F_m was selected to measure the harmonic means of TPrate and precision, whereas AG_m was selected to measure the increase of TPrate rates without decreasing TNrate.

3.6. Experimental Design

In this study, MLP, RF, and LR were compared over six different structures (original, balanced with ROS, balanced with RUS, balanced with SMOTE, balanced with SMOTE ENN, and balanced with SMOTE Tomek) using stratified 10-fold cross validation. The datasets were alienated in training, validation, and testing by 60%, 20%, and 20%, respectively, to minimize sampling bias. The methodology used to conduct this study is summarized in Figure 3.

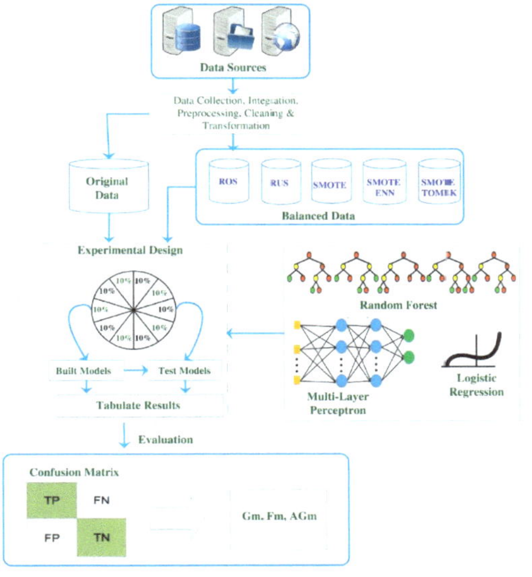

Figure 3. Overview of the experimental design.

Data balancing techniques for predicting student dropout using machine learning can help identify the key determinants of dropout more accurately. This can help schools and other educational institutions better understand the factors that lead to student dropout and take appropriate measures to prevent it. In addition to that, educational institutions can anticipate when students are at risk of dropping out and intervene early to provide the necessary support. This can help reduce the rate of student dropout and improve educational outcomes. By understanding the key determinants of student dropout and intervening early, educational stakeholders can provide targeted interventions to improve educational outcomes. This can help improve student success and reduce the overall dropout rate. Furthermore, data balancing techniques can also help identify disparities in educational outcomes among different groups of students, such as those from different backgrounds or those with different levels of academic achievement. This can help identify and address disparities in educational outcomes and promote equity in education.

4. Results and Discussion

The study used two datasets to compare data balancing techniques. The datasets used were highly imbalanced due to the fact that there are still many students in school compared to students who drop out, which makes balancing the data very important in this study because the focus was primarily on the minority class, in this case dropouts. The results showed that the SMOTE ENN data balancing technique had very good solutions for achieving greater performance, followed by SMOTE TOMEK and RUS on the Uwezo datasets. For the Indian dataset, the SMOTE ENN data balancing technique performed better, followed by SMOTE TOMEK and ROS (Table 1).

The SMOTE ENN data balancing technique has shown very good solutions for achieving greater performance due to its ability to provide in-depth data cleaning. Similar results were reported by [78] when assessing a number of methods to balance machine learning data. Furthermore, [79] stressed the techniques and importance of handling data imbalance when developing training sets from a machine learning model, and [80] emphasized the good performance of hybrid data balancing techniques such as SMOTE-RSB, SMOTE-TOMEK, and SMOTE ENN when dealing with highly imbalanced data like in the case of student dropout.

Table 1. Comparison of data balancing techniques (Uwezo and India datasets).

Preprocessing	Models	G_m	F_m	AG_m	G_m	F_m	AG_m
	Uwezo dataset					India dataset	
None	LR	0.000	0.000	0.000	0.000	0.000	0.000
	MLP	0.011	0.002	0.012	0.000	0.000	0.000
	RF	0.004	8.32×10^{-5}	0.004	0.031	0.002	0.031
ROS	LR	0.536	0.547	1.010			
	MLP	0.499	0.438	0.920	0.524	**0.450**	0.957
	RF	0.293	0.270	0.449	0.707	0.667	1.207
RUS	LR	0.548	0.546	1.042	0.582	0.570	1.085
	MLP	0.512	0.332	1.031	0.515	0.139	0.925
	RF	0.624	0.561	1.192	0.711	0.667	1.210
SMOTE	LR	0.551	0.556	1.034	0.648	0.603	1.190
	MLP	0.525	0.475	0.967	0.555	0.410	1.032
	RF	0.661	0.645	1.138	0.707	0.667	1.207
SMOTE ENN	LR	**0.562**	**0.572**	**1.079**	**0.722**	**0.638**	**1.343**
	MLP	**0.577**	0.491	**1.104**	**0.791**	0.438	**1.531**
	RF	**0.676**	**0.666**	1.176	**0.738**	**0.706**	**1.283**
SMOTE Tomek	LR	0.550	0.556	1.032	0.655	0.605	1.201
	MLP	0.546	**0.508**	1.015	0.735	0.441	1.390
	RF	0.663	0.646	1.140	0.707	0.667	1.206

On the contrary, the RUS technique performed the worst in the study's experiment evaluating data sampling techniques. This could be due to the nature of the loss of certain potential information that could have an impact on the learning process. Similar results were reported by [81,82] when assessing multiple approaches to managing imbalanced datasets. However, it was reported that this approach improved predictive performance in other studies compared with the lack of data sampling techniques [83,84]. Most datasets in the real world are not balanced, i.e., there is a majority and minority class, and if data balancing is ignored when training the machine learning model, it may lead to bias towards one class, and the model will learn more about the majority class and learn less about or ignore the minority class. Hence, handling unbalanced data is very important when developing a machine learning model.

Models Performance

Three machine learning models used in data balancing techniques were evaluated, and the findings showed that LR was the best model to correctly classify the highest number of student dropouts and misclassify the lowest, followed by MLP and RF in the Uwezo (Figure 4) and Indian datasets (Figure 5).

Similar metrics (G_m, F_m, and AG_m) were used by [41,75–89] in evaluating the performance of the developed models in order to take the class distribution into account. In addition, accuracy has been reported as a common metric for measuring the degree of correctness of machine learning models [66,67]. However, its limitations in the imbalanced domain make it unsuitable for evaluating models with imbalanced data [2,72].

Moreover, this study found that LR and MLP were the best models to correctly classifying the highest number of student dropouts and misclassifying the lowest. This may be due to the ability of LR to model the probability of binary results and the power of MLP to produce satisfactory results for nonlinear relationships. Similar results were reported by [42,90] when determining the accuracy of their predictive models for the early prediction of stroke and student dropout, respectively. Both studies indicated that LR was the best-performing classification model relative to the others. These results, however, contradict what was reported by [91] in their study of evaluating the performance of supervised machine learning models in healthcare, where K-Nearest Neighbor and Random Forest were reported to outperform other models such as Logistic Regression and Naive Bayes.

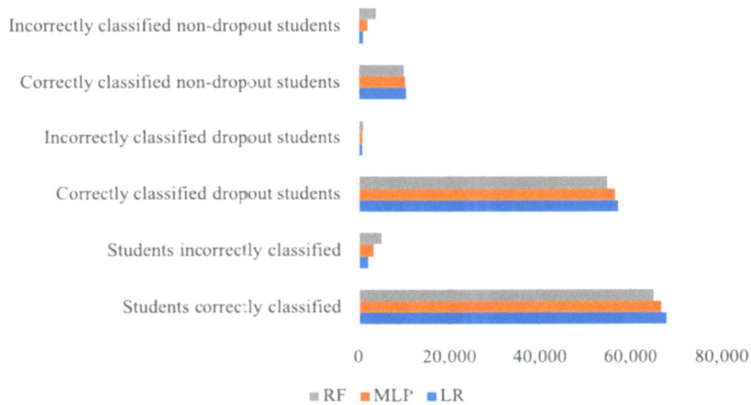

Figure 4. Comparison of models' performance in terms of numbers of correctly and incorrectly classified students (the Uwezo dataset).

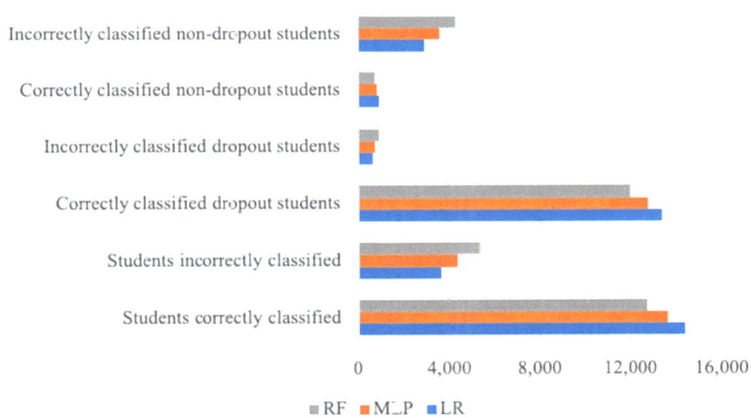

Figure 5. Comparison of models' performance in terms of numbers of correctly and incorrectly classified students (the India dataset).

The issue of predicting student dropout using a machine learning model is an important one, and it's been addressed by many different approaches. Data balancing is one of the most promising of these methods. Data balancing techniques are designed to identify the key determinants of student dropout and then use machine learning to develop a model that can accurately predict dropout rates. Data balancing techniques involve creating a data set that is as balanced as possible. This means that the data must be stratified to ensure that the populations being compared are equal in terms of key attributes. By ensuring that the data is balanced in terms of key attributes, it allows the machine learning model to accurately predict dropout rates. The machine learning solution presented in this study can be used to accurately predict students at risk of dropping out of school and provide early measures for intervention.

5. Conclusions

Based on the analysis of the results, the study concluded that the SMOTE ENN balancing technique provides a good solution for achieving superior performance. Furthermore LR has been considered a potential model for the type of data used due to its high accuracy in classifying the dropout class, which is the focus of this study. The study also concluded that the use of data balancing techniques before model development helps to improve the performance of the predictive results when measured by the G_m, F_m, and AG_m. In other words, predictive outcomes were improved by comparing original (unbalanced) data with data that were collected using sampling techniques. In a real-world environment, most datasets are imbalanced and contain a large number of anticipated examples with only a small number of unexpected examples. Most of the interest is in the predictions of the unexpected examples. Machine learning models are not as precise for predicting the minority class in unbalanced datasets. Therefore, a data balancing task is required as part of the pre-processing phase to deal with this situation. This study is limited to the application of data sampling techniques to address the problem of student dropout. Prospective future directions will focus on alternative methods, including algorithmic modification and cost-sensitive learning, with the aim of improving the predictive power of the machine learning model.

Funding: This work was carried out with the aid of a grant from the Artificial Intelligence for Development in Africa Program, a program funded by the Canada's International Development Research Centre, Ottawa, Canada and the Swedish International Development Cooperation Agency, grant number 109704-001/002.

Institutional Review Board Statement: Not applicable.

Informed Consent Statement: Not applicable.

Data Availability Statement: The data presented in this study are publicly available at http://www.twaw37e1za.o datasets and https://www.kaggle.com/imrandude/studentdropindia2016 (accessed on 30 January 2017).

Acknowledgments: The author would like to thank the UNESCO-L'Oreal Foundation for supporting this study.

Conflicts of Interest: The author declares no conflict of interest.

Abbreviations

The following abbreviations are used in this manuscript:

RUS	Random Under Sampling
ROS	Random Over Sampling
SMOTE	Synthetic Minority Over Sampling Technique
SMOTE ENN	Synthetic Minority Over Sampling Technique with Edited Nearest Neighbor
SMOTE TOMEK	Synthetic Minority Over Sampling Technique with Tomek links
LR	Logistic Regression
RF	Random Forest
MLP	Multi-Layer Perceptron
G_m	Geometric Mean
F_m	F-measure
AG_m	Adjusted Geometric Mean

Notes

1 http://www.twaweza.org/go/uwezo-datasets (accessed on 30 January 2017).

2 https://www.kaggle.com/imrandude/studentdropindia2016 (accessed on 30 January 2017)

References

1. Lin, W.J.; Chen, J.J. Class-imbalanced classifiers for high-dimensional data. *Brief. Bioinform.* **2013**, *14*, 13–26. [CrossRef] [PubMed]
2. López, V.; Fernández, A.; García, S.; Palade, V.; Herrera, F. An insight into classification with imbalanced data: Empirical results and current trends on using data intrinsic characteristics. *Inf. Sci.* **2013**, *250*, 113–141. [CrossRef]
3. Krawczyk, B. Combining One-vs-One Decomposition and Ensemble Learning for Multi-class. In *Proceedings of the 9th International Conference on Computer Recognition Systems CORES 2015*; Springer International Publishing: Berlin/Heidelberg, Germany, 2015; pp. 27–36. [CrossRef]
4. Galar, M.; Fernández, A.; Barrenechea, E.; Bustince, H.; Herrera, F. New Ordering-Based Pruning Metrics for Ensembles of Classifiers in Imbalanced Datasets. In *Proceedings of the 9th International Conference on Computer Recognition Systems CORES 2015*; Springer International Publishing: Berlin/Heidelberg, Germany, 2016. [CrossRef]
5. Krawczyk, B. Learning from imbalanced data: Open challenges and future directions. *Prog. Artif. Intell.* **2016**, *5*, 221–232. [CrossRef]
6. Borowska, K.; Topczewska, M. New Data Level Approach for Imbalanced Data Classification Improvement. In *Proceedings of the 9th International Conference on Computer Recognition Systems CORES 2015*; Springer International Publishing: Berlin/Heidelberg, Germany, 2016; pp. 283–294. [CrossRef]
7. Mazumder, R.U.; Begum, S.A.; Biswas, D. Rough Fuzzy Classi fi cation for Class Imbalanced Data. In *Proceedings of Fourth International Conference on Soft Computing for Problem Solving*; Springer: Delhi India, 2015. [CrossRef]
8. Abdi, L.; Hashemi, S. An Ensemble Pruning Approach Based on Reinforcement Learning in Presence of Multi-class Imbalanced Data. In *Proceedings of the Third International Conference on Soft Computing for Problem Solving*; Springer: Delhi, India, 2014. [CrossRef]
9. Sonak, A.; Patankar, R.A A Survey on Methods to Handle Imbalance Dataset. *Int. J. Comput. Sci. Mob. Comput.* **2015**, *4*, 338–343.
10. Ali, H.; Salleh, M.N.M.; Saedudin, R.; Hussain, K.; Mushtaq, M.F. Imbalance class problems in data mining: A review. *Indones. J. Electr. Eng. Comput. Sci.* **2019**, *14*, 1552–1563. [CrossRef]
11. Realinho, V.; Machado, J.; Baptista, L.; Martins, M.V. Predicting Student Dropout and Academic Success. *Data* **2022**, *7*, 146. [CrossRef]
12. Thammasiri, D.; Delen, D.; Meesad, P.; Kasap, N. A critical assessment of imbalanced class distribution problem: The case of predicting freshmen student attrition. *Expert Syst. Appl.* **2013**, *41*, 321–330. [CrossRef]
13. UNESCO. *Estimation of the Numbers and Rates of Out-of-school Children and Adolescents Using Administrative and Household Survey Data*; UNESCO Institute for Statistics: Montreal, QC, Canada, 2017; pp. 1–33. [CrossRef]
14. Valles-coral, M.A.; Salazar-ram, L.; Injante, R.; Hernandez-torres, E.A.; Ju, J ; Navarro-cabrera, J.R.; Pinedo, L.; Vidaurre-rojas, P. Density-Based Unsupervised Learning Algorithm to Categorize College Students into Dropout Risk Levels. *Data* **2022**, *7*, 165. [CrossRef]
15. Mduma, N. Data Driven Approach for Predicting Student Dropout in Secondary Schools. Ph.D. Thesis, NM-AIST, Arusha, Tanzania, 2020.
16. Gao, T. Hybrid Classification Approach of SMOTE and Instance Selection for Imbalanced Datasets. Ph.D. Thesis, Iowa State University, Ames, IA, USA, 2015.
17. Hoens, T.R.; Chawla, N V. Imbalanced Datasets: From Sampling to Classifiers. In *Imbalanced Learning: Foundations, Algorithms, and Applications*; John Wiley & Inc.: Hoboken, NJ, USA, 2013; pp. 43–59. [CrossRef]
18. Elhassan, T.; Aljurf, M.; Shoukri, M. Classification of Imbalance Data using Tomek Link (T-Link) Combined with Random Under-sampling (RUS) as a Data Reduction Method. *J. Inform. Data Min.* **2016**, *1*, 1–12.
19. Santoso, B.; Wijayanto, H.; Notodiputro, K.A.; Sartono, B. Synthetic Over Sampling Methods for Handling Class Imbalanced Problems: A Review. In *IOP Conference Series: Earth and Environmental Science*; IOP Publishing: Bristol, UK, 2017. [CrossRef]
20. Skryjomski, P. Influence of minority class instance types on SMOTE imbalanced data oversampling. *Proc. Mach. Learn. Res.* **2017**, *74*, 7–21.
21. Yu, X.; Zhou, M.; Chen, X.; Deng, L.; Wang, L. Using Class Imbalance Learning for Cross-Company Defect Prediction. *Int. Conf. Softw. Eng. Knowl. Eng.* **2017**, 117–122. [CrossRef]
22. Douzas, G.; Bacao, F. Geometric SMOTE: Effective oversampling for imbalanced learning through a geometric extension of SMOTE. *arXiv* **2017**, arXiv:1709.07377.
23. Shilbayeh, S.A. *Cost Sensitive Meta Learning Samar Ali Shilbayeh School of Computing, Science and Engineering*; University of Salford: Salford, UK, 2015.
24. Kumar, M.; Singh, A.; Handa, D. Literature Survey on Educational Dropout Prediction. *Int. J. Educ. Manag. Eng.* **2017**, *7*, 8–19. [CrossRef]
25. Siri, A.; Siri, A. Predicting Students' Dropout at University Using Artificial Neural Networks. *Ital. J. Sociol. Educ.* **2015**, *7*, 225–247.
26. Oancea, B.; Dragoescu, R.; Ciucu, S. Predicting Students' Results in Higher Education Using Neural Networks. In Proceedings of the International Conference on Applied Information and Communication Technologies, Baku, Azerbaijan, 23–25 October 2013; pp. 190–193.
27. Saranya, A.; Rajeswari, J. Enhanced Prediction of Student Dropouts Using Fuzzy Inference System and Logistic Regression. *ICTACT J. Soft Comput.* **2016**, *6*, 1157–1162. [CrossRef]

28. Fei, M.; Yeung, D.Y. Temporal Models for Predicting Student Dropout in Massive Open Online Courses. In Proceedings of the 2015 IEEE International Conference on Data Mining Workshop (ICDMW), Atlantic City, NJ, USA, 14–17 November 2015; pp. 256–263. [CrossRef]
29. Goga, M.; Kuyoro, S.; Goga, N. A Recommender for Improving the Student Academic Performance. *Procedia Soc. Behav. Sci.* **2015**, *180*, 1481–1488. [CrossRef]
30. Sales, A.; Balby, L.; Cajueiro, A. Exploiting Academic Records for Predicting Student Drop Out: A case study in Brazilian higher education. *J. Inf. Data Manag.* **2016**, *7*, 166–180.
31. Nagrecha, S.; Dillon, J.Z.; Chawla, N.V. MOOC Dropout Prediction: Lessons Learned from Making Pipelines Interpretable. In *Proceedings of the 26th International Conference on World Wide Web Companion*; ACM: New York, NY, USA, 2017; pp. 351–359. [CrossRef]
32. Aulck, L.; Velagapudi, N.; Blumenstock, J.; West, J. Predicting Student Dropout in Higher Education. ICML Workshop on #Data4Good: Machine Learning in Social Good Applications 2016. *arXiv* **2017**, 16–20. [CrossRef]
33. Halland, R.; Igel, C.; Alstrup, S. High-School Dropout Prediction Using Machine Learning: A Danish Large-scale Study. In Proceedings of the 23rd European Symposium on Artificial Neural Networks, Bruges, Belgium, 22–23 April 2015; pp. 22–24.
34. Kemper, L.; Vorhoff, G.; Wigger, B.U. Predicting student dropout: A machine learning approach. *Eur. J. High. Educ.* **2020**, *10*, 28–48. [CrossRef]
35. de Oliveira Durso, S.; da Cunha, J.V. Determinant Factors for Undergraduate Student's Dropout in Accounting Studies Department of A Brazilian Public University. *Fed. Univ. Minas Gerais* **2018**, *34*, 186332. [CrossRef]
36. Nath, S.R.; Ferris, D.; Kabir, M.M.; Chowdhury, T.; Hossain, A. Transition and Dropout in Lower Income Countries: Case Studies of Secondary Education in Bangladesh and Uganda. *World Innov. Summit Educ.* **2017**. Available online: https://www.wise-qatar.org/app/uploads/2019/04/rr.3.2017_brac.pdf (accessed on 1 January 2023).
37. Wang, X.; Schneider, H. *A Study of Modelling Approaches for Predicting Dropout in a Business College*; Louisiana State University: Baton Rouge, LA, USA, 2018; pp. 1–8.
38. Franklin, B.J.; Trouard, S.B. An Analysis of Dropout Predictors within a State High School Graduation Panel. *Schooling* **2014**, *5*, 1–8.
39. Helou, I. Analytical and experimental investigation of steel friction dampers and horizontal brake pads in chevron frames under cyclic loads. *Issues Inf. Sci. Inf. Technol. Educ.* **2018**, *15*, 249–278.
40. Aguiar, E.; Dame, N.; Miller, D.; Yuhas, B.; Addison, K.L. Who, When, and Why: A Machine Learning Approach to Prioritizing Students at Risk of not Graduating High School on Time. *ACM* **2015**, 93–102. [CrossRef]
41. Rovira, S.; Puertas, E.; Igual, L. Data-driven System to Predict Academic Grades and Dropout. *PLoS ONE* **2017**, *12*, e0171207. [CrossRef]
42. Mgala, M.; Mbogho, A. Data-driven Intervention-level Prediction Modeling for Academic Performance. In Proceedings of the Seventh International Conference on Information and Communication Technologies and Development, Singapore, 15–18 May 2015; pp. 2:1–2:8. [CrossRef]
43. Voyant, C.; Paoli, C.; Nivet, M.l.; Notton, G. Multi-layer Perceptron and Pruning. *Turk. J. Forecast.* **2017**, *1*, 1–6.
44. Ramchoun, H.; Amine, M.; Idrissi, J.; Ghanou, Y.; Ettaouil, M. Multilayer Perceptron: Archi-tecture Optimization and Training. *Int. J. Interact. Multimed. Artif. Intell.* **2016**, *4*, 26. [CrossRef]
45. Fesghandis, G.S. Comparison of Multilayer Perceptron and Radial Basis Function Neural Networks in Predicting the Success of New Product Development. *Eng. Technol. Appl. Sci. Res.* **2017**, *7*, 1425–1428. [CrossRef]
46. Rani, K.U. Advancements in Multi-Layer Perceptron Training to Improve Classification. *Int. J. Recent Innov. Trends Comput. Commun.* **2017**, *5*, 353. [CrossRef]
47. Ahmed, K.; Shahid, S.; Haroon, S.B.; Xiao-jun, W. Multilayer perceptron neural network for downscaling rainfall in arid region: A case study of Baluchistan, Pakistan. *J. Earth Syst. Sci.* **2015**, *124*, 1325–1341. [CrossRef]
48. Taravat, A.; Proud, S.; Peronaci, S.; Del Frate, F.; Oppelt, N. Multilayer perceptron neural networks model for meteosat second generation SEVIRI daytime cloud masking. *Remote Sens.* **2015**, *7*, 1529–1539. [CrossRef]
49. Wu, Z.; Lin, W.; Zhang, Z.; Wen, A.; Lin, L. An Ensemble Random Forest Algorithm for Insurance Big Data Analysis. In Proceedings of the 2017 IEEE International Conference on Computational Science and Engineering and IEEE/IFIP International Conference on Embedded and Ubiquitous Computing, CSE and EUC 2017, Guangzhou, China, 21–24 July 2017; Volume 1, pp. 531–536. [CrossRef]
50. Compo, P.; Pca, E.; Variances, A.U.; Analysis, B.S. Submitted to the Annals of Statistics. *Ann. Stat.* **2017**, *45*, 1–37.
51. Biau, G.; Scornet, E. A Random Forest Guided Tour. *TEST* **2015**, *25*, 197–227. [CrossRef]
52. Prajwala, T.R. A Comparative Study on Decision Tree and Random Forest Using R Tool. *Ijarcce* **2015**, *4*, 196–199. [CrossRef]
53. Ibrahim, M. Scalability and Performance of Random Forest based Learning-to-Rank for Information Retrieval. In *ACM SIGIR Forum*; ACM: New York, NY, USA, 2017; Volume 51, pp. 73–74.
54. Kulkarni, A.D.; Lowe, B. Random Forest for Land Cover Classification. *Int. J. Recent Innov. Trends Comput. Commun.* **2016**, *4*, 58–63.
55. Fabris, F.; Doherty, A.; Palmer, D.; de Magalhães, J.P.; Freitas, A.A. A new approach for interpreting Random Forest models and its application to the biology of ageing. *Bioinformatics* **2018**, *34*, 2449–2456. [CrossRef]
56. Goel, E.; Abhilasha, E. Random Forest: A Review. *Int. J. Adv. Res. Comput. Sci. Softw. Eng.* **2017**, *7*, 251–257. [CrossRef]

57. Aydın, C. Classification of the Fire Station Requirement with Using Machine Learning Algorithms. *I.J. Inf. Technol. Comput. Sci.* **2019**, *11*, 24–30. [CrossRef]
58. Klusowski, J.M. *Complete Analysis of a Random Forest Model*; Rutgers University: New Brunswick, NJ, USA, 2018.
59. Tyralis, H.; Papacharalampous, G. Variable selection in time series forecasting using random forests. *Algorithms* **2017**, *10*, 114. [CrossRef]
60. Ahmadlou, M.; Delavar, M.R.; Shafizadeh-Moghadam, H.; Tayyebi, A. Modeling urban dynamics using random forest: Implementing Roc and Toc for model evaluation. *Int. Arch. Photogramm. Remote Sens. Spat. Inf. Sci. ISPRS Arch.* **2016**, *41*, 285–290. [CrossRef]
61. Genuer, R.; Poggi, J.M.; Tuleau-Malot, C.; Villa-Vialaneix, N. Random Forests for Big Data. *Big Data Res.* **2015**, *9*, 28–46. [CrossRef]
62. Kudakwashe, M.; Mohammed Yesuf, K. Application of Binary Logistic Regression in Assessing Risk Factors Affecting the Prevalence of Toxoplasmosis. *Am. J. Appl. Math. Stat.* **2014**, *2*, 357–363. [CrossRef]
63. Sperandei, S. Understanding logistic regression analysis. *Biochem. Med.* **2014**, *24*, 12–18. [CrossRef] [PubMed]
64. Park, H.A. An introduction to logistic regression: From basic concepts to interpretation with particular attention to nursing domain. *J. Korean Acad. Nurs.* **2013**, *43*, 154–164. [CrossRef] [PubMed]
65. Shu, D.; He, W. A New Method for Logistic Model Assessment. *Int. J. Stat. Probab.* **2017**, *6*, 120. [CrossRef]
66. Ameri, S.; Fard, M.J.; Chinnam, R.B.; Reddy, C.K. Survival Analysis based Framework for Early Prediction of Student Dropouts. *ACM* **2016**, 903–912. [CrossRef]
67. Lakkaraju, H.; Aguiar, E.; Shan, C.; Miller, D.; Bhanpuri, N.; Ghani, R.; Addison, K.L. A Machine Learning Framework to Identify Students at Risk of Adverse Academic Outcomes. In Proceedings of the 21th ACM SIGKDD International Conference on Knowledge Discovery and Data Mining, Sydney, NSW, Australia, 10-13 August 2015; pp. 1909–1918. [CrossRef]
68. Susheel Kumar, S.M.; Laxkar, D.; Adhikari, S.; Vijayarajan, V. Assessment of various supervised learning algorithms using different performance metrics. *IOP Conf. Ser. Mater. Sci. Eng.* **2017**, *263*, 042087. [CrossRef]
69. Maggo, S.; Gupta, C. A Machine Learning based Efficient Software Reusability Prediction Model for Java Based Object Oriented Software. *I.J. Inf. Technol. Comput. Sci.* **2014**, 1–13. [CrossRef]
70. Liang, J.; Li, C.; Zheng, L. Machine learning application in MOOCs: Dropout prediction. In Proceedings of the ICCSE 2016 11th International Conference on Computer Science and Education, Nagoya, Japan, 23–25 August 2016; pp. 52–57. [CrossRef]
71. Longadge, R.; Dongre, S.S.; Malik, L. Class imbalance problem in data mining: Review. *Int. J. Comput. Sci. Netw.* **2013**, *2*, 83–87. [CrossRef]
72. Yilmaz, D.; Boz, H.; Yücel, M.; Günay, E. Prediction of student dropout from a university in Turkey using data balancing techniques. *Comput. Educ.* **2020**, *108*, 11–29. [CrossRef]
73. Mesut, G.; Demir, I.; Batur, K.; Sahin, F. Applying data balancing techniques to predict student dropout using machine learning. *Int. J. Adv. Comput. Technol.* **2017**, *5*, 1–8.
74. Antar, K.; Al-Dmour, R.; Zbaidieh, M.; Al-Kabi, M. Prediction of Student Dropouts Using Machine Learning Techniques. *Int. J. Comput. Appl.* **2020**, *5*, 1–8. [CrossRef]
75. Jain, A.; Singh, U.; Kumar, S. Application of data balancing techniques to predict student dropout using machine learning. *Int. J. Comput. Appl.* **2018**, *11*, 430–439.
76. Barros, T.M.; Neto, P.A.; Silva, I.; Guedes, L.A. Predictive models for imbalanced data: A school dropout perspective. *Educ. Sci.* **2019**, *9*, 275. [CrossRef]
77. Kotsiantis, S.B. Supervised Machine Learning: A Review of Classification Techniques. *Informatica* **2007**, *31*, 249–268.
78. Batista, G.E.A.P.A.; Prati, R.C.; Monard, M.C. A Study of the Behavior of Several Methods for Balancing Machine Learning Training Data. *SIGKDD Explor. Newsl.* **2004**, *6*, 20–29. [CrossRef]
79. Farquad, M.A.; Bose, I. Preprocessing Unbalanced Data Using Support Vector Machine. *Decis. Support Syst.* **2012**, *53*, 226–233. [CrossRef]
80. Ramentol, E.; Caballero, Y.; Bello, R.; Herrera, F. SMOTE-RSB *: A Hybrid Preprocessing Approach Based on Oversampling and Undersampling for High Imbalanced Data-sets Using SMOTE and Rough Sets Theory. *Knowl. Inf. Syst.* **2012**, *33*, 245–265. [CrossRef]
81. Yen, S.J.; Lee, Y.S. Cluster-based Under-sampling Approaches for Imbalanced Data Distributions. *Expert Syst. Appl.* **2009**, *36*, 5718–5727. [CrossRef]
82. Wang, S.; Yao, X. Using Class Imbalance Learning for Software Defect Prediction. *IEEE Trans. Reliab.* **2013**, *62*, 434–443. [CrossRef]
83. Burez, J.; Van den Poel, D. Handling Class Imbalance in Customer Churn Prediction. *Expert Syst. Appl.* **2009**, *36*, 4626–4636. [CrossRef]
84. Prusa, J.; Khoshgoftaar, T.M.; DIttman, D.J.; Napolitano, A. Using Random Undersampling to Alleviate Class Imbalance on Tweet Sentiment Data. In Proceedings of the IEEE 16th International Conference on Information Reuse and Integration, IRI 2015, San Francisco, CA, USA, 13–15 August 2015; pp. 197–202. [CrossRef]
85. Aulck, L.; Aras, R.; Li, L.; Heureux, C.L.; Lu, P.; West, J. STEM-ming the Tide: Predicting STEM Attrition Using Student Transcript Data. *arXiv* **2017**, arXiv:1708.09344. [CrossRef]
86. Batuwita, R.; Palade, V. Adjusted Geometric-mean: A Novel Performance Measure for Imbalanced Bioinformatics Datasets Learning. *J. Bioinform. Comput. Biol.* **2012**, *10*, 1250003. [CrossRef]

87. Kim, M.J.; Kang, D.K.; Kim, H.B. Geometric mean based boosting algorithm with over-sampling to resolve data imbalance problem for bankruptcy prediction. *Expert Syst. Appl.* **2015**, *42*, 1074–1082. [CrossRef]
88. Mgala, M. Investigating Prediction Modelling of Academic Performance for Students in Rural Schools in Kenya. Ph.D. Thesis, University of Cape Town, Cape Town, South Africa, 2016.
89. Kuncheva, L.I.; Arnaiz-González, Á.; Díez-Pastor, J.F.; Gunn, I.A. Instance Selection Improves Geometric Mean Accuracy: A Study on Imbalanced Data Classification. *Prog. Artif. Intell.* **2019**, *8*, 215–228. [CrossRef]
90. Hakim, A. Performance Evaluation of Machine Learning Techniques for Early Prediction of Brain Strokes. Ph.D. Thesis, United International University, Dhaka, Bangladesh, 2019.
91. Amin, M.Z.; Ali, A. Performance Evaluation of Supervised Machine Learning Classifiers for Predicting Healthcare Operational Decisions. *Tech. Rep.* **2017**. [CrossRef]

data

MDPI

Article

Multi-Level Analysis of Learning Management Systems' User Acceptance Exemplified in Two System Case Studies †

Parisa Shayan [1,*], Roberto Rondinelli [2], Menno van Zaanen [3] and Martin Atzmueller [4,5]

1. School of Humanities and Digital Sciences, Cognitive Science and Artificial Intelligence, Tilburg University, 5037 AB Tilburg, The Netherlands
2. Department of Economics and Statistics, University of Naples Federico II, Via Cintia 26, 80126 Naples, Italy
3. South African Centre for Digital Language Resources, North-West University, Potchefstroom 2520, South Africa
4. Semantic Information Systems Group, Osnabrück University, 49090 Osnabrück, Germany
5. German Research Center for Artificial Intelligence (DKFI), 49090 Osnabrück, Germany

* Correspondence: p.shayan@tilburguniversity.edu

† This paper is an extended version of our paper published in ABIS'19: Proceedings of the 23rd International Workshop on Personalization and Recommendation on the Web and Beyond; ACM: Boston, MA, USA, 2019; pp. 7–13.

Abstract: There has recently been an increasing interest in Learning Management Systems (LMSs). It is currently unclear, however, exactly how these systems are perceived by their users. This article analyzes data on user acceptance for two LMSs (Blackboard and Canvas). The respective data are collected using a questionnaire modeled after the Technology Acceptance Model (TAM); it relates several variables that influence system acceptability, allowing for a detailed analysis of the system acceptance. We present analyses at two levels of the questionnaire data: questions and constructs (taken from TAM) as well as on different analysis levels using targeted methods. First, we investigate the differences between the above LMSs using statistical tests (*t*-test). Second, we provide results at the question level using descriptive indices, such as the mean and the Gini heterogeneity index, and apply methods for ordinal data using the Cumulative Link Mixed Model (CLMM). Next, we apply the same approach at the TAM construct level plus descriptive network analysis (degree centrality and bipartite motifs) to explore the variability of users' answers and the degree of users' satisfaction considering the extracted patterns. In the context of TAM, the statistical model is able to analyze LMS acceptance on the question level. As we are also very much interested in identifying LMS acceptance at the construct level, in this article, we provide both statistical analysis as well as network analysis to explore the connection between questionnaire data and relational data. A network analysis approach is particularly useful when analyzing LMS acceptance on the construct level, as this can take the structure of the users' answers across questions per construct into account. Taken together, these results suggest a higher rate of user acceptance among Canvas users compared to Blackboard both for the question and construct level. Likewise, the descriptive network modeling for Canvas indicates a slightly higher concordance between Canvas users than Blackboard at the construct level.

Keywords: Learning Management System; Technology Acceptance Model; Cumulative Link Mixed Model; descriptive network analysis

Citation: Shayan, P.; Rondinelli, R.; van Zaanen, M.; Atzmueller, M. Multi-Level Analysis of Learning Management Systems' User Acceptance Exemplified in Two System Case Studies. *Data* **2023**, *8*, 45. https://doi.org/10.3390/data8030045

Academic Editors: Antonio Sarasa Cabezuelo and Ramón González del Campo Rodríguez Barbero

Received: 5 December 2022
Revised: 15 February 2023
Accepted: 16 February 2023
Published: 22 February 2023

1. Introduction

Within the context of higher education, Learning Management Systems (LMSs) are often used to support learning processes. LMSs are software frameworks that provide functionality that helps to share information from the instructor to users, e.g., via catalogs, instructional content, such as learning objectives, assignments, lecture slides, or other course content. Additionally, the system can collect information on the behavior of users and their interaction with the system. This includes, for example, managing registered users' logins, observing interactions with the provided course material, and in general monitoring users' activities with the system [1].

The effectiveness of the use of LMSs depends heavily on whether users (and instructors) are willing to use the system. However, not everybody may accept the use of LMSs on the same level. In this article, a significantly adapted and extended revision of [2], we specifically aim to investigate how users (in this case, students) perceive and accept the use of two LMSs: Blackboard and Canvas. These LMSs are used at Tilburg University (Tilburg, The Netherlands), where recently, the transition from Blackboard to Canvas took place. With respect to [2], where only Blackboard was analyzed, the research in this article provides a more extensive comparison and an in-depth understanding of the user acceptance of the two systems: Blackboard versus Canvas.

We apply the Technology Acceptance Model (TAM), which was introduced by [3]. TAM is an information systems theory that is commonly used to model users' acceptance of (novel) technologies. The model is adapted from the Theory of Reasoned Action (TRA) [4], which is specifically designed to model user acceptance of information systems [5]. TAM (see Figure 1) consists of five constructs (in addition to variables external to the model), which contain aspects that influence the actual use of the technology under consideration:

- External Variables (EV) represent contextual information from users and the environment.
- Perceived Usefulness (PU) is described as the extent to which users confirm that using the system improves their job performance.
- Perceived Ease of Use (PEU) is the extent to which users confirm that using the system would be free of corporeal and cerebral exertion.
- Attitude Towards Using the Technology (ATUT) relates to the users' perceptions of using the system, i.e., what is their attitude toward actually using the system (in all means).
- Behavioral Intention to use the Technology (BIT) denotes the users' intention of using the system.
- Actual Technology Use (ATU) assesses the system's performance and the extent to which it can meet the users' requirements [6].

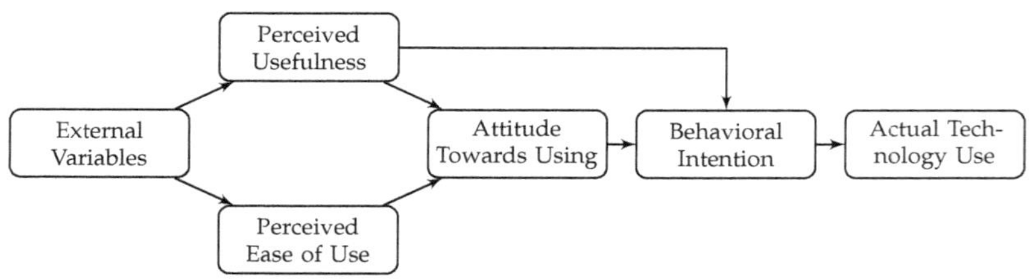

Figure 1. Technology Acceptance Model (adopted from [3]).

As shown in Figure 2, in this article, we extend the results presented in [2] in different ways. The first difference is related to the way data are considered. In [2], the answers to the questions could be selected from a 5-point Likert scale (ranging from strongly disagree (1) to strongly agree (5)). Therefore, [2] contemplated three different views of the data set by analyzing users' acceptance of Blackboard LMS specifically focusing on questions answered with scores "less than 3", "equal to 3", and "more than 3", whereas here, the data set is considered analyzed in its entirety.

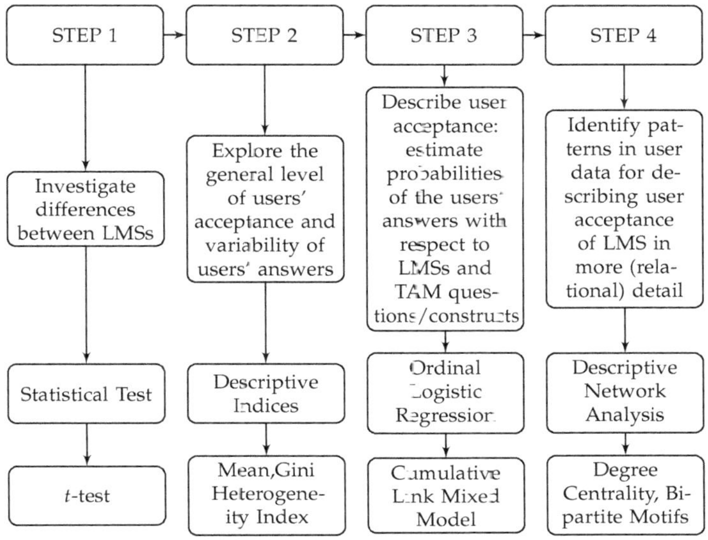

Figure 2. Summarizing the multi-level approach with the respective analysis methods.

We structure the analysis of the acceptance of the above LMSs according to the Technology Acceptance Model (TAM) [3], organizing this in two levels: questions and constructs. For both levels, first, we make use of descriptive statistics to explore the users' acceptance, showing general trends as well as variation between the answers provided by the users. Second, we employ a Cumulative Link Mixed Model, which describes user acceptance by estimating the probability of the users' answers considering the different LMSs and the questions/constructs. Third, we apply the descriptive network modeling to the obtained answers on TAM constructs (which is a similar technique as [2]), but for the two LMSs, i.e., Blackboard and Canvas) used on questionnaire data in which questions are organized in constructs. It represents the information provided by users for each construct of TAM as a network relating users based on their answers. Essentially, the network analysis approach identifies interesting patterns in the participant data. These patterns describe user acceptance of LMS in a more fine-grained manner (compared to the statistics approach). Overall, this enables different analysis levels using the respective targeted methods. Therefore, the main purpose is to demonstrate the data science techniques for data collection, processing, evaluation, and analysis, which are used in the context of the two LMSs.

Summing up, in this article, we target the following research questions, which extend those investigated in [2] (e.g., statistical and network analysis) to additional LMSs plus descriptive indices, e.g., the mean and the Gini heterogeneity index as well as the Cumulative Link Mixed Model, which enables a more comprehensive discussion:

1. What is the users' level of LMS acceptance for the following two LMSs: Blackboard versus Canvas?
2. What is the level of concordance for the users' acceptance?
3. How can we provide patterns for the users' LMS acceptance?

The rest of the article is structured as follows. In Section 2, we start with a brief introduction of the background related to LMSs and TAM. Next, Section 3 provides a description of the information coming from the questionnaire (for the two LMSs) and the methods that we employed for our investigations, e.g., the descriptive statistical indices and the statistical model. Additionally, in Section 4, we show the results of our analyses by comparing the data of the two LMSs (Blackboard versus Canvas). We provide results on both question and construct levels as follows:

1. We examine the results of statistical tests to show if there are significant differences between the two LMSs.
2. In addition, we make use of descriptive statistical indices, e.g., the mean and the Gini heterogeneity index, where the latter helps us to investigate the users' acceptance and the fluctuations amongst the answers provided by the users on a question-by-question basis.
3. Furthermore, we use the Cumulative Link Mixed Model (which corresponds well to the ordinal data collected from the questionnaire on a 5-point Likert scale ranging from 1 to 5) to see the differences in the probability of answering questions/constructs while comparing the two LMSs.
4. Finally, we apply descriptive network analysis approaches to provide patterns of users' LMS acceptance as well as the concordance in answering compared to [2].

This way, the Cumulative Link Mixed Model as well as network analysis can provide interesting insights for our data by measuring the effect of the questions and constructs plus their interactions with LMSs on the users' answers. This can be a complement to the general overview of the two LMSs, which is provided using descriptive statistical analyses. With respect to the TAM, the above approaches can be extended using descriptive network modeling, which results in patterns of users' LMS acceptance. Ultimately, we conclude with a summary and discussion of interesting future research directions.

2. Background

In this section, we first provide a brief overview of LMSs in general as well as some background information on the specific LMSs used in this study: Blackboard and Canvas. After that, we provide a short overview of the Technology Acceptance Model (TAM), which forms the basis for the design of the questionnaire, which is used to measure LMS acceptance.

2.1. LMS (Blackboard and Canvas)

An LMS is an electronic framework that allows for the creation, storage, reuse, management, and delivery of learning content. Most current LMSs are online, web-based systems that provide different interfaces for different functionalities or for different stakeholders. From the user perspective, an LMS provides learning content, such as lecture slides, instructional videos, and assessments, including exams or assignments (for online/offline use, and for local/distant learning). An LMS may also provide interaction with the instructor, e.g., by facilitating the submission of worked-out assignments, or through the use of forums, but also potentially with other users when working in groups [5]. From the instructor's perspective, an LMS allows for the easy distribution of learning material but also deals with user registration, user progress, and user results [7]. In general, it collects data to manage the learning and teaching process [1]. These data can be made available through reports that help instructors manage users better. As an example, they can organize users into groups to centralize reports and assignments. Using more advanced reports, it is also possible to follow the progress of large groups of users [8].

Various LMSs exist, with overall similar functionality, but also specific variations in user interfaces or in the level of functionality. In particular, Blackboard and Canvas are web-based LMSs that support both on-campus and online courses to plan, perform, and appraise learning processes. Blackboard is a popular LMS in the US, which is mostly aimed at colleges and universities (although other school types also use it) [9]. Canvas is another LMS specifically created for the academic environment and educational institutions [10].

The main differences between these LMSs can be found in two areas:

1. Implementation and integration: Blackboard is originally designed for universities that want to host their own data. However, for universities that do not have enough resources or do not want to host the data, both Blackboard and Canvas allow for cloud deployment. Moreover, Canvas has a wide variety of tools to choose from, whereas

the Blackboard LMS only integrates with Dropbox, PowerSchool, and OneDrive (although these functionalities may change over time).

2. Features: The two LMSs all have basic functionality: namely, Blackboard and Canvas both have a number of features in common, such as multi-user support, configurable learning portals, user-friendly design, and powerful user management capabilities. However, there are differences in additional features. For example, Blackboard users have to purchase modules that allow for specialized collaboration, such as the web conferencing function, whilst one primary feature of Canvas is the use of video as a source of content and collaboration.

2.2. Technology Acceptance Model (TAM)

As mentioned above, the TAM model describes that once users are provided with a new technology, several factors can influence their decision to use this technology. This process takes place within a certain environment, which is described using EV. These can be social factors (e.g., facilitating conditions, skills, and language), cultural factors (such as the perceived effect of using the technology within a social group), or political factors (e.g., the influence of technology on a political crisis). Within this environment, the directly important factors are PU and PEU. As improving the PU and PEU will also lead to an improvement of the ATUT and the BIT, developers need to realize the importance of the perceived system's usefulness and its ease of use [11]. The ATUT relates to the user's perception of the desirability of using the system, whereas BIT is the likelihood of the user actually using the system [12]. The ATU is now directly affected by the user's BIT.

Although numerous models have been proposed to describe the acceptance of systems, TAM describes this from a situational perspective and as such fits well to describe LMSs and e-learning [13–15]. Most other models aim to provide a detailed account, but they typically specialize due to the added complexity. For example, TAM2 [6] and the unified theory of acceptance and use of technology (UTAUT) [16], which are direct extensions of the standard TAM, focus on specifying new variables describing the EV in more detail. Alternatively, the valence model has a major focus on organizational aspects [17].

TAM has also been extended to include other types of information. For example, these models include two types of perceived usefulness (short-term and long-term) [18] or add a construct of compatibility [19]. The TAM3 version [20] includes constructs describing trust and risk. In addition, [21] examined individual acceptance and website usage and added two new structures to TAM: the value of perceived fun and the appeal of perceived presentation. Ref. [22] added playfulness constructs to analyze the World Wide Web acceptance. Several other publications show the usefulness of (extensions of) TAM in specific situations, such as the online shopping acceptance model (OSAM) to study online shopping behavior [23], whereas [24] used TAM to understand RFID acceptance and [25] investigated mobile service acceptance with perceived usefulness as the most important indicated factor.

Some studies relate the TAM to psychological models such as the Theory of Planned Behavior (TPB) and a decomposed TPB model. For instance, the study of [26] applies these models in Hong Kong's healthcare setting. The results highlight the superiority of TAM over TPB in explaining physicians' intention to use telemedicine technology.

More relevant to our study, some previous research investigated the impact of demographics (e.g., gender, age, field of study) on TAM constructs. The results show that there is not a substantial connection between perceived usefulness and users' demographics [27,28]. However, older users seem to better comprehend the usefulness of the system under consideration [29]. Additionally, the users' level of education seems to play a crucial role in the perceived usefulness [30,31]. Similar results have been obtained for other constructs, e.g., BIT, ATUT, and ATU [29,32]. Previous work emphasized statistical descriptions within the model [11,33,34]. Similarly, [35] confirmed the relationship amid PEU, PU, ATUT, and the overall impact on BIT. In addition, they showed that the external variables, e.g., job relevance, have a robust association with TAM constructs such that job relevance can have a positive effect on LMS usefulness (PU). Furthermore, [36] measured the usability of three open-source LMSs: Moodle, ILIAS, and Atutor. According to the results, Moodle, due to the attractive interface, was most easy to use compared to the other two LMSs. Considering this overview, we believe that TAM is a good fit and a proper framework to predict the behavioral intention to use a system.

Summarizing, TAM delivers a concrete (and simple) model to define users' acceptance of novel technologies and can be successfully applied in an LMS context. The model consists of five constructs and their interconnections. As such, they describe how we can expect users to accept LMSs, e.g., Blackboard and Canvas. To investigate acceptance, TAM provides us with specific areas that influence acceptance, which can be used to ask the users specifically about their perception of these areas. This means that the model can help us to structure a questionnaire for the investigation into the acceptance of the LMSs. This provides information on two levels: the individual questions as well as their organization in the TAM constructs. Exactly how this is accomplished is described in Section 3 and 4.

2.3. Statistical Analysis

To investigate the LMS users' acceptance as measured by the questionnaire structured according to TAM, we perform some statistical analyses on the data. This allows us to observe the acceptance per system and to compare the considered LMSs. For this, we use a statistical model based on the questionnaire values (per question and construct). In addition, we look at the Gini heterogeneity index [37], which has been applied to questionnaire data in order to evaluate whether answers are concentrated mostly in only one category (i.e., potential answer) or whether they are mostly equally distributed along all answers of a question. In other words, this evaluates the level of accordance between individuals.

Furthermore, a mixed model can tell us whether there is a difference in the probability of selecting a value (answer) for each variable (question) when comparing the datasets or not. This means that by allowing for an interaction between question and interface, this model makes a big assumption; i.e., the probability of each answer can be modeled as a linear combination of the likelihood of each answer under each interface as well as the likelihood of each answer for each question [38].

2.4. Network Analysis (Centrality and Motifs)

Network analysis methods in general are almost exclusively used to analyze relational data. In relational data, the links (relationships) between actors (users, objects, companies, etc.) matter to explain some phenomena in the data. This type of relational data can be modeled using a network, which consists of a set of nodes (also called vertices), which represent the actors or objects, and a set of edges (or arcs or links), which describe the relationships. Networks, which can also be represented as graphs, can be *directed* if the edges only run in one direction (from one node to another) or *undirected* or *biunivocal* if the edges run in both directions between two nodes. Additionally, if the edges of a network merely depict the absence/presence of a relationship between the nodes, it is called *unweighted*, whereas when the strength of a link is provided, the network is called *weighted*. A specific type of network is the *2-mode* or *bipartite* network (in contrast to the usual one-mode network). These networks are made up of two distinct types of nodes, and the edges only exist between the different types (connecting one of each type). Networks may also be even more complicated. For example, multi-layer networks comprise multiple, dissimilar kinds of nodes and edges. This allows for many global systems (e.g., social networks) to be represented as networks [39].

The advantage of viewing data as a network is that a range of networks analysis methods may be used to extract additional information (compared to "regular" statistical analysis methods, i.e., descriptive statistics and inferential statistics) as these methods can focus on the inherent relational aspects of the data, which may provide additional information that is closely related to the research questions. Ref. [40] discussed the use of network approaches on questionnaire data; an additional discussion on the different network analyses can be found in [41], stressing that the network analysis approach focuses on properties of pairs of users (i.e., dyadic relationships), thus also providing information on a more relational level.

Furthermore, [42] proposed a network analysis model from a Likert-scale survey. They created a bipartite network from users' answers based on Likert-scale selections. To present the number of users selecting similar answers, they used the edge weights. In other words, using the edge weight in the network, the similarity of the Likert-scale selections could be presented. They were also able to find the advantages of this approach by comparing network analysis and principal component analysis (PCA) so they construct a meaningful network based on the similarities and differences of answering between users. According to [42], this proposed methodology can be generalized to any set of Likert-scale surveys for network-based modeling. Likewise, in [2], the users' answers were based on Likert-scale selection (from 1 to 5). We thus considered three different views of the data set, focusing on questions answered with scores "less than 3", "equal to 3", and "more than 3". We applied the frequency distributions through descriptive network analysis tools, namely degree centrality and bipartite motifs.

Centrality measurement provides information about the importance of a node in the graph. There are four main centrality criteria: degree centrality, closeness centrality, betweenness centrality, and eigenvector centrality. Degree centrality for a node is basically defined via its degree; thus, the greater the degree, the more central the node will be. In other words, a high degree of centrality implies that a node includes more connections than the average graph. There are two types of criteria for directed graphs: in-degree and out-degree. The former is considered to be the number of edges that point toward the node and the latter is considered to be the number of nodes directed away from the given node [40]. The closeness measures the average distance between one node and other nodes, so the more central a node is, the closer the node is to the other nodes. The betweenness is defined as the number of shortest paths in which a node is located which is commonly used to see the information flow in the graph. The higher the betweenness, the more information flows within the graph. The eigenvector is about a node's relative impact within the network or how connected a node is to other highly connected nodes [39].

Whereas the basic network analysis metrics, e.g., degree centrality provides information of the overall structure of the network, it is possible to gain further insight into the structure of the network using more advanced, structural analyses by applying a motif extraction approach, c.f., [43,44].

Motifs are particular subgraphs of bipartite networks considered as the basic "building blocks" of networks that include both types of nodes [45]. As shown in Figure 3, you may observe two nodes in the top set (A) and three nodes in the lower set (B) in motifs 14, 15, and 16. The product of binomial coefficients, selecting two nodes from A and three nodes from B, thereby gives the maximum number of node combinations that could exist in these patterns: $\binom{A}{2}\binom{B}{3}$ [46]. With respect to our data set, this indicates that motif configurations include one or two users and many questions (3, 4, or 5) or many users (3, 4, or 5) and one or two questions. They represent patterns of questions receiving the same answer from a user or patterns of users responding to a question in the same way.

In general, the sizes of motifs can be varied from two to six nodes (larger is possible, but depending on the size of the bipartite network, these may hardly ever occur) and include all the isomorphism classes. Bipartite motifs can be used in different ways, for example, to compute the number of repetitions of different motifs in a network [47]. Likewise, they can be helpful while quantifying the role of nodes in a group by counting the number of nodes that appear in various positions of motifs [48]. The benefit of motifs is that with respect to traditional indices, they are much more sensitive to changes within the network. This means that while many network configurations have similar index values, a small number of network configurations have the same motif structure [43]. Furthermore, bipartite motifs are well suited to represent the relationships (answers) between one user and a group of questions, a group of users and one question, or a group of users and a group of questions.

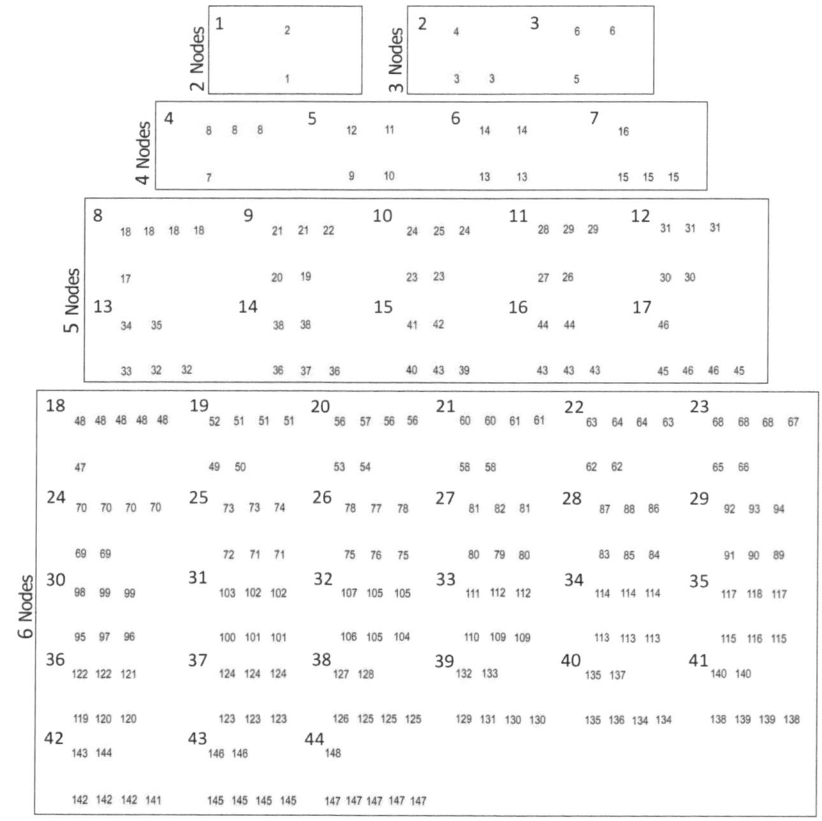

Figure 3. All possible bipartite motifs from two to six nodes.

3. Methodology

To collect information on the acceptance of LMSs by users, we used a questionnaire [49]. The types of questions and their answers allowed us to perform quantitative analyses.

As explained in Section 1, at the question and the construct level, we investigate statistical tests. This helps to see whether there are significant differences between the LMSs. Furthermore, we make use of descriptive statistical analysis (e.g., mean and Gini heterogeneity index) as well as statistical modeling (Cumulative Link Mixed Model), where the former shows the level of user acceptance and their concordance, while the latter helps us to estimate the answers given by the users considering the distinction between LMSs, questions/constructs, and their interaction. Finally, we apply descriptive network modeling (e.g., degree centrality and bipartite motifs) at the construct level to examine the variability of users' answers and the patterns of user satisfaction in different networks accordingly.

Below, we provide an overview of the approach. First, we describe the design of the questionnaire and, second, we provide an overview of the methods used for the descriptive as well as the statistical and network-based modeling analyses.

3.1. Material

For the current study, we collected data on the acceptance of two LMSs at one university using the same questionnaire. We will first provide information on the questionnaire that was used followed by a short description of the data set for each LMS. Based on the answers to the questions, we provide a short analysis of the reliability of the questionnaire.

The questionnaire consists of two parts. The first part is a small set of demographic questions. The second part consists of 30 questions taken from [49], which together measure the five TAM constructs. Perceived usefulness (PU) is measured in questions 1–6, perceived ease of use (PEU) is measured in questions 7–11, behavioral intention (BIT) is measured in questions 12–15, attitude toward using (ATUT) is measured in questions 16–23, and actual technology use (ATU) is measured in questions 24–30. Table 1 provides an overview of the questions. Note that the answers to the questions in the second part can be selected from a 5-point Likert scale (ranging from strongly disagree (1) to strongly agree (5)).

Table 1. List of questions in the questionnaire (where "LMS" is replaced by the name of the LMS under consideration).

Q	Description
PU	
1	LMS helps me to increase my learning productivity
2	LMS helps me to find the course materials
3	LMS helps me to submit the assignments
4	LMS increases my academic performance
5	LMS helps me in the learning process
6	LMS helps me to ask and discuss some topics with the lecturer
PEU	
7	LMS is easy to operate
8	LMS uses understandable language
9	LMS uses the appropriate background color and font
10	LMS has a systematic menu
11	LMS is accessible from within and outside of the university
BIT	
12	I have the intention to use LMS every day
13	I have the intention to check the latest materials on LMS
14	I have the intention to check my grade through LMS
15	I have the intention to encourage my fellow users to use LMS
ATUT	
16	I use LMS without any compulsion from anyone
17	I need LMS
18	I am happy when I use LMS
19	Using LMS to submit the assignment is an innovative idea
20	Using LMS to download the course materials is an innovative idea
21	Using LMS to discuss with lecturer/fellow users is a positive idea
22	Using LMS is a good and wise decision
23	I am going to encourage my fellow users to use LMS
ATU	
24	I use LMS to support the learning activities
25	I always access LMS every day
26	I get the course materials from LMS
27	I download and upload assignments through LMS
28	I use LMS to check my grades
29	I am satisfied using LMS
30	I tell my fellow users about my satisfaction using LMS

The data of the first LMS, Blackboard, describes answers to the questions in the questionnaire from 51 pre-master LMS users (out of a total of 118 people registered as pre-master students in the full academic year; note that the pre-master program is only one semester, but student registration is captured per academic year) from the School of

Humanities and Digital Sciences School at Tilburg University (Tilburg, The Netherlands). These were collected during the spring (i.e., last) semester in the academic year 2018–2019. Of the 51 users, 25 (49.0%) were female and 26 (51.0%) were male. The age of the users was distributed as follows: 44 (86.3%) were in the age range between 20 and 30, six (11.8%) were in the age range between 31 and 40, and one (2.0%) user was over 40 years old. This part of the data set has also been used in a previous study [2].

For the second LMS, Canvas, answers from 49 pre-master users (out of a total of 95 people registered as pre-master students that academic year) from the School of Humanities and Digital Sciences at Tilburg University were collected during the fall (i.e., first) semester in the academic year 2019–2020. Out of 49 users, 27 (55.1%) were female and 22 (44.9%) were male. Most users (46, 93.9%) were in the age range between 20 and 30, two (4.1%) were in the age range between 31 and 40, and one (2.0%) was over 40 years old.

Note that both Blackboard and Canvas users have relatively similar experiences of LMS (e.g., submit the assignments, check the course materials/grades, and discuss with lecturer/fellow users through the LMS).

To illustrate the reliability of the questionnaire (both for the overall questionnaire as well as for the individual TAM constructs separately), we compute the Average Variance Extracted (AVE), Composite Reliability (CR), R-squared (R^2), and the respective Cronbach αs [50]. AVE is used to measure the variance degree for a construct; the values of greater than 0.5 indicate that the reliability of the result is more acceptable. CR is an indicator to measure the internal integrity in which the values should be higher than 0.6. R^2 measures the proportion of the variance for each construct such that the values greater than zero are acceptable. Table 2 shows the different values for the two LMSs. Previous research has already shown the reliability of the questionnaire with Cronbach α-values above 0.7 [49], and here, we observe similar results with Cronbach $\alpha > 0.7$ for all constructs in LMSs. The overall Cronbach αs for the LMSs are larger than 0.9, which means that the questionnaire's reliability is excellent. Additionally, the reliability of the questionnaire for each construct is considered acceptable.

Table 2. Average Variance Extracted (AVE), Composite Reliability (CR), R^2, and Cronbach α of the results from the questionnaire for each TAM construct and total (combined) LMS (for Blackboard and Canvas).

Blackboard	AVE	CR	R^2	α
PU	0.401	0.763	0.145	0.822
PEU	0.477	0.972	0.106	0.706
BIT	0.552	0.831	0.142	0.701
ATUT	0.859	0.687	0.265	0.855
ATU	0.352	0.754	0.149	0.836
Total	0.128	0.737	0.559	0.942
Canvas	**AVE**	**CR**	**R^2**	**α**
PU	0.414	0.776	0.167	0.813
PEU	0.721	0.927	0.156	0.891
BIT	0.494	0.791	0.236	0.744
ATUT	0.541	0.903	0.465	0.876
ATU	0.578	0.904	0.238	0.866
Total	0.161	0.789	0.417	0.944

3.2. Statistical Analyses

To investigate the LMS users' acceptance as measured by the questionnaire structured according to TAM, we perform statistical analysis on the data, using descriptives and modeling methods. This allows us to observe the acceptance per system and to compare the two LMSs considered in this study at the question and construct level.

For the descriptive statistical tools, we use the mean, standard deviation, and the Gini heterogeneity index [37] for both the question and construct level. The Gini heterogeneity index indicates how far answers in Likert-scale questionnaire data are concentrated mostly in only one specific answer (i.e., value) or whether they are more equally distributed over all answers to a question. In other words, this evaluates the variability of each question, namely the level of accordance among individuals. While for the question level, the Gini heterogeneity index is a natural index, due to the reliability observed in Cronbach's α, we can also apply it at the construct level. Each construct is then considered to be a unique block (same distributions of questions) and can be vectorized.

The Gini heterogeneity index was proposed by Corrado Gini [37,51] as one instantiation of statistical inequality measures. Here, we investigate a specific ordinal variable with its associated set of categories with the Gini heterogeneity index G defined as:

$$G = \frac{m}{m-1}(1 - \sum_{j=1}^{m} {p_j}^2), \tag{1}$$

where m is the number of categories described by the ordinal variable (in our case $m = 5$ as we deal with 5-point Likert scales), and p_j is the relative frequency of each category, $j = 1, \ldots, m$.

If we observe only occurrences in one category (i.e., the relative frequency for one category $p_j = 1$), then the heterogeneity is minimal ($G = 0$) (and for the questionnaire data, the concordance of users' answers is maximal). If the relative frequencies are the same for all categories $p_j = \frac{1}{m}$ for $j = 1, \ldots, m$, then the heterogeneity is maximal ($G = 1$), but in the questionnaire case, the concordance in answers between users is at the minimum.

The Cumulative Link Mixed Model (CLMM) [38,52,53] is a statistical modeling approach that can tell us whether there is a question or construct effect on the probability of selecting a value (answer) by a user when comparing the LMSs. Similar to the mean and Gini heterogeneity index, we apply this model to both the question and construct levels.

We can organize the questionnaire answer data for different users in a table with five columns. Each row represents the answer to a particular question by a particular user. Additional information on the LMS and the corresponding construct is also added. Therefore, we can represent the LMS data in a matrix with users as rows and questions (per construct) as columns; however, to apply the CLMM, it is required to re-organize the data.

The main goal of the new organization is to put the two LMSs together, considering that in each row, one answer given by a user to a question from one LMS is repeated (i.e., one user will be repeated 30 times in 30 rows). The final matrix is composed of four variables: the first is the answer given by the user, the second indicates the LMS, the third is the corresponding question, and the fourth is the corresponding construct. We can thus model the probability of answers by including LMSs and the question/construct effect as well as the interaction between them (i.e., LMS and question/construct) by adding a grouping factor effect for the users. Allowing for interaction between the question and the LMS, this model makes an important assumption (i.e., the probability of each answer (1–5) is modeled as a linear combination of the likelihood of each answer under each LMS as well as the likelihood of each answer for each question/construct).

3.3. Network Analysis

Before we can apply network analysis measures to investigate relational information in the data, the data set will need to be converted into a network. Networks can be represented as graphs, which are typically used to visualize them, but they can also be represented using adjacency matrices (which are called affiliation matrices for two-mode networks) or using edge lists. Adjacency matrices consist of n rows and n columns that are labeled with identifiers for the n nodes, and each entry (i, j) in the $n \times n$ matrix (each cell) represents the value of the link between the two nodes i and j. Unweighted graphs may have true/false or 1/0 values in their matrices, whereas weighted graphs may contain real numbers indicating the weights of edges in the respective entries of the adjacency matrix. In the case of questionnaire data, there are three types of information: users, questions, and answers given by users to the questions. From this, we can construct a weighted bipartite network for each TAM construct with the set of users U and the set of questions Q as nodes. The edges between elements from the sets U and Q indicate that a user (from U) provides an answer to a question (from Q) and the answer information can then be encoded as a weight on each edge. The same as previous study [2], for each TAM construct, we create three unweighted bipartite graphs with the same nodes as the original graph (representing users and questions), but only those edges that adhere to a particular criterion: edges that have the answer (weights). 1. less than three, 2. equal to three, 3. more than three (see Figure 4).

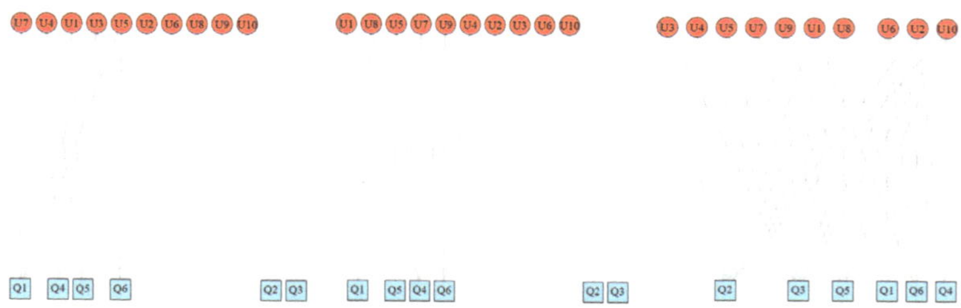

Figure 4. From the left to the right, an example of "less than 3", "equal to 3", and "more than 3" networks (User/Question) within the PU construct of the Blackboard questionnaire.

3.3.1. Degree Centrality

For network characterization and the identification of interesting properties, we can apply descriptive network analysis methods or apply more complex models. These analyses provide information on the overall shape or other properties of the network. Here, we are interested in studying and comparing the users' LMSs acceptance at the construct level. This is well measured by the variability of the users' degree centrality distribution in their unipartite weighted networks, which are obtained as projections from the constructed bipartite networks.

Considering an adjacency matrix A describing the LMS acceptance results for one construct, with the main diagonal equal to zero (so that there are no self-loops for the set of nodes), the formulation of the weighted degree is defined as

$$d_i = \sum_{j=1}^{n} a_{ij}, \qquad (2)$$

where n is the number of (participant) nodes and a_{ij} represents the entries of the adjacency matrix A. The degree centrality d_i can be seen as the level of concordance of each participant with respect to the other participants. The average value of the degree distribution is the average level of concordance between participants, so it describes a similar measure to the Gini heterogeneity index but is now on the construct level. In addition, the variance of the composite degree centrality indicates the consistency of users' acceptance within the network. Focusing on it and allowing for comparison between constructs and LMSs, we make use of the normalized coefficient of variation.

3.3.2. Bipartite Motifs

As mentioned above, motifs are specific subgraphs of bipartite networks that include two sets of nodes. Given this, we represent our data set as a bipartite network with users as a set of nodes U and questions as a set of nodes Q to indicate the patterns between users and questions.

For instance (as displayed in Figure 3), you may observe two nodes in the top set (A) and three nodes in the lower set (B) in motifs 14, 15, and 16. The product of binomial coefficients, selecting two nodes from A and three nodes from B, thereby gives the maximum

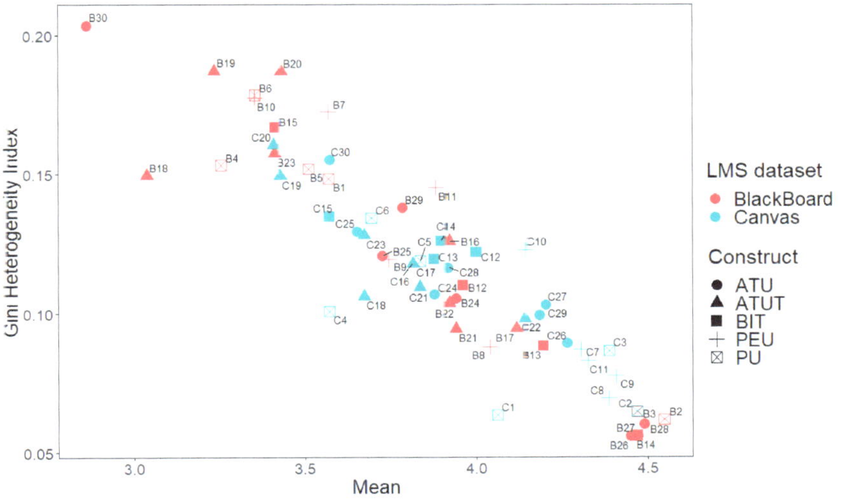

Figure 5. Descriptive summary from the answers to the TAM questions and the variability between the answers per question for Blackboard and Canvas.

Second, for both systems, there is a trend to have higher Gini values if the scores are lower. This can (partially) be expected. If we consider the extremes, e.g., mean scores close to either the maximum or minimum values, then this means that users generally need to select the extreme value (otherwise, the mean would be further away from the extreme value). Similarly, the standard deviation would be small. If we now consider mean scores closer to the middle of the possible scores, there may be more variation in the scores to end up with such a mean. Still, one may expect that if some Blackboard users, with mean scores around two, give a score of two to the questions, then the Gini index would be lower, indicating higher internal coherence. In this case, the mean scores for the Blackboard users are established due to some variation in the scores provided by the users, which is also reflected in the higher standard deviations in Table 3.

The figure also indicates a few outlier questions. Some Blackboard questions show lower scores. In particular, question 30 has a low score (and high Gini value). This is the last question in the questionnaire (see Table 1) and deals with whether a user would tell other users about their satisfaction. The corresponding values for Canvas are also on the lower side, but these are not as extreme. The other questions that could be seen as outliers (low values compared to Canvas) are mostly related to ATUT questions (18, 19, 20) and PU (4, 6). These will be discussed below (in Section 4.3).

4.2. Cumulative Link Mixed Model (Question Level)

To investigate the relationships between LMSs and the answers to the questions, we build a Cumulative Link Mixed Model. This model predicts the most likely answer (out of values 1–5) for each question while considering the LMS. The model relies on a linear combination of the weights of the LMS and the question. This model makes the assumption that the likelihood of an answer can be modeled as a linear combination of information from the LMS and the question (allowing for an interaction between the question and LMS).

The model fits the answers given by the participants based on the LMS and question variables where we also look at potential interactions. The participant is incorporated in the model as a random variable. The weights of the LMS and question variables can be found in Table 4 with Blackboard and Q1 as a reference. The interaction effects between LMS and questions can be found in Table 5 (due to space limitations).

Table 4. Weights and p-values for the two variables (LMS and questions) indicating the significant influences of the values for the variables in the Cumulative Link Mixed Model, where "***" indicates $p < 0.001$, "**" indicates $p < 0.01$, "*" indicates $p < 0.05$, and "." indicates $p < 0.1$.

Coefficients	Weight	p		Coefficients	Weight	p	
Canvas	−0.611	0.022	*	Q16	0.322	0.263	
Q2	2.363	<0.001	***	Q17	0.473	0.100	
Q3	2.105	<0.001	***	Q18	−1.274	<0.001	***
Q4	−1.187	<0.001	***	Q19	−1.303	<0.001	***
Q5	−0.473	0.093	.	Q20	−1.024	<0.001	***
Q6	−0.750	0.009	**	Q21	0.198	0.489	
Q7	0.538	0.065	.	Q22	0.598	0.034	*
Q8	1.282	<0.001	***	Q23	−0.810	0.004	**
Q9	0.875	0.002	**	Q24	0.286	0.312	
Q10	−0.018	0.950		Q25	−0.356	0.203	
Q11	1.086	<0.001	***	Q26	1.745	<0.001	***
Q12	0.486	0.092	.	Q27	1.735	<0.001	***
Q13	0.752	0.009	**	Q28	1.284	<0.001	***
Q14	1.242	<0.001	***	Q29	0.541	0.057	.
Q15	−0.926	<0.001	***	Q30	−1.554	<0.001	***

Table 5. Weights and p-values between two variables (LMS and questions) indicating the significant influences of the values for the variables in the Cumulative Link Mixed Model, where "***" indicates $p < 0.001$, "**" indicates $p < 0.01$, "*" indicates $p < 0.05$, and "." indicates $p < 0.1$.

Coefficients	Weight	p		Coefficients	Weight	p	
Canvas-Q2	0.803	0.007	**	Canvas-Q17	1.060	<0.001	***
Canvas-Q3	0.705	0.018	*	Canvas-Q18	−0.188	0.487	
Canvas-Q4	0.260	0.341		Canvas-Q19	0.359	0.198	
Canvas-Q5	0.214	0.447		Canvas-Q20	0.655	0.021	*
Canvas-Q6	0.291	0.311		Canvas-Q21	0.716	0.012	*
Canvas-Q7	−0.459	0.115		Canvas-Q22	0.212	0.453	
Canvas-Q8	0.053	0.855		Canvas-Q23	0.236	0.395	
Canvas-Q9	−0.479	0.097		Canvas-Q24	0.716	0.011	*
Canvas-Q10	−0.616	0.033	*	Canvas-Q25	0.744	0.007	**
Canvas-Q11	−0.023	0.938		Canvas-Q26	0.934	<0.001	***
Canvas-Q12	0.484	0.092	.	Canvas-Q27	1.023	<0.001	***
Canvas-Q13	1.001	<0.001	***	Canvas-Q28	1.568	<0.001	***
Canvas-Q14	1.526	<0.001	***	Canvas-Q29	−0.005	0.986	
Canvas-Q15	0.407	0.144		Canvas-Q30	−0.342	0.220	
Canvas-Q16	0.755	0.009	**				

According to the results, there are significant differences between the questions (in contrast to the t-tests per question). Looking at the model, we observe that Blackboard and Canvas users do not agree with finding course materials (Q2) or submitting assignments (Q3) through LMS. Therefore, they do not believe that the above LMSs can increase their academic performance (Q4). In addition, Blackboard and Canvas users do not agree with the understandability (Q8) and accessibility (Q11) of the LMSs. Additionally, Blackboard and Canvas users have no intention to check their grades through LMS (Q14) and to encourage their peers to use LMS (Q15). Furthermore, Blackboard and Canvas users are strongly uncertain about accepting LMS as a system that makes them happy (Q18). So, they cannot consider the LMSs as an innovative idea to download course materials (Q20/Q26), submit assignments (Q19/Q27), and check their grades (Q28). That is why they do not intend to tell their fellows about their satisfaction with using LMS (Q30).

4.3. Descriptive Analysis (Construct Level)

Table 6 and Figure 6 provide information similar to Table 3 and Figure 5. The mean, standard deviation, and Gini heterogeneity index values are provided per construct (instead of per question). Furthermore, the *t*-tests are applied to each construct to examine the statistical relationship between LMSs for each of the constructs. (Note that the values in Table 2 showed that the questions per construct provide consistent results, which allows us to combine these values and analyze constructs).

Table 6. Mean (*M*), standard deviation (*SD*), and Gini heterogeneity index (Gini) values for each TAM construct for Blackboard and Canvas LMSs. In addition, the *t*-values of the *t*-tests comparing the results per construct between the systems and the corresponding *p*-values are provided. At the bottom of the table, the mean and standard deviations over all questions are provided.

	Blackboard			Canvas				
	M	*SD*	Gini	*M*	*SD*	Gini	*t*	*p*
PU	3.784	(0.664)	0.095	4.003	(0.535)	0.075	0.157	0.876
PEU	3.718	(0.687)	0.103	4.314	(0.628)	0.080	0.150	0.881
BIT	4.009	(0.587)	0.081	3.837	(0.630)	0.092	0.140	0.889
ATUT	3.627	(0.666)	0.104	3.724	(0.636)	0.094	0.215	0.830
ATU	3.960	(0.565)	0.079	3.953	(0.538)	0.090	0.176	0.861

The *t*-test results indicate that there are no significant differences between the systems for each construct ($p > 0.5$).

Looking at Figure 6, we see a relatively similar picture to Figure 5. Again, some Blackboard constructs are in the top left corner, indicating that the mean values for the constructs are lower than that of Canvas, but also the Gini heterogeneity index values are higher, indicating a less consistent answer selection by the Blackboard users.

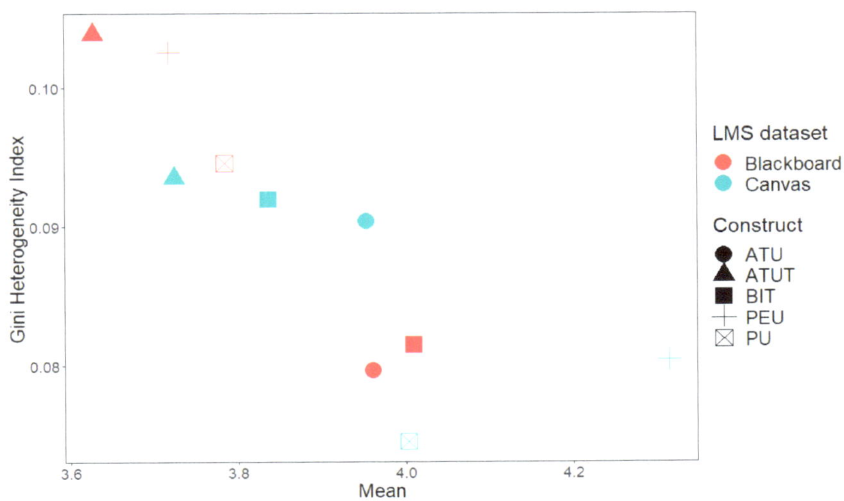

Figure 6. Descriptive summary from the answers to the TAM constructs and the variability between the answers per construct for Blackboard and Canvas.

Considering the distribution of the constructs, we see that for Blackboard and Canvas, the results for ATUT are the lowest. BIT is the highest value for Blackboard, but it is (after ATUT) the lowest for Canvas. This shows that the acceptance of the systems is different based on different properties.

4.4. Cumulative Link Mixed Model (Construct Level)

Similar to the analysis performed on a question basis, we here investigate the relationships between LMSs and the TAM constructs. Again, we build a Cumulative Link Mixed Model, which predicts the most likely answer. However, here, the answers are grouped per TAM construct, but we keep the LMS into account. Note that this can be accomplished as the questions related to the different TAM constructs show consistent behavior. This CLMM results in a linear model using the weights of the LMS and the TAM constructs.

Similar to the model that fits the answers based on the questions (and the LMS), here, we also consider possible interactions between the LMS and TAM constructs. Again, the participant variable is a random variable in the model. Table 7 provides all the weights and their p-values of the model. Blackboard and ATU are taken as the reference values.

Table 7. Weights and p-values for the two variables (LMS and TAM constructs) indicating the significant influences of the values for the variables in the Cumulative Link Mixed Model, where "***" indicates $p < 0.001$.

Coefficients	Weight	p	
Canvas	0.031	<0.001	***
PU	−0.205	<0.001	***
PEU	0.219	<0.001	***
BIT	−0.111	<0.001	***
ATUT	−0.741	<0.001	***
Canvas-PU	−0.275	<0.001	***
Canvas-PEU	−0.823	<0.001	***
Canvas-BIT	0.169	<0.001	***
Cavas-ATUT	−0.169	<0.001	***

According to the model, there are significant differences between all constructs (in contrast to the t-tests per construct). Considering the ATU construct as a reference and the CLMM results for the questions, you can see a significant difference in particular between the PU and BIT constructs. This means that the Blackboard and Canvas users do not agree with the LMSs' usefulness (PU). That is why they do not intend to use the systems or encourage their peers to use the LMSs (BIT).

4.5. Descriptive Network Analysis (Construct Level)

4.5.1. Degree Centrality

According to the normalized coefficient of variation of the degree distribution depicted in Figure 7, there is a clear trend of low variability for "more than 3" networks compared with "equal to 3" and "less than 3" networks for both Blackboard and Canvas users. In other words, they tend to have a similar perception about the acceptance of the LMSs mostly for the aspects they answered with a high score. This is also due to the high rate of 4 and 5 answers, which increases the likelihood that users give the same answers to the same questions.

What is interesting for Blackboard is the low variability of almost all networks except for BIT and PU with variability of over 20% for the "less than 3" networks. Canvas, on the other hand, has higher variability on average for each structure and network type. This is especially noticeable for the PEU construct of the "less than 3" and the "equal to 3" networks. This means that the perception of the acceptance is more variable with the use of Canvas than with Blackboard; the users' low answers make its definition even more evident.

Therefore, in general, the low variability of degree centrality distribution (concerning 4 and 5 answers, which are the more frequent answers) indicates a certain consistency in users' answers: namely, both Blackboard and Canvas users have the same understanding of different features of the LMSs.

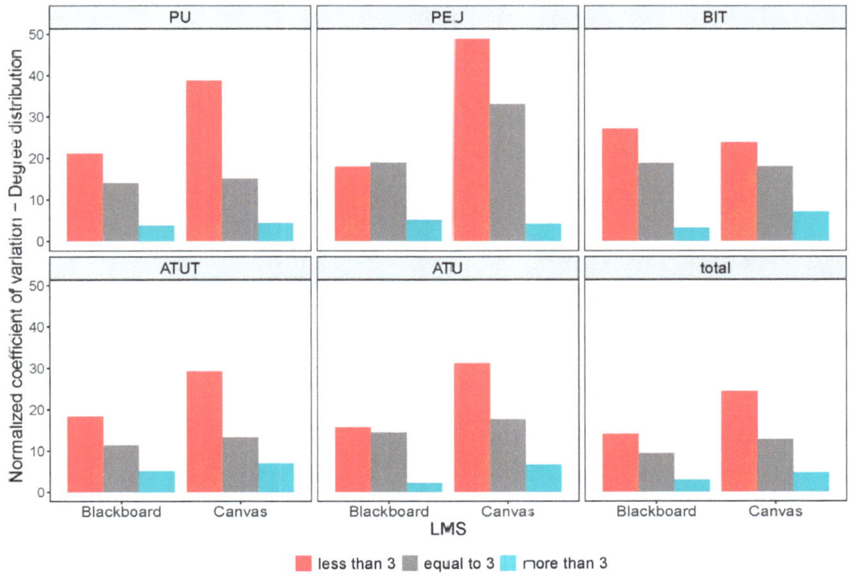

Figure 7. Normalized coefficient of variation of the degree centrality distribution for Blackboard and Canvas users.

4.5.2. Bipartite Motifs

As shown in Figures 8 and 9, the results prove the frequency of high answers for motifs for the two LMSs. As answers are related to the ratio of nodes, we can then compare answers on the basis of the network types. According to the details provided in the paragraphs covered in Section 3, the most intriguing extreme theme combinations are those with one or two users and a large number of questions (3, 4, or 5), or numerous users (3, 4, or 5) with a small number of questions (one or two). They outline patterns of questions that elicit the same answers from users or groups of users who answer similarly to the questions. This is consistent with the following motifs: 4, 7, 8, 12, 16, 17, 23, and 44 (see Figure 3).

Looking at the "less than 3" networks, we observe a global different behavior for both LMSs, specifically with Blackboard having higher relative frequencies of PU, PEU, and ATUT constructs than Canvas. This corresponds to the motifs 1, 3, 5, 7, 17, 38, 40, 42, and 44. This means that the above motifs are very important for PU, PEU, and ATUT compared to the other two TAM constructs. While the above TAM constructs (i.e., PU, PEU, and ATUT) for the first seven motifs with less than five nodes (Figure 8) have high frequencies, the graph shows lower frequencies for the remaining motifs (Figure 9). This means that low scores are fairly evenly distributed across both LMSs; however, the similarity between PU (Q1–6), PEU (Q7–11), and ATUT (Q16–23) is interesting with respect to the TAM (see Figure 1), and the ATUT construct derives directly from the PU and PEU constructs. With respect to the questions (Q1–6, Q7–11, and Q16–23), this means that both Blackboard and Canvas users do not concur that the above LMSs are helpful to their learning process. In addition, they do not consider the LMS as an innovative idea to submit their assignment or to download their course materials. In addition, the users do not accept the usability of the LMSs in terms of easy operation, understandability, and accessibility as well as the system interface. Therefore, they do not intend to use the system frequently or to encourage their peers to do so.

According to the "equal to 3" networks, for Blackboard, for the first seven motifs with less than five nodes (Figure 8), we observe the same behavior as "less than 3" networks with the high proportion of neutral answers for PU and ATUT than other constructs in Blackboard. While for Canvas, you can see the higher peaks at motifs 5, 13, 17, 38, and 40 for BIT, ATU, and ATUT which are the patterns of users that have given a similar answer to a question or the patterns of questions that receive the similar answers by a user. In the meantime, for the remaining motifs (Figure 9), the two LMSs demonstrate the lower frequencies with a steady decline for Blackboard. With respect to the TAM (see Figure 1) and ATUT being derived directly from the PU construct, the resemblance between PU (Q1–6) and ATUT (Q16–23) constructs for Blackboard can be intriguing. This indicates that the Blackboard users do not care that the Blackboard LMS is beneficial to their learning process. Furthermore, they do not consider the Blackboard LMS to be a novel way to submit assignments or download course materials. Therefore, they might or might not encourage fellows to use the system. In the meantime, the resemblance between BIT, ATUT, and ATU constructs can be thought-provoking, since BIT derives directly from ATUT and also from ATU from BIT constructs. In light of the questions (Q12–15, Q16–23, Q24–30), this suggests that the Canvas users are neutral about using the system frequently or encouraging their peers to do so because it makes no difference to them that the Canvas LMS is a novel idea of submitting assignments or downloading course materials. That is why they might or might not be satisfied with the system.

Finally, for the "more than 3" network, what is striking is a sharp increase for all constructs and high frequencies for almost all motifs for both LMSs. This depicts the patterns of questions that receive similar answers from a user or patterns of users that give similar answers to a question for the above motifs. In addition, the overlap between ATU (Q24–30) and BIT (Q12–15) in Blackboard as well as the high frequency of PU (1–6) and PEU (Q7–11) in Canvas is very interesting. Considering the TAM (see Figure 1), ATU derives directly from the BIT construct and PU moves exactly in line with PEU. According to the questions in Blackboard (Q24–30, Q12–15), the Blackboard users are happy with the LMS and have the intention to use it regularly. Regarding the questions (Q16–23, Q7–11) in Canvas, this suggests that the Canvas users think choosing the LMS is a smart choice in terms of usability and accessibility. They believe that Canvas LMS is useful for increasing learning productivity as well as submitting assignments or downloading course materials. In the meantime, PEU in Canvas has higher frequencies than other constructs. This indicates that Canvas users have fully accepted the usability of the LMS compared to Blackboard users.

As a result, what is interesting once looking at the above figures is the similar behavior of Blackboard and Canvas for the "more than 3" network. This can be immediately observed in the "more than 3" networks: the motif answers are always higher than the "less than 3" and "equal to 3" networks for the two LMSs. This is the pattern of users that give a similar answer (high scores) to a question or the patterns of questions that receive similar answers (high scores) by a user. Furthermore, it seems that the most important constructs are ATU and BIT for Blackboard plus PU and PEU for Canvas. As mentioned above, with respect to the TAM, this makes sense. Since PU moves along the PEU construct and ATU extracts from the BIT construct, this means that the system's usability and ease of use interact in some way (for Canvas users). Meanwhile, users' intention to use the LMS can have an impact on their actual use of the system (for Blackboard users). Furthermore, with respect to the questions, the Blackboard users are delighted with the system and want to use the Blackboard LMS frequently. Likewise, Canvas users see LMS as a sensible and reasonable choice with regard to usefulness and usability. They believe it would be wise to discuss this with the instructor and peers. Nevertheless, both Blackboard and Canvas users do not consider using the LMS as an innovative idea and a wise decision. In relation to the motif, this means that for both Blackboard and Canvas users, the "more than 3" answers are higher than the "less than 3" and "equal to 3" answers. While the most important constructs for Blackboard users are their intention to use LMS and its actual use, the most essential constructs for Canvas users are their perception of the LMS usability and the system's ease of use.

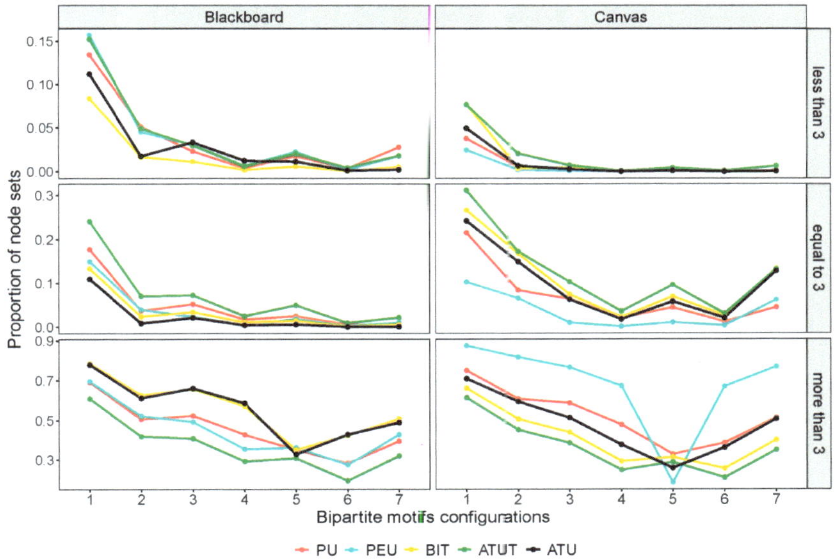

Figure 8. Relative motif frequencies on user/construct for "less than 5 nodes" configurations (Blackboard and Canvas).

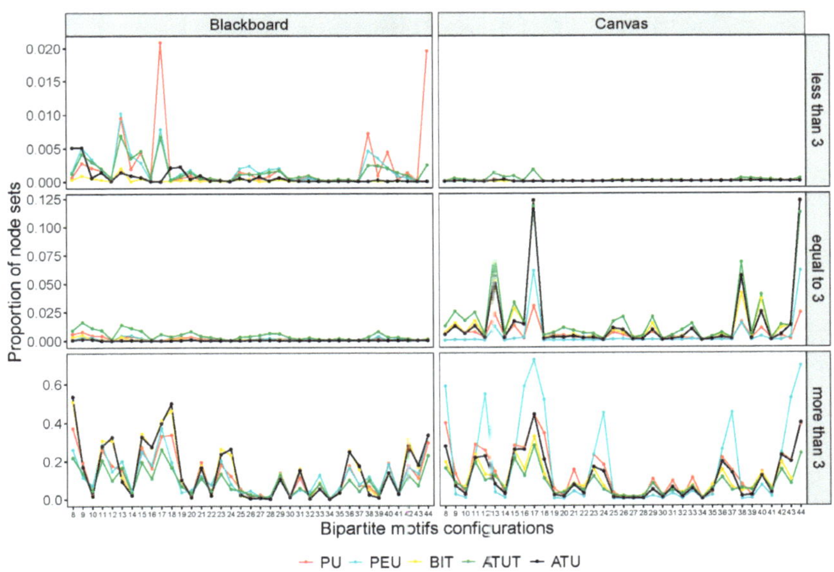

Figure 9. Relative motif frequencies on user/construct for "more than 4 nodes" configurations (Blackboard and Canvas).

Although the frequencies of the motifs configurations depend on the number of questions contained in a construct, we can conclude that the use of motifs highlights the following: while for the "more than 3" networks, both Blackboard and Canvas show a global similar behavior (similar variations in the observed motifs configurations), for the "less than 3" and "equal to 3" networks, their behavior is reversed (evident from Figure 9).

5. Discussion

The aim of this article is to demonstrate the data science techniques and approaches for data collection, processing, evaluating, and analyzing the users' acceptance of the two LMSs: Blackboard and Canvas. We do this on two levels (both individual questions as well as TAM constructs) and through using different techniques. First, we applied statistical analysis, i.e., *t*-tests, to investigate whether there are significant differences between the two LMSs (or not). Second, we provide results utilizing descriptive indices, e.g., the mean, the standard deviation, and the Gini heterogeneity index to assess the general level of users' acceptance and the variability of their answers. Third, we estimated the effect of the questions and constructs as well as their interactions with the above LMSs on the users' answers through the Cumulative Link Mixed Model. Fourth, we experimented with descriptive network analysis, e.g., degree centrality and bipartite motifs to see the variation of users' answers and the patterns of users' satisfaction within TAM constructs.

On the question level, for the *t*-tests, we see that no significant differences can be found between the two LMSs. Similar results are found when analyzing the differences on the TAM construct level as well. There may be two reasons for this: Firstly, Blackboard and Canvas are both developed in the US with approximately similar capabilities. Secondly, the user groups are relatively similar, namely both systems were evaluated in the [2], mostly by users who studied at the School of Humanities and Digital Sciences with somewhat similar experiences of LMSs. Regarding the consistency of answering the questions, we observe that overall, the participants are very consistent (when looking at the low Gini values). Additionally, the standard deviations are relatively small, indicating that not only did participants typically select the same answers, but the answers they select are also close together. For instance, if participants select mostly either answer 1 or answer 5, this would lead to a relatively low Gini score but a relatively high standard deviation. In the data, both Gini and standard deviation values are low.

When comparing consistency between LMSs, we see that Blackboard has slightly larger standard deviations and Gini scores than Canvas. For Blackboard, participants are somewhat less consistent in answering their questions compared to Canvas. While considering the results from the CLMM, we see that Blackboard and Canvas behave somewhat differently. There are some differences between the questions, but it is difficult to find clear patterns. Some questions are functionality specific (such as Q2 and Q3, which deal with finding course materials and submitting assignments). However, most questions of the PU and BIT TAM constructs show significant differences (both on the question and construct level, with Q1 and the ATU construct as a reference). In addition, the weight is negative for both constructs, but it is slightly lower for the PU.

For the network analysis, we examined bipartite motifs and degree centrality based on TAM constructs in the unweighted bipartite networks and weighted unipartite networks, respectively, which are derived from the three networks: "less than 3", "equal to 3", and "more than 3". As estimated, according to the motifs, network "more than 3" presented a rising trend and higher consistency compared to the other two networks. What stands out was the high frequencies of ATU and BIT for Blackboard as well as PE and PEU for Canvas. This means that most users gave high answers to the above constructs with respect to the other TAM constructs. Meanwhile, according to the TAM (see Figure 1), PU moves exactly along the PEU construct and ATU derives directly from the BIT construct. This striking resemblance between the pattern of motifs and TAM structure is very interesting. This proves how the LMS usability and the system's ease of use are related. At the same time, the actual use of the LMS by users originates from their intention to use the system. For the degree centrality distribution, the surprising result was the higher variability of BIT and PU for Blackboard (the same way as CLMM) as well as the PEU for Canvas. This striking similarity between the BIT and PU constructs is interesting. Looking at the TAM model, the BIT directly drives from the PU construct, meaning that Blackboard users do not accept the usability of the LMS. That is why they do not intend to use the system frequently or encourage their colleagues to do so. Furthermore, the remarkable variability of the PEU construct for Canvas is intriguing. This indicates that Canvas users have a more variable perception of the LMS ease of use than Blackboard users. These results seem to suggest that most Canvas users find the system easy to use, but they are not necessarily fully satisfied with the system in its actual usage.

Overall, this study extends the previous work [2] (where only Blackboard was analyzed) to a larger context (Blackboard plus Canvas) focusing more on extra methodologies. In the previous study [2], basic statistical analysis, as well as network analysis, was applied to investigate the acceptance of the Blackboard LMS. Descriptive network analysis showed the consistency of users' perspectives toward the system; then, the descriptive statistics results enabled the extraction of actions related to the Blackboard LMS. In the current study, we provided a broader comparison and in-depth comprehension of user acceptance of the two systems (Blackboard versus Canvas). Therefore, we experimented with the general level of acceptance and its heterogeneity plus the Cumulative Link Mixed Model (CLMM) at the question and construct level through descriptive analysis as well as network analysis at the construct level. The results confirmed the higher acceptance and consistency among Canvas users compared to Blackboard, which helped the Tilburg University LMS group while switching LMSs from Blackboard to Canvas.

In comparison to the previous study [2], the strengths of the current study were employing a Cumulative Link Mixed Model to describe user acceptance by estimating the probability of the users' answers with respect to the different LMSs and the questions/constructs. Additionally, the network analysis approaches used in the current study revealed interesting patterns in the participant data to more precisely describe the user acceptance of LMSs. Nevertheless, the generalizability of these results is subject to certain limitations. For instance, no major strategic conclusions can be made based on the results from this study, as these results are based on a small sample from one university in The Netherlands. Another limitation of this study is that demographic information has not been taken into account. Previous work [2] has shown that gender may have an impact on the results (although that study did not show a considerable influence).

6. Conclusions

In this article, we presented several analyses of questionnaire results that investigate the acceptance of two learning management systems (LMSs), namely Blackboard and Canvas. The analyses were performed on two levels: questions and constructs that stem from the Technology Acceptance Model (TAM), which also formed the basis for the questions in the questionnaire.

We compared the LMSs using statistical properties. We provided descriptive measures, e.g., mean, standard deviations, and Gini heterogeneity index results for both levels. The means provide overall values for the answers to the questions in the questionnaire, whereas the standard deviation shows the spread. The Gini heterogeneity index indicates the consistency in the answers. We also applied the Cumulative Link Mixed Model (CLMM) to both levels. The Cumulative Link Mixed Model examines the impact of the questions and TAM constructs with the LMSs on the users' answers. Finally, we experimented with descriptive network analysis, e.g., degree centrality and bipartite motifs to see the variability of users' answers and extract the patterns of user satisfaction across TAM constructs.

The results showed that overall, participants were very consistent in providing their answers. Both standard deviations and Gini heterogeneity scores were low for both questions and TAM constructs. The overall scores were high, indicating that participants seem to accept the use of LMSs. The Blackboard system, however, showed slightly lower scores compared to the Canvas system. We propose that this may be due to the differences in functionality of the two LMSs: namely, Canvas is somewhat better and more innovative in design than Blackboard.

Investigating the combination of metrics provides a fine-grained analysis of the results. Not only a statistical model is built which shows differences: the model is applied to two levels, which illustrates the differences in the construct as well as individual questions. The use of the Gini heterogeneity index provides additional information on the consistency of answers between the participants (which may be different from the spread measured by the standard deviation). Finally, through descriptive network analysis, for both Blackboard and Canvas, we observed high equilibrium, which was due to a large proportion of satisfaction among users. For the Canvas users, however, the perception of the LMS acceptance was higher than the Blackboard users per construct.

As mentioned above, the reason for choosing the two LMSs in our data analysis of user acceptance is due to the academic context. The context for both LMS groups was almost the same: namely, the same university, the same educational program, and the same environment (both LMS users studied at the School of Humanities and Digital Sciences, and the LMSs were evaluated in The Netherlands).

Overall, the main goal of the current study was to illustrate how data science methodologies were applied for data collection, processing, assessment, and analysis in a particular context, namely the LMSs: Blackboard and Canvas. The empirical findings contribute in several ways to a new understanding of the LMSs and provide a basis for the analysis of LMSs' user acceptance. Firstly, the t-tests indicate that there are no significant differences between the two LMSs. Secondly, what is interesting is the relatively higher standard deviations and Gini scores for Blackboard than Canvas, meaning that Blackboard users are less consistent in answering their questions. Thirdly, looking at the CLMM results, you can see that Blackboard and Canvas users behave differently in answering some functional questions and constructs, i.e., Q2 and Q3 (finding course materials and submitting assignments) as well as PU and BIT (LMSs usefulness and users' intention to use the LMSs). Lastly, according to network analysis, namely bipartite motifs, the high frequencies of ATU and BIT for Blackboard, as well as PE and PEU for Canvas, are particularly noticeable. This indicates that compared to the other TAM constructs, the above constructs received high answers from the majority of users. Additionally, this proves the relationship between LMSs' usability and ease of use. Furthermore, the surprising result for the degree centrality distribution is the higher variability of BIT and PU for Blackboard as well as PEU for Canvas. This indicates that Blackboard users have no intention of using the system frequently or encouraging their colleagues to do so. The considerable variability of the PEU construct for Canvas, however, is unexpected, meaning that Canvas users have a more varied perception of the LMS's usability than Blackboard users. These findings appear to suggest that while the majority of Canvas users regard the system as easy to use, they are not always completely satisfied with it in practice or in actual use.

For future work, we would like to further investigate the underlying reasons for the differences between the acceptance of the LMS systems. This can be completed by separately querying students from different educational backgrounds and comparing these results. Additionally, the cultural (geographic) differences can be investigated further as well.

Author Contributions: The authors' contributions are as follows: conceptualization: P.S., R.R., M.v.Z. and M.A.; methodology: P.S., R.R., M.v.Z. and M.A.; software: P.S. and R.R.; validation, P.S., R.R., M.v.Z. and M.A.; formal analysis: P.S., R.R., M.v.Z. and M.A.; investigation: P.S. and R.R.; resources: P.S.; data curation: P.S.; writing—original draft preparation: P.S. and R.R.; writing—review and editing: P.S., R.R., M.v.Z. and M.A.; visualization: P.S. and R.R.; supervision: M.v.Z. and M.A.; project administration: M.A.; funding acquisition: M.A. All authors have read and agreed to the published version of the manuscript.

Funding: This research was funded by the German Research Foundation (DFG) project "MODUS" grant number AT 88/4-1.

Institutional Review Board Statement: The study was conducted in accordance with the Declaration of Tilburg, The Netherlands, and approved by the Ethics Committee of the School of Humanities and Digital Sciences School at Tilburg University (protocol code REC # 2018/60 and date of approval 18 January 2019).

Informed Consent Statement: Informed consent was obtained from all subjects involved in the study.

Data Availability Statement: Data supporting reported results can be found in the link below: https://github.com/p877/MDPI (accessed on 5 December 2022).

Acknowledgments: The research leading to this work has partially been funded by the German Research Foundation (DFG) project "MODUS" under grant AT 88/4-1.

Conflicts of Interest: The authors declare no conflict of interest. The funders had no role in the design of the study; in the collection, analyses, or interpretation of data; in the writing of the manuscript; or in the decision to publish the results.

Abbreviations

The following abbreviations are used in this manuscript:

LMSs	Learning Management Systems
TAM	Technology Acceptance Model
CLMM	Cumulative Link Mixed Model
EV	External Variables
PU	Perceived Usefulness
PEU	Perceived Ease of Use
ATUT	Attitude Toward Using the Technology
BIT	Behavioral Intention to use the Technology
ATU	Actual Technology Use
UTAUT	Unified Theory of Acceptance and Use of Technology
TPB	Theory of Planned Behavior
OSAM	Online Shopping Acceptance Model
PCA	Principal Component Analysis
AVE	Average Variance Extracted
CR	Composite Reliability
SD	Standard Deviation
M	Mean

References

1. Szabo, M.; Flesher, K. CMI Theory and Practice: Historical Roots of Learning Management Systems. In Proceedings of the E-Learn: World Conference on E-Learning in Corporate, Government, Healthcare and Higher Education, Montreal, QC, Canada, 15–19 October 2002; Association for the Advancement of Computing in Education: Montreal, QC, Canada, 2002.
2. Shayan, P.; Rondinelli, R.; Zaanen, M.; Atzmueller, M. Descriptive Network Modeling and Analysis for Investigating User Acceptance in a Learning Management System Context. In Proceedings of the ABIS '19: 23rd International Workshop on Personalization and Recommendation on the Web and Beyond, Hof, Germany, 17 September 2019; ACM: Boston, MA, USA, 2019; pp. 7–13. [CrossRef]
3. Davis, F.D. Perceived usefulness, perceived ease of use, and user acceptance of information technology. *MIS Q.* **1989**, *13*, 319–340. [CrossRef]
4. Hill, R.J.; Fishbein, M.; Ajzen, I. Belief, Attitude, Intention and Behavior: An Introduction to Theory and Research. *Contemp. Sociol.* **1977**, *6*, 244. [CrossRef]
5. Milošević, M.; Zećirović, E.; Krneta, R. Technology Acceptance Models And Learning Management Systems: Case Study. In Proceedings of the Fifth International Conference on e-Learning, Belgrade, Serbia, 22–23 September 2014 ; University of Belgrade: Belgrade, Serbia, 2014.
6. Venkatesh, V.; Davis, F.D. A Theoretical Extension of the Technology Acceptance Model: Four Longitudinal Field Studies. *Manag. Sci.* **2000**, *46*, 186–204. [CrossRef]
7. Karrer, T. Understanding E-Learning 2.0: Introduction to Tools and Technologies. In Proceedings of the Elearning Development Conference and Expo, San Francisco, CA, USA, 10–13 October 2006 ; TechEmpower, Inc.: San Jose, CA, USA, 2007.
8. Caballero, J.A.; Palomo, M.; Dodero, J.M.; Rodríguez, G.; Ibarra, M.S. Integrating External Evidence of Skill Assessment In Virtual Learning Environments. In Proceedings of the Fifth International Conference on e-Learning, Belgrade, Serbia, 22–23 September 2014 ; University of Belgrade: Belgrade, Serbia, 2014; pp. 70–80.
9. Bradford, P.; Porciello, M.; Balkon, N.; Backus, D. The Blackboard learning system: The be all and end all in educational instruction? *J. Educ. Technol. Syst.* **2007**, *35*, 301–314. [CrossRef]
10. Baepler, P. *The Canvas Learning Management System: Instructor And Student Experience*; Technical Report; Center for Educational Innovation (CEI), University of Minnesota: Minneapolis, MN, USA, 2017.
11. Cowen, R. The transfer, translation, and transformation of educational processes: And their shape-shifting. *Comp. Educ.* **2009**, *45*, 315–327. [CrossRef]

12. Surendran, P. Technology Acceptance Model: A Survey of Literature. *Int. J. Bus. Soc. Res. (IJBSR)* **2012**, *2*, 175–178.
13. Fathema, N.; Sutton, K.L. Factors influencing faculty members' Learning Management Systems adoption behavior: An analysis using the Technology Acceptance Model. *Int. J. Trends Econ. Manag. Technol. (IJTEMT)* **2013**, *2*, 20–28.
14. Fathema, N.; Ross, M.; Witte, M. Student acceptance of university web portals: A quantitative study. *Int. J. Web Portals* **2014**, *6*, 42–58. [CrossRef]
15. Fathema, N.; Shannon, D.; Ross, M. Expanding The Technology Acceptance Model (TAM) to Examine Faculty Use of Learning Management Systems (LMSs) In Higher Education Institutions. *J. Online Learn. Teach.* **2015**, *11*, 210–233.
16. Venkatesh, V.; Morris, M.G.; Davis, G.B.; Davis, F.D. User Acceptance of Information Technology: Toward a Unified View. *MIS Q.* **2003**, *27*, 425–478. [CrossRef]
17. Stieninger, M.; Nedbal, D. Diffusion and Acceptance of Cloud Computing in SMEs: Towards a Valence Model of Relevant Factors. In Proceedings of the 2014 47th Hawaii International Conference on System Sciences, Washington, DC, USA, 6–9 January 2014; IEEE: Waikoloa, HI, USA, 2014; pp. 3307–3316.
18. Chau, P.Y.K. An Empirical Assessment of a Modified Technology Acceptance Model. *J. Manag. Inf. Syst.* **1996**, *13*, 185–204. [CrossRef]
19. Agarwal, R.; Prasad, J. A Conceptual and Operational Definition of Personal Innovativeness in the Domain of Information Technology. *Inf. Syst. Res.* **1998**, *9*, 204–215. [CrossRef]
20. Venkatesh, V.; Bala, H. Technology Acceptance Model 3 and a Research Agenda on Interventions. *Decis. Sci.* **2008**, *39*, 273–315. [CrossRef]
21. Van der Heijden, H. *Using the Technology Acceptance Model to Predict Website Usage: Extensions and Empirical Test*; Technical Report, Serie Research Memoranda, Research Memorandum; University of Amsterdam: Amsterdam, The Netherlands, 2000.
22. Moon, J.W.; Kim, Y.G. Extending the TAM for a World-Wide-Web context. *Inf. Manag.* **2001**, *38*, 217–230. [CrossRef]
23. Zhou, L.; Dai, L.; Zhang, D. Online shopping acceptance model—A critical survey of consumer factors in online shopping. *J. Electron. Commer. Res.* **2007**, *8*, 41–62.
24. Müller-Seitz, G.; Dautzenberg, K.; Creusen, U.; Stromereder, C. Customer acceptance of RFID technology: Evidence from the German electronic retail sector. *J. Retail. Consum. Serv.* **2009**, *16*, 31–39. [CrossRef]
25. Ervasti, M.; Helaakoski, H. Case study of application-based mobile service acceptance and development in Finland. *Int. J. Inf. Technol. Manag.* **2010**, *9*, 243–259. [CrossRef]
26. Chau, P.Y.K.; Jen-Hwa Hu, P. Information Technology Acceptance by Individual Professionals: A Model of Comparison Approach. *Decis. Sci.* **2001**, *32*, 699–719. [CrossRef]
27. Shen, H.; Luo, L.; Sun, Z. What Affects Lower Grade Learner's Perceived Usefulness and Perceived Ease of Use of Mobile Digital Textbook Learning System? An Empirical Factor Analyses Investigation in China. *Int. J. Multimed. Ubiquitous Eng.* **2015**, *10*, 33–46. [CrossRef]
28. Arumugam, R. The usage of technology among education students in university Uttara Malaysia: An application of extended technology acceptance model. *Int. J. Educ. Dev. Using Inf. Commun. Technol. (IJEDICT)* **2011**, *7*, 4–17.
29. Claar, C.; Dias, L.P.; Shields, R. Student acceptance of learning management systems: A study on demographics. *Issues Inf. Syst.* **2014**, *15*, 409–417.
30. Cakir, O. The factors that affect online learners' satisfaction. *Anthropologist* **2014**, *17*, 895–902. [CrossRef]
31. Dahlstrom, E.; Brooks, D.C.; Bichsel, J. *The Current Ecosystem of Learning Management Systems in Higher Education: Student, Faculty, and Its Perspectives*; Technical Report; EDUCAUSE Center for Analysis and Research: Boulder, CO, USA, 2014.
32. Kurkinen, E. The effect of age on technology acceptance among field police officers. In Proceedings of the 10th International ISCRAM Conference, Baden-Baden, Germany, 12–15 May 2013.
33. Chua, C.; Montalbo, J. Assessing students' satisfaction on the use of virtual learning environment (VLE): An input to a campus-wide e-learning design and implementation. *Inf. Knowl. Manag.* **2014**, *3*, 108–115.
34. Marmon, M.; Vanscoder, J.; Gordesky, J. Online student satisfaction: An examination of preference, asynchronous course elements and collaboration among online students. *Curr. Issues Educ. (CIE)* **2014**, *17*, 1–12.
35. Alharbi, S.; Drew, S. Using the Technology Acceptance Model in Understanding Academics' Behavioural Intention to Use Learning Management Systems. *Int. J. Adv. Comput. Sci. Appl.* **2014**, *5*, 13. [CrossRef]
36. Hock, S.Y.; Omar, R.; Mahmud, M. Comparing the Usability and Users Acceptance of Open Sources Learning Management System (LMS). *Int. J. Sci. Res. Publ.* **2015**, *5*, 1–5.
37. Ceriani, L.; Verme, P. The origins of the Gini index: Extracts from Variabilità e Mutabilità (1912) by Corrado Gini. *J. Econ. Inequal.* **2012**, *10*, 421–443. [CrossRef]
38. Field, A.; Miles, J.; Field, Z. *Discovering Statistics Using R*; SAGE Publications Ltd.: London, UK, 2012; Volume 81. [CrossRef]
39. Newman, M.E. The structure and function of complex networks. *SIAM Rev.* **2003**, *45*, 167–256. [CrossRef]
40. Clemente, S.P.; Everett, M.G. Network analysis of 2-mode data. *Soc. Netw.* **1997**, *19*, 243–269.
41. Wellman, B.; Berkowitz, S.D. *Social Structures: A Network Approach*; Cambridge University Press: Cambridge, UK; New York, NY, USA, 1988.
42. Dalka, R.; Sachmpazidi, D.; Henderson, C.; Zwolak, J. Network analysis approach to Likert-style surveys. *Phys. Rev. Phys. Educ. Res.* **2022**, *18*, 1–15. [CrossRef]

43. Simmons, B.I.; Sweering, M.J.M.; Schillinger, M.; Dicksand, L.V.; Sutherland, W.J.; Di Clemente, R. Bmotif: A package for motif analyses of bipartite networks. *Methodol. Ecology Evol.* **2018**, *10*, 695–701. [CrossRef]
44. Simmons, B.I.; Cirtwill, A.R.; Baker, N.J.; Wauchope, H.S.; Dicks, L.V.; Stouffer, D.B.; Sutherland, W.J. Motifs in bipartite ecological networks: Uncovering indirect interactions. *Oikos* **2019**, *128*, 154–170. [CrossRef]
45. Milo, R.; Shen-Orr, S.; Itzkovitz, S.; Kashtan, N.; Chklovskii, D.; Alon, U. Network Motifs: Simple Building Blocks of Complex Networks. *Science* **2002**, *298*, 824–827. [CrossRef]
46. Poisot, T.; Stouffer, D. How ecological networks evolve. *bioRxiv* **2016**, bioRxiv: 071993.
47. Isla, J.; Jácome-Flores, M.; Pareja, D.; Jordano, P. Drivers of individual-based, antagonistic interaction networks during plant range expansion. *J. Ecol.* **2022**, *110*, 1266–1276. [CrossRef]
48. Baker, N.J.; Kaartinen, R.; Roslin, T.; Stouffer, D.B. Species' roles in food webs show fidelity across a highly variable oak forest. *Ecography* **2015**, *38*, 130–139. [CrossRef]
49. Siang, J.J.; Santoso, H.B. Students' perspective of learning management system: An empirical evidence of technology acceptance model in emerging countries. *Res. World* **2015**, *6*, 1–14.
50. Cronbach, L.J. Coefficient alpha and the internal structure of tests. *Psychometrika* **1951**, *16*, 297–334. [CrossRef]
51. Gini, C. *Variabilità e Mutabilità: Contributo allo Studio delle Distribuzioni e delle Relazioni Statistiche*; Studi Economici-Giuridici della Regia Facoltà di Giurisprudenza, Anno III, Parte II; Tipogr. di P. Cuppini: Bologna, Italy 1912.
52. Bruin, J. *Newtest: Command to Compute New Test @ONLINE*; Technical Report; UCLA: Statistical Consulting Group: Los Angeles, CA, USA, 2006.
53. Christensen, R. Cumulative Link Models for Ordinal Regression with the R Package ordinal. In *Proceedings of the Cumulative Link Models*; Technical University of Denmark: Lyngby, Denmark, 2018.

data

MDPI

Article

Federated Learning for Data Analytics in Education

Christian Fachola [1,†], Agustín Tornaría [2,†], Paola Bermolen [2,†], Germán Capdehourat [3,4,†], Lorena Etcheverry [1,*,†] and María Inés Fariello [2,†]

[1] Instituto de Computación, Facultad de Ingeniería, Universidad de la República, Montevideo 11300, Uruguay
[2] Instituto de Matemática, Facultad de Ingeniería, Universidad de la República, Montevideo 11300, Uruguay
[3] Instituto de Ingeniería Eléctrica, Facultad de Ingeniería, Universidad de la República, Montevideo 11300, Uruguay
[4] Ceibal, Montevideo 11500, Uruguay
* Correspondence: lorenae@fing.edu.uy; Tel.: +598-2714-2714 (ext. 12148)
† These authors contributed equally to this work.

Abstract: Federated learning techniques aim to train and build machine learning models based on distributed datasets across multiple devices while avoiding data leakage. The main idea is to perform training on remote devices or isolated data centers without transferring data to centralized repositories, thus mitigating privacy risks. Data analytics in education, in particular learning analytics, is a promising scenario to apply this approach to address the legal and ethical issues related to processing sensitive data. Indeed, given the nature of the data to be studied (personal data, educational outcomes, and data concerning minors), it is essential to ensure that the conduct of these studies and the publication of the results provide the necessary guarantees to protect the privacy of the individuals involved and the protection of their data. In addition, the application of quantitative techniques based on the exploitation of data on the use of educational platforms, student performance, use of devices, etc., can account for educational problems such as the determination of user profiles, personalized learning trajectories, or early dropout indicators and alerts, among others. This paper presents the application of federated learning techniques to a well-known learning analytics problem: student dropout prediction. The experiments allow us to conclude that the proposed solutions achieve comparable results from the performance point of view with the centralized versions, avoiding the concentration of all the data in a single place for training the models.

Keywords: federated learning; learning analytics

Citation: Fachola, C.; Tornaría, A.; Bermolen, P.; Capdehourat, G.; Etcheverry, L.; Fariello, M.I. Federated Learning for Data Analytics in Education. *Data* **2023**, *8*, 43. https://doi.org/10.3390/data8020043

Academic Editors: Antonio Sarasa Cabezuelo, Ramón González del Campo and Rodríguez Barbero

Received: 30 December 2022
Revised: 8 February 2023
Accepted: 14 February 2023
Published: 20 February 2023

1. Introduction

Education systems are usually composed of different educational centers (kindergartens, high schools, etc.). Each center involves students and teachers who interact daily in various learning activities, generating valuable information, both locally for the individual school and globally for the whole educational system. When these interactions occur through digital educational platforms, a potentially massive volume of data is generated that can be harnessed for various academic and pedagogical purposes.

Regardless of the governance and organization of each country's education system, there are usually government entities above the schools. One of their main tasks is collecting and analyzing data on the education system. In its traditional approach, building, training, and deploying machine learning (ML) models and artificial intelligence (AI) techniques involve simple data-sharing models. Data must be fused, cleaned, and integrated and then used to train and test the models. This procedure faces challenges related to individuals' privacy and personal data protection. These privacy and ethical issues are essential in learning analytics (LA), which is the application of quantitative techniques to educational data to help solve problems such as the design of teaching trajectories or the development of early dropout alerts. In the latter case, the prediction result would be significant mostly for the community of the analyzed individuals, and the prediction should be treated as personal data. The privacy issues and the ethical use of data in LA applications have been widely documented in the literature [1–3].

There are two ways of ensuring privacy in LA. On the one hand, the privacy-preserving data-publishing approach, which consists of applying data de-identification and anonymization techniques (e.g., satisfying the definition of k-anonymity [4]) and then using conventional ML methods [5,6]. On the other hand, in the privacy-preserving data mining or statistical disclosure control approach, the analyst does not directly access the data but uses a query mechanism that adds statistical noise to the response, implementing differential privacy [7]. The latter strategy may be more robust and scalable, but some authors suggest that it may be challenging to implement in practice [8].

Another way to tackle the privacy-preserving data issue is to use a decentralized approach such as federated learning (FL), initially proposed by Google [9] to build ML models using distributed datasets across multiple devices. Its main goal is to train ML models on remote devices or isolated data centers without transferring the data to centralized repositories. FL incorporates ideas from multiple areas, including cryptography, ML, heterogeneous computing, and distributed systems. In recent years, the concept has been growing and consolidating, along lines ranging from improvements in security aspects and the study of statistical problems that arise in the distributed scenario to the extension of the concept to cover collaborative learning scenarios between organizations [10].

In the context of LA, FL provides mechanisms that allow fitting models based on the data generated by a set of schools but avoid the concentration of raw data generated at each school. This scheme improves data management regarding privacy preservation but uses more information for training models than independently fitting a model for each school. It also avoids storage duplication in a central server and each school. In this context, we see a clear opportunity to capitalize on the benefits of FL.

There are two main variants of FL: horizontal and vertical [11], typically associated with the two different use cases called cross-device and cross-silo. In the first case, the data are horizontally partitioned since the data structure in the different devices is the same. Each device has its own data set, but all the sets share the same attributes or variables. The records in each data set have the same fields but are for different participants. A known example is the one that originated federated learning: smartphones' predictive keyboard. Communication problems play a relevant role in this case, as devices are only sometimes available, hindering machine learning models' training rounds. In contrast, in the vertical case, the partitioning corresponds to where different data sets share common identifiers (e.g., information from the same users). Still, each data set includes different fields in its records. This case corresponds to the exchange of information between different institutions, which usually involves data communication between well-established data centers. This second scenario is usually applied to integrating data from different sources without gathering all the data in one server. For example, a typical case could be a cross-silo scheme in which different government agencies share information on their citizens.

Our Proposal and Related Works

Our work uses horizontal FL techniques to apply LA to data distributed across different educational institutions. This scenario presents specific characteristics that differ from the ones observed in typical cross-device FL systems. In our proposal, each educational center manages all the information related to its teachers and students. This situation does not necessarily imply that each educational center has its on-premises data center. We could also have educational platforms in the cloud, where data are hosted on third-party servers. However, we assume that each educational center has the administration rights to all data on its teachers and students. Thus, each institution in the educational system can be seen as a silo in the proposed federated scheme.

In Figure 1, we illustrate the proposed cross-silo scheme for the educational system. The goal is to use the federated learning paradigm to enable a centralized analysis of education system data while avoiding the corresponding centralization of raw data. The proposed approach would allow higher government institutions in charge of the education system to conduct data analytics using ML models while preserving the teachers' and students' privacy. A similar scenario has been extensively studied in the context of healthcare applications [12–14]. scheme. The main difference is that the proposed education cross-silo scheme uses horizontal rather than vertical partitioning. In our case, the different educational institutions share the same data schema, with all of them sharing the same attributes for students and courses.

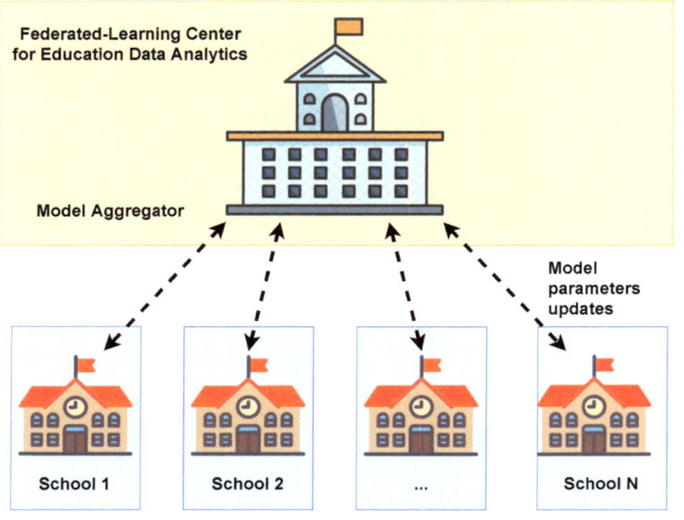

Figure 1. Our proposal for a cross-silo federated learning scheme for centralized data analysis of the educational system.

To evaluate the proposed education federated scheme, we analyzed a well-known learning analytics problem: student dropout prediction through a neural network model. We assume a scenario in which a global analysis is required without centralizing student data stored in different educational centers. To validate the proposed scheme, we compared the accuracy of the neural network model in an FL framework with the centralized case to determine if we lose performance compared to gathering all the data together. We also compared the accuracy obtained for each school after training the models under a federated scheme with the case where each school trains an individual model separately (i.e., each school uses only its data for dropout prediction). Finally, we extended the analysis to the case of the non-homogeneous distribution of dropout student rates among the different institutions. In cases where the data between clients are heterogeneous, algorithms can have good accuracy when considering all the data together. Still, they can be unfair to schools with a different data distribution than the rest. In the case of FL, this means that the schools can have very different sizes and also that there can be different biases in the schools, so the algorithms can be unfair with some of the schools and have a lower accuracy for those cases [15,16].

The LA problem we address has been studied before. In particular, the development of dropout-prediction systems is a relevant concern in many educational communities, and different proposals have been devised in this regard. In particular, our approach is based on the work presented in [17], where the dropout detection problem is addressed centrally in the context of online learning platforms. Finally, very few papers present the application of FL in the context of LA. A framework for educational data analysis is described in [18], introducing a similar education federated scheme to our proposal.

However, it does not emphasize the evaluation of the obtained results or the discussion of how different parameters affect the convergence of the solutions.

The rest of the document is organized as follows. First, in Section 2, we describe the main FL concepts and present the dataset and models used to evaluate the framework proposed. Next, in Section 3, we describe the experiments carried out on federated dropout prediction and present the corresponding results. Finally, in Section 4, we discuss over the insights observed, while Section 5 concludes the paper, commenting on future research lines.

2. Methods

In this section, we describe the main concepts of FL; we then present the dataset that was used to evaluate the performance of the models and, finally the network architecture.

2.1. Main Concepts of Federated Learning

In the FL setting, the participating entities are usually classified as servers and clients; the server is the one that orchestrates the model training, and the clients are the ones who store the data and also run the models locally. In LA, the clients would be the schools and the server would be a governmental entity.

For computing the model's parameters, an iterative process between model parameter estimation within the clients and actualization of the parameters in the server is carried out. In each iteration, specific clients are chosen to train the model with its own data locally. Then, the server aggregates all clients' results to update the model's state, which will be deployed on new clients in the next iteration, repeating the process [19].

Figure 2 shows the steps necessary to train a model using the FL scheme. First, the central server sends the last model parameters to the nodes or initial parameters in case it is the first run (step 1). Then, in step 2, data are selected at each node, and each local model is trained based on the last parameters (step 3). At the end of the local training, each client communicates the updated parameters of the local model to the global model (step 4), where the updates of each model are combined, generating a new model (step 5). Finally, the process is restarted from step 1 (step 6). The model developed in step 5 can then be put into production.

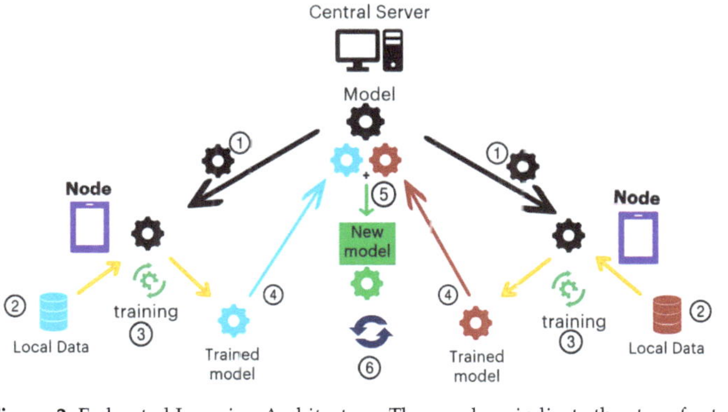

Figure 2. Federated Learning Architecture. The numbers indicate the steps for training a model in an FL scheme. Reprinted/adapted with permission from Ref. [20]. 2020, Faisal Zaman.

Each deployment, local training on the selected clients, and update of the server model cycle constitutes a round. In the case of neural networks (NN), each client trains the model independently using classical gradient descent, and sends the computed weights to the server. The server updates the weights using federated averaging algorithm [21]. This algorithm averages the received client's weights. In the federated case for neural networks, one has to consider the parameters that define the behavior of the models on the clients epochs and batch-size) but also those specific to the federation, specifically, the number of rounds, the number of clients chosen per round, the total number of clients, and how the data are distributed among them. The parameters mentioned above may influence, a priori, the performance of the models obtained. Therefore, one of our goals is to experiment in this direction to understand the effects of each of these parameters.

2.2. Dataset Description and Pre-Processing

As already mentioned, our work focuses on studying the applicability of FL to student dropout prediction. For this purpose, we use the KDDCup2015 dataset, which contains activity logs from XuetangX, a Chinese MOOC (Massive Online Open Course) learning platform [22]. Data are provided about the student activity on each course over time. Student information includes a record of participation in several activities of each course (discussion forum, quiz, media usage, etc.). There are 21 activities, and their availability varies across courses. We can calculate metrics, such as dropout, for a particular course or student across all the courses it takes.

The logs have 42 M individual entries and have a total size of approximately 2.1 GB. There are 77,083 students and 247 courses. On the one hand, there are typically many students per course, as is expected from a MOOC platform, and a count of how many courses have what amount of students can be seen as a histogram in Figure 3. On the other hand, the vast majority of students only enroll in a few courses, with 46% of them enrolling in just 2. Table 1 shows how many courses students tend to enroll in, with percentages.

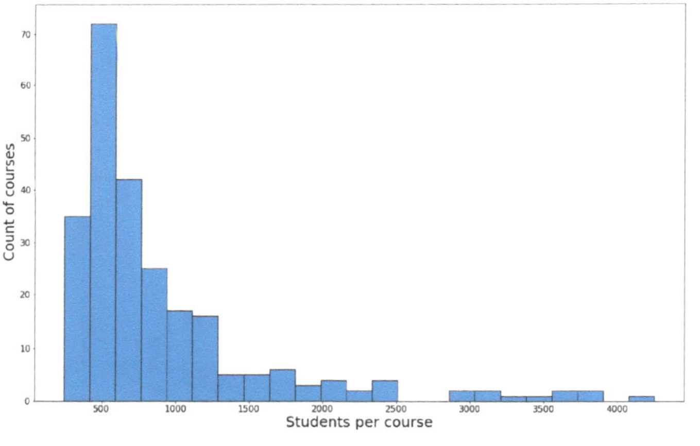

Figure 3. Histogram showing the count of courses with a certain number of students. About 70 courses have about 500 students.

Table 1. Percentage of students taking a certain number of courses. The vast majority of students take only a few courses.

% of All Students	# Courses	Students Taking # Courses
46%	2	35,683
17%	3	13,271
16%	1	12,411
8%	4	6277
4%	5	3212
9%	>5	6229

From the individual entries of the raw activity logs, we group data by course and student, counting the number of entries per activity. The final dataset has 225,642 entries of 21 features, where each entry corresponds to a distinct pair $(course\_id, student\_id)$, which is also identified by a key type column called *enroll_id* number. The features are the activity counts. For instance, for the entry of *enroll_id K* corresponding to the enrollment of student S in course C, one feature is the number of times that S reproduced a video featured on the course's C web page. Another feature is the number of times S deleted a comment in the forum of course C. A complete list of features can be found in the Appendix A. The data-preparation code is available at our repository [23].

2.3. Network Architecture

We used a fully connected NN architecture consisting of the input layer, three hidden layers of size 100, and an output layer of 1 neuron with a sigmoid as an activation function. In addition, we used the Adam optimizer [24] and binary cross-entropy as a loss function. This architecture was used across all experiments. In Section 3, we compare its performance for different training schemes (federated and centralized) and training parameters. The centralized NN will be trained using Tensorflow. For federating, we use the Federated Averaging algorithm [21] implemented in Tensorflow Federated.

3. Experimental Results

This section presents our implementation of different scenarios using Federated Learning frameworks and the experiments carried out in each case using a public data set from KDDCup2015 [22]. As mentioned in Section 1, we used the approach presented in [17] to predict student dropout. For every *enroll_id*, there is a label saying whether the student dropped out of the course. We then used these labels to train and test a deep learning model that predicts dropout.

The experiments had four main goals: (1) to evaluate the influence of the training parameters on the accuracy and total training time of the federated models, (2) to assert whether the federated models can reach the accuracy of the centralized setting or not, (3) to evaluate the performance of the federation compared to training models locally on each institution, and (4) to compare performance when data distribution varies across institutions. These objectives are crucial to understanding when a federated scheme is a good alternative and how best to implement it. Objective (1) is addressed in Section 3.1, testing different parameters; then, in Section 3.2, we explore our second goal by increasing the number of rounds. Finally objectives (3) and (4) are addressed in Section 3.3, where we compare different schemes for performing training and evaluation (e.g., where each client trains its model separately, where all data are centralized, and a federated version), also varying the data distribution approach.

3.1. Federated Learning Parameters

In addition to the usual local parameters of each client, such as the epochs and the batch size, in the federated version, we need to deal with additional ones, such as the number of training rounds, the number of clients chosen in each round, the total number of clients, and how the data are distributed among them.

3.1.1. Experiment

A centralized model was trained for 20 epochs. The number of epochs was chosen empirically; we searched for a number large enough to allow for parameter tuning in the federated approach that also maintained an acceptable level of accuracy without much overfitting. We sampled 70% of all students and collect their data to build the training dataset, using the remaining to build the test set. The model was evaluated using 50 different random splits of the students.

In the federated model, each client represents one school and comprises samples of 1000 students, totaling 77 schools (clients). Clients did not share students but could share courses. The training-evaluation process consisted of 50 different random splits in a 70/30 proportion. Using 1000 students per client, the training was performed for 54 clients (%70). The remaining data were used for testing; this is carried out in a centralized manner where a model with the same architecture was initialized with weights from the federated model at the end of each round.

We used different combinations on the number of rounds (R) and local epochs (E), leaving a fixed number of total epochs $R \times E = 20$. This number is not arbitrary; it is the same number of training epochs as for the centralized model, so experiments are comparable. It also has a fair number of divisors, which allow us to play with different values of E and R. We vary the number of clients used per round using 1 client (minimum availability of clients), 14 clients (25% availability), 27 clients (50%), 43 clients (75%), and 54 clients (maximum availability).

3.1.2. Results

The centralized version achieves a mean accuracy of 81.7% $\pm$ 0.07% across the 50 different splits and a mean running time of 105 $\pm$ 2 s. This case is our baseline. Figure 4 shows the results in terms of accuracy for the federated case, where each point represents the result of one of the 50 executions. Increasing the number of clients from 1 to 14 causes a leap of 2–3% in the mean accuracy. At the same time, additional increments in the number of clients only cause a marginal increase in the mean accuracy but also produce a rise in the execution time (see Figure A1 in Appendix A). If the number of clients is fixed, we can see that favoring the number of rounds R over local epochs E tends to yield a better accuracy overall (boxes in each group go up), but again, this will cause an increment in time. It is also worth noting that variance decreases as we increase the clients and the rounds.

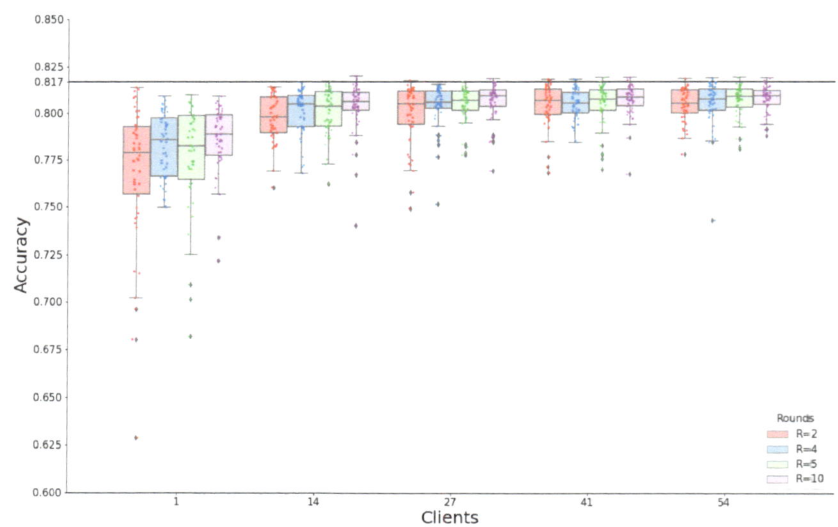

Figure 4. Accuracy results of dropout prediction (Federated version), averaging over 50 random executions with different number of clients per round (C), number of rounds (R), and local epochs of clients (E), where $R \times E = 20$. Each box contains 50 points. The black line marks accuracy averaged by the centralized model.

Finally, Figure 4 also shows that the performance in the federated setting is close to the one found in the centralized model, with a mean accuracy larger than 76% in every experiment (which goes up to 80% when excluding the experiments with one client per round), a top accuracy of 82% (reached on the run with 14 clients and ten rounds) and with around 63% of all individual runs, across all experiments, with more than 80% accuracy. However, some executions still have relatively low accuracy.

3.2. Further Tuning of the Federation

In this section we present our experiments to assert whether the federated models can reach the accuracy of the centralized setting or not.

3.2.1. Experiment

We determine whether it is possible to consistently reach the results of the centralized environment. Therefore, we repeated the experiments, running as many rounds as needed to reach 81.7% accuracy, the top accuracy of the centralized model. This evaluation scheme is inspired by the method's comparison presented in [21].

3.2.2. Results

Figure 5 shows our results; we can see that it is possible to reach the accuracy of the centralized model in every case, with the caveat that many rounds may be needed. The maximum number of rounds is needed when training with one client per round, and the resulting accuracy presents a significant variance. From 14 clients onward, the results do not vary significantly; that is to say, increasing the number of clients does not necessarily improve convergence. Increasing E lowers the average R necessary to reach our baseline accuracy (81.7%) in every case. However, there is no 1:1 inverse relationship; for instance, if E = 2, an average of 16 rounds is needed, but if E = 10, we need six rounds. Thus an ×5 increase ratio in E but only an ×2.6 decrease ratio on R.

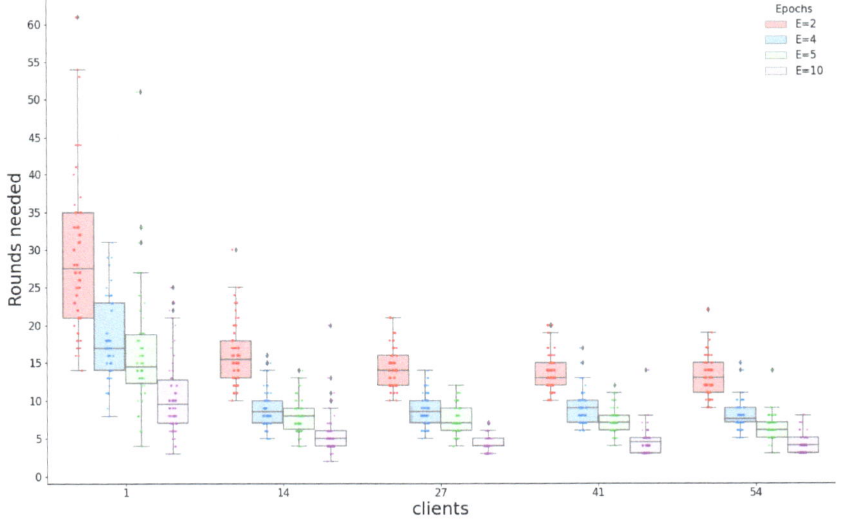

Figure 5. Number of rounds R needed to reach the centralized accuracy baseline (81.7%), averaging over 50 random executions with different number of clients per round (C) and local epochs at clients (E).

3.3. Federated Learning Performance

The experiments described in Section 3.1 test the interaction between different parameters and how they affect performance, given a fixed experimental setup. In this section, we select the parameters but vary the setups. We compare three training schemes: (1) each institution trains a model using only its local data, (2) a federated setting, and (3) a centralized approach, training on data collected from all institutions.

This experiment differs from Section 3.1 because the training parameters are fixed on a batch size equal to 32 in every case, 20 epochs for schemes 1 and 3, 10 rounds and 2 local epochs on scheme 2, and 50% of the clients on each round. We also introduced a new contender, which is models trained on each client, representing the case of institutions only using their data for an in-house model to be used for themselves, possibly only occurring with large and well-funded institutions, since they would need enough data (enough students) and resources.

To perform an experiment, we first have to simulate the clients. Since simply sampling from the original dataset of (students, courses) pairs could result in two clients having the same student assigned to different courses on each client, we first select a fixed number of students and define the client based on them. Then, for each student S selected to be a part of the client, we generate all the pairs (S, course), which finally constitute the institution's data. This emulates how, in the real world, each student would typically take all of their courses at the same institution. For clarity, let us explain what one execution or run of an experiment in this section consists of: first, we sample a fixed number of students and define a client with their corresponding data (e.g., all the entries (student, course) for every student in each client). We do this to form each client until we run out of students; secondly, we partition each client, where 70% of the data are reserved for training and the other 30% for testing, and the training and testing are performed on each proposed scheme. This process (training and testing) must be performed to ensure a fair comparison between the three schemes. This is why we have data reserved from each client, so we always use the same data for testing. In scheme (1), each institution trains and tests on its own. In scheme (2), training is federated (using that 70% of data from each client), and testing is performed on each client on the remaining 30%. We test it with a model of the same architecture but with weights resulting after the training (in a fashion similar to Section 3.1). Finally, in scheme (3), all training data from institutions are merged into a single training dataset, but testing is performed separately on each client's held-out data. For each scheme, we then report the average accuracy across all institutions.

The second purpose of this section is to assess the performance of the federated version on different data-distribution scenarios. This is important because, in real life, institutions come in all shapes and sizes. Therefore, we will present the scenarios in the following.

3.3.1. Homogeneous Data Distribution

The first scenario is defined by what we call a homogeneous data distribution; that is, each simulated institution (client) is generated by sampling randomly from the original dataset without any bias other than the one already present on the dataset (which favors the positive class, e.g., cases of dropout). This scenario replicates the case in real life where all institutions participating in the federation are comparable, at least when it comes to what is being predicted, in this case, early dropout of students (we could say they are equally engaging). The whole dataset has 76% dropout cases and 24% non-dropout cases; In this scenario, in the first phase of the execution, where we select the students, we sample randomly from the dataset. Therefore the label distribution mentioned will be approximately the same on each of the clients.

3.3.2. Heterogeneous Data Distribution

In the second scenario, the clients are generated by sampling from subsets of the original data, which are partitioned into three parts, using criteria based on the dropout rate. Given a student S, we define their dropout rate as the ratio of courses where they dropped out, out of all the courses in which they are enrolled. Figure 6 shows the distribution of this number across all students. We can see that most students have a dropout rate of 1; they dropped out of all their courses. Others have a dropout rate based around 0.5, so they dropped out from about half of their courses, and the rest have dropped out from no courses, so their dropout rate is 0. Based on this insight, we define three intervals: students with dropout rates from 0 to 0.2, from 0.2 to 0.8, and from 0.8 to 1. The size of the categories is shown in Table 2. Therefore, now with the defined categories, on the first phase of the execution of the experiment, we again sample 1000 students to define each client, generating its pairs as explained before, but with the condition that all students must be part of the same category of dropout rate. Based on Table 2, this process yields 9 clients with students having a low dropout rate (8 in 1000 students, 1 in 723), therefore having a label distribution skewed towards the negative class (e.g., cases of no dropout); 21 clients with a medium dropout rate, so a neutral label distribution; and 7 clients with a high dropout rate and therefore with a label distribution skewed towards the positive class. This heterogeneous scenario tries to emulate the case in real life where different institutions may have different levels of overall engagement (e.g., different overall dropout rates), either because of their teaching methods, the socio-economic background of their students, or any other reason. With our experiments, we hope to see how this affects each scheme's performance and what scheme works better for each kind of institution.

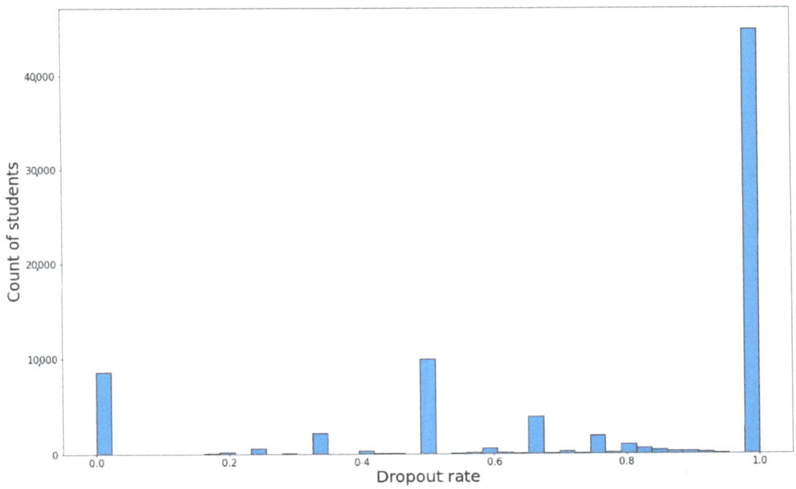

Figure 6. Dropout rate distribution over all students.

Table 2. Student categories according to dropout rate.

Students with a low dropout rate (lower than 0.2):	8895	11.54%
Students with a medium dropout rate (between 0.2 and 0.8):	20,567	26.68%
Students with a high dropout rate (higher than 0.8):	47,621	61.78%

3.3.3. Results

Figure 7 shows the results of 50 independent runs of the experiment in the first scenario, all schemes. By an independent run, we mean executing the experiment from the first step, randomly generating the clients. In each run, the clients are different; therefore, the results vary accordingly. Each point in this figure represents the average accuracy across clients achieved on each of their test data. The figure features the results of each scheme separately (histograms on diagonal), and one versus another, so it shows at the same time how they perform individually but also how they compare to one another. It clearly shows which scheme performs better by counting the number of dots above or below the drawn $y = x$ lines. More points above the line means the scheme referenced on *y-axis* performs better.

If we focus on the histograms, we can observe that the results have different variances. When the institutions train separately, they all have an accuracy of around 81%, which is very focused. The centralized version has a more significant dispersion, and the federation has an even bigger one.

The cases in which each institution trains separately have better results on average than cases with the federated version; see quadrant at mid-left. This could be because the federation needs more rounds. However, the institutions training alone tend to perform worse than the centralized scheme, so sharing data proves beneficial. Finally, the federated version performs worse than the centralized model, but it could be because of a lack of rounds. Under the hypothesis of homogeneous data distribution, federating the training does not increase accuracy. However, given enough rounds, it could equal it, based on what we showed in Section 3.1.

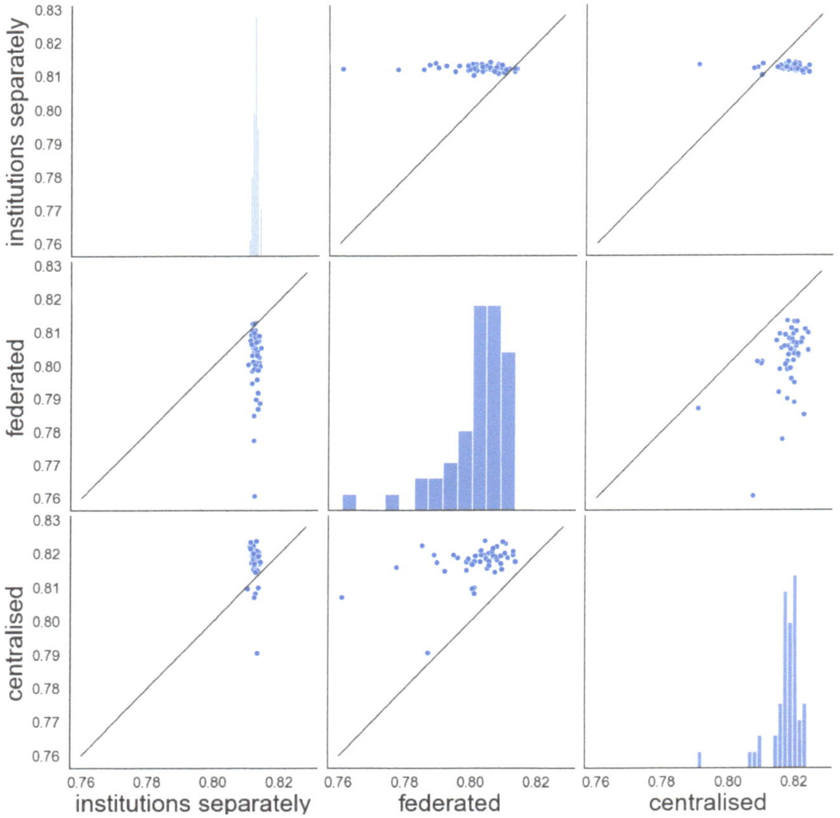

Figure 7. Mean accuracy comparison of 50 independent runs using three training schemes: institutions alone, federated, and centralized. Homogeneous scenario (clients made sampling randomly from the data set). Figure is symmetrical.

Moving on to the heterogeneous scenario, Figure 8 shows the mean accuracy of 50 independent runs for each scheme. This figure is similar to Figure 7, but here we indicate the type of institution with colors according to the categories defined before.

Since fewer students have a low dropout rate (11.54% of the whole population), they are underrepresented in the total dataset, so they perform the worst in the centralized scheme (quadrant at bottom-right). Furthermore, this category only features nine institutions out of the 77 available; therefore, they are sampled less frequently during the federated averaging algorithm in the federated scheme and tend to perform poorly, seeing that they also have the worst performance for this scheme (middle quadrant). However, when the training is carried out separately, the models fit nicely to each institution's data, regardless of their low proportion on the whole federation. The models never see the data on the rest of the institutions since they train and test in isolation. In this case, we achieve better performance (top-left, where orange and blue bins are overlapped) than in centralized and federated schemes (see orange dots on bottom-left and middle-left).

Interestingly, in institutions with medium dropout rates (green), the scheme of separated clients has its worst performance. This could be caused by the fact that here, the labels on each client are more balanced since the students in this category tend to have a 50/50 dropout rate, as opposed to the other categories where institutions have mostly only positive (high dropout) or only negative (low dropout) labels. This setup makes it harder for a model to generalize, hence the poorer performance The federated and centralized schemes perform similarly.

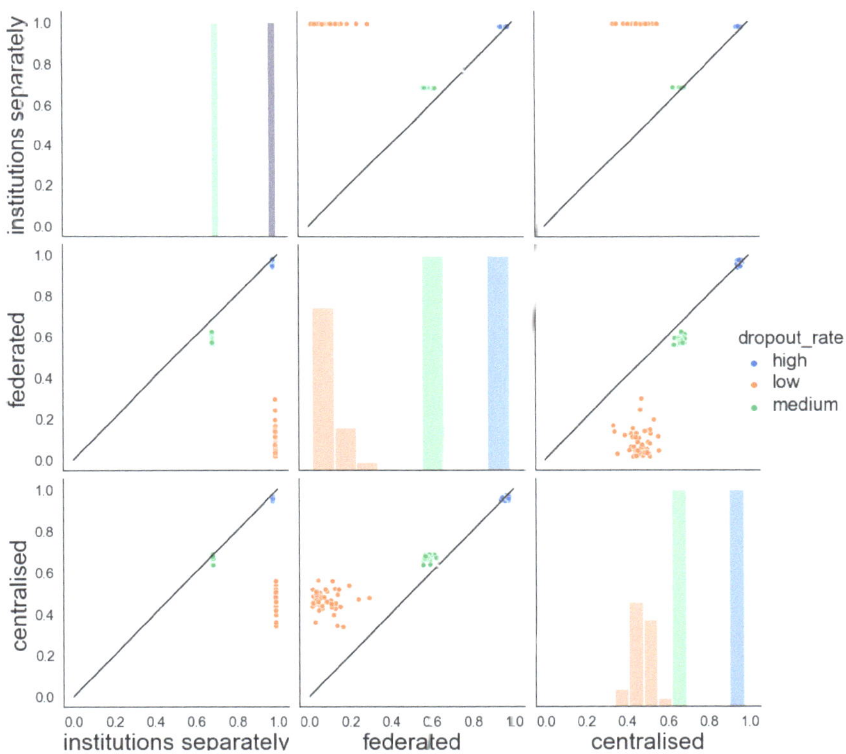

Figure 8. Mean accuracy comparison of 50 independent runs using three training schemes: institutions alone, federated, and centralized. Dropout rate varies between clients according to the categories defined in Table 2.

Lastly, in terms of the high dropout rate, the largest category, each scheme has its best result, with the institutions separated also achieving it with the low dropout rate. Federation performs just as well as the centralized (mid-bottom) and just as well as with separate institutions (mid-left). This class is the biggest, so the centralized model generalizes easily; correspondingly, it is the one with the most clients, and the federation learns from it more often (these clients are selected for the most rounds, on average). These clients are highly unbalanced, so it is also easy for the isolated models to learn to predict there (just as in the low dropout rate).

We can conclude that if the institution belongs to a "favoured class", i.e., those for which there are more data, then there is not much difference between using one scheme or another. If the institution has a random distribution, it is better to use approaches that leverage data from other clients, such as centralized or federated schemes. In this case, there is no evidence of performance loss using federation. To complete this analysis, if the institution belongs to one of the categories with fewer data, it is more convenient to use a customized model trained only with its data. This makes sense, because both the centralized model and the federated model will not be trained with this outlier category enough. As we saw on the results, this makes them perform poorly on the outlier institutions.

4. Discussion

The results show that increasing the number of clients and favoring more rounds result in higher accuracy. Concentrating resources on more rounds than local epochs of clients without network constraints brings better results. If time and connectivity are not an issue, using as many clients as possible per round is also optimal. However, the gain is not substantial, and reasonable results could be achieved using much fewer data (as in our experiment, using 25% and 50% of all clients). FL has the potential to achieve the same results as traditional ML in real-world settings, as it does in our experiments. However, testing in a non-experimental setting is needed to confirm this.

However, it is crucial to remember that this is a simulation, and we have yet to consider the problems involved in transferring information over the network in the real world. For example, when connectivity is an issue (such as in institutions in rural areas), it may only be possible to use some clients simultaneously in the same round. Latency may also be a factor to consider. For example, an increase in communication time could make it prohibitive to run many rounds. In these cases, it would be advisable to favor local epochs, even though the experiment in Figure 2 showed that it is not optimal in terms of accuracy. It is also worth noting that we have yet to consider privacy-preserving schemes (e.g., differential privacy [25]) in our experiments.

It is essential to remember that all experiments are based on MOOCs data; this should be kept in mind when extrapolating the results to the context of a physical institution. Some variables have an equivalent (number of courses taken, for example), but others certainly differ. On this note, the number of total courses, students, and especially students per course may not be typical of physical institutions (see Section 2).

It is crucial to notice that many potential issues would make this type of model unfeasible in the real world, such as heterogeneity in sampling and data storage across institutions, lack of processing capacity of underfunded institutions that could lead to discrimination, sampling bias of institutions in different parts of the territory, etc. All these aspects deserve thorough analysis and discussion before adopting this type of solution, as well as further experimentation to gauge the possible limitations of the federated approach.

Finally, we have focused on assessing whether a model can yield similar results in federated and centralized training settings. We also have explored the extent to which each client benefits from the federation, depending mostly on data distribution patterns. We showed that in some cases it is feasible to benefit from the patterns learned by the model at other institutions and that the obtained results are better than simply training a model of its own. Further experiments are needed to extend the observations to other scenarios, for example, considering the relative size of the institutions, their hardware and connection capabilities, etc.

5. Conclusions

In this paper, we evaluate the application of federated learning for learning analytics, specifically for student dropout prediction based on students' activities. We implemented a neural network model to predict students' behavior, and we explored different training scenarios (centralized and federated under various data-distribution hypotheses). In addition, we evaluated the influence on the prediction results of parameters such as the number of clients, the data distribution, the batch size, and the number of epochs. Although more exhaustive evaluations of the approach are still to be carried out, the results are auspicious. Our future work includes using real data and studying the possible repercussions that enabling mechanisms such as differential privacy could have. In all cases, interesting conclusions are reached, which demonstrate the feasibility of this approach and allow for envisioning its application at institutional and industrial levels in many scenarios.

Author Contributions: Conceptualization, P.B., G.C., L.E. and M.I.F.; software, C.F. and A.T.; validation, G.C. and M.I.F.; investigation, P.B., G.C., L.E., M.I.F., C.F. and A.T.; writing—original draft preparation, P.B., G.C., L.E. and M.I.F.; writing—review and editing, L.E.; visualization, M.I.F.; supervision, P.B., G.C., L.E. and M.I.F. All authors have read and agreed to the published version of the manuscript.

Funding: This research was funded by Agencia Nacional de Innovación e Investigación (ANII) Uruguay, Grant Number FMV_3_2020_1_162910.

Institutional Review Board Statement: Not applicable.

Informed Consent Statement: Not applicable.

Data Availability Statement: Not applicable.

Conflicts of Interest: The authors declare no conflict of interest. The funders had no role in the design of the study; in the collection, analyses, or interpretation of data; in the writing of the manuscript; or in the decision to publish the results.

Appendix A

Features vector

There are different actions tracked on the platforms: video actions (seek, play, pause, stop, load), problem actions (get the problem, check, reset), forum actions (create a thread, comment, delete thread, delete comment), click actions, and closing the page. Therefore, the entire feature vector has 21 features, each corresponding to one action, counting the number of times the student performed the action during their enrollment in the course. Programatically, it is a list such as the following:

```
['seek_video#num', 'play_video#num', 'pause_video#num', 'stop_video#num',
'load_video#num', 'problem_get#num', 'problem_check#num',
'problem_save#num', 'reset_problem#num', 'problem_check_correct#num',
'problem_check_incorrect#num', 'create_thread#num',
'create_comment#num', 'delete_thread#num', 'delete_comment#num',
'click_info#num', 'click_courseware#num', 'click_about#num',
'click_forum#num', 'click_progress#num', 'close_courseware#num']
```

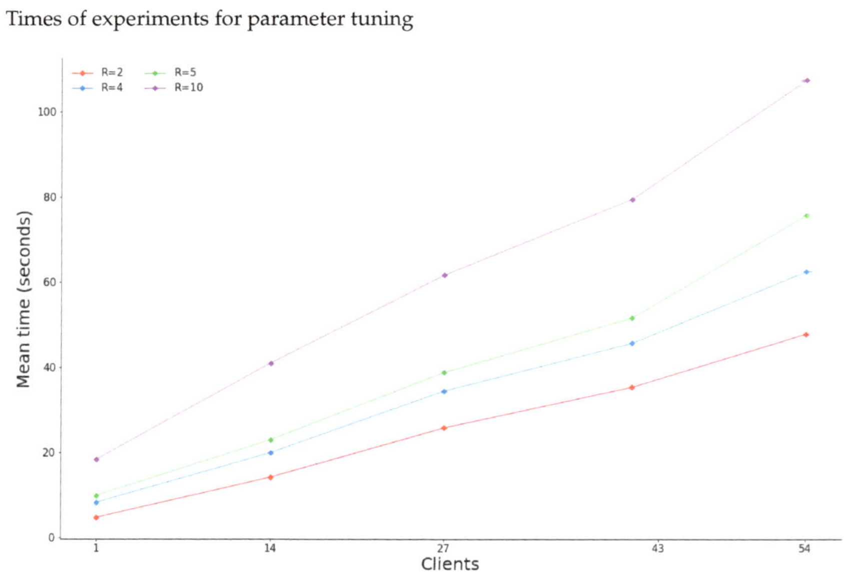

Figure A1. Mean time results of federation applied to dropout prediction, averaging over 50 random executions with different amount of clients per round (C), number of rounds (R), and local epochs of clients (E),s where $R \times E = 20$.

Hardware specifications for dropout experiments:

- CPU 2.3 GHz Quad-Core Intel Core i7.
- 16 GB of RAM.

References

1. Drachsler, H.; Kismihók, G.; Chen, W.; Hoel, T.; Berg, A.; Cooper, A.; Scheffel, M.; Ferguson, R. Ethical and privacy issues in the design of learning analytics applications. In *ACM International Conference Proceeding Series*; Association for Computing Machinery: New York, NY, USA, 2016; Volume 25–29, pp. 492–493. [CrossRef]
2. Banihashem, S.K.; Aliabadi, K.; Pourroostaei Ardakani, S.; Delaver, A.; Nili Ahmadabadi, M. Learning Analytics: A Systematic Literature Review. *Interdiscip. J. Virtual Learn. Med. Sci.* **2018**, *9*, 63024. [CrossRef]
3. Mangaroska, K.; Giannakos, M. Learning Analytics for Learning Design: A Systematic Literature Review of Analytics-Driven Design to Enhance Learning. *IEEE Trans. Learn. Technol.* **2019**, *12*, 516–534. [CrossRef]
4. Sweeney, L. k-anonymity: A model for protecting privacy. *Int. J. Uncertain. Fuzziness Knowl.-Based Syst.* **2002**, *10*, 557–570. [CrossRef]
5. Khalil, M.; Ebner, M. De-Identification in Learning Analytics. *J. Learn. Anal.* **2016**, *3*, 129–138. [CrossRef]
6. Kyritsi, K.H.; Zorkadis, V.; Stavropoulos, E.C.; Verykios, V.S. The pursuit of patterns in educational data mining as a threat to student privacy. *J. Interact. Media Educ.* **2019**, *2019*, 2. [CrossRef]
7. Dwork, C. Differential privacy: A survey of results. In *Proceedings of the International Conference on Theory and Applications of Models of Computation*; Springer: Berlin/Heidelberg, Germany, 2008; pp. 1–19.
8. Gursoy, M.E.; Inan, A.; Nergiz, M.E.; Saygin, Y. Privacy-Preserving Learning Analytics: Challenges and Techniques. *IEEE Trans. Learn. Technol.* **2017**, *10*, 68–81. [CrossRef]
9. Konečný, J.; McMahan, H.B.; Ramage, D.; Richtarik, P. Federated Optimization: Distributed Machine Learning for On-Device Intelligence. *arXiv* **2016**, arXiv:1610.02527. https://doi.org/10.48550/arXiv.1610.02527
10. Li, T.; Sahu, A.K.; Talwalkar, A.; Smith, V. Federated Learning: Challenges, Methods, and Future Directions. *IEEE Signal Process. Mag.* **2020**, *37*, 50–60. [CrossRef]
11. Yang, Q.; Liu, Y.; Chen, T.; Tong, Y. Federated Machine Learning: Concept and Applications. *ACM Trans. Intell. Syst. Technol.* **2019**, *10*, 1–19. [CrossRef]
12. Hakak, S.; Ray, S.; Khan, W.Z.; Scheme, E. A framework for edge-assisted healthcare data analytics using federated learning. In Proceedings of the 2020 IEEE International Conference on Big Data (Big Data), Atlanta, GA, USA, 10–13 December 2020; pp. 3423–3427.

13. Nguyen, D.C.; Pham, Q.V.; Pathirana, P.N.; Ding, M.; Seneviratne, A.; Lin, Z.; Dobre, O.; Hwang, W.J. Federated learning for smart healthcare: A survey. *ACM Comput. Surv. (CSUR)* **2022**, *55*, 1–37. [CrossRef]
14. Rieke, N.; Hancox, J.; Li, W.; Milletari, F.; Roth, H.R.; Albarqouni, S.; Bakas, S.; Galtier, M.N.; Landman, B.A.; Maier-Hein, K.; et al. The future of digital health with federated learning. *NPJ Digit. Med.* **2020**, *3*, 1–7. [CrossRef]
15. Divi, S.; Lin, Y.S.; Farrukh, H.; Celik, Z.B. New Metrics to Evaluate the Performance and Fairness of Personalized Federated Learning. *arXiv* **2021**, arXiv:2107.13173. https://doi.org/10.48550/ARXIV.2107.13173.
16. Shi, Y.; Yu, H.; Leung, C. A Survey of Fairness-Aware Federated Learning. *arXiv* **2021**, arXiv:2111.01872.
17. Feng, W.; Tang, J.; Liu, T.X. Understanding dropouts in MOOCs. In *Proceedings of the AAAI Conference on Artificial Intelligence*; AAAI: Palo Alto, CA, USA, 2019; Volume 33, pp. 517–524.
18. Guo, S.; Zeng, D. Pedagogical Data Federation toward Education 4.0. In *Proceedings of the 6th International Conference on Frontiers of Educational Technologies*; Association for Computing Machinery: New York, NY, USA, 2020; pp. 51–55. [CrossRef]
19. Kairouz, P.; McMahan, B.; Song, S.; Thakkar, O.; Thakurta, A.; Xu, Z. Practical and Private (Deep) Learning without Sampling or Shuffling. *arXiv* **2021**, arXiv:2103.00039. https://doi.org/10.48550/arXiv.2103.00039.
20. Zaman, F. Instilling Responsible and Reliable AI Development with Federated Learning. 2020. Available online: https://medium.com/accenture-the-dock/instilling-responsible-and-reliable-ai-development-with-federated-learning-d23c366c5efd (accessed on 3 January 2023).
21. McMahan, B.; Moore, E.; Ramage, D.; Hampson, S.; y Arcas, B.A. Communication-efficient learning of deep networks from decentralized data. In Proceedings of the Artificial Intelligence and Statistics, Fort Lauderdale, FL, USA, 20–22 April 2017; pp. 1273–1282.
22. KDD. KDDCup. 2015. Available online: http://moocdata.cn/challenges/kdd-cup-2015 (accessed on 3 January 2023).
23. FLEA. FLEA Project Public Repository. 2022. Available online: https://gitlab.fing.edu.uy/lorenae/flea (accessed on 3 January 2023).
24. Kingma, D.P.; Ba, J. Adam: A method for stochastic optimization. *arXiv* **2014**, arXiv:1412.6980.
25. Liu, B.; Ding, M.; Shaham, S.; Rahayu, W.; Farokhi, F.; Lin, Z. When Machine Learning Meets Privacy: A Survey and Outlook. *ACM Comput. Surv.* **2021**, *54*, 1–36.

data

MDPI

Data Descriptor

Data from Zimbabwean College Students on the Measurement Invariance of the Entrepreneurship Goal and Implementation Intentions Scales

Takawira Munyaradzi Ndofirepi

Faculty of Management Sciences, Central University of Technology, Bloemfontein 9300, South Africa

Abstract: This article analyses primary data on the entrepreneurship intentions of selected Zimbabwean college students. The goal of this study was to examine the measurement invariance of the entrepreneurship goal and implementation intention scales across gender groups in a higher education setting. Entrepreneurship goal intentions (EGI) and entrepreneurship implementation intentions (EII) are examined as separate but related constructs. To address the research goal, a positivist philosophy and quantitative research approach were used. A cross-sectional survey was used to collect data from a convenient sample of 262 college students in Zimbabwe. A researcher-administered questionnaire, written in English, was distributed to the respondents and collected after completion. Multi-group confirmatory analysis was performed on the dataset using JASP computer software. The results obtained confirmed all four levels of measurement invariance, namely configural, metric, scalar, and strict invariance. The pattern of the results validates the consistency of the measurement properties of the entrepreneurial intention instruments designed in developed countries across different contexts of use. Researchers, entrepreneurship educators, and policymakers in Zimbabwe can use the results of this analysis to quantify potential entrepreneurs among young adults and to come up with intervention measures to support future entrepreneurship.

Dataset: https://doi.org/10.17632/74nhxtmrzx.1.

Dataset License: CC BY 4.0.

Keywords: entrepreneurial intentions; measurement invariance; multigroup analysis; gender; Zimbabwe

Citation: Ndofirepi, T.M. Data from Zimbabwean College Students on the Measurement Invariance of the Entrepreneurship Goal and Implementation Intentions Scales. *Data* **2022**, *7*, 172. https://doi.org/10.3390/data7120172

Academic Editor: Antonio Sarasa Cabezuelo

Received: 26 October 2022
Accepted: 26 November 2022
Published: 29 November 2022

Publisher's Note: MDPI stays neutral with regard to jurisdictional claims in published maps and institutional affiliations.

1. Summary

The entrepreneurship intention construct is an important component in understanding the entrepreneurial mindset. From a cognitive perspective, the concept of entrepreneurial intentions sheds some light on why some people seek out opportunities to set up and manage business ventures, while others do not [1]. According to [2], entrepreneurial intent is "a self-acknowledged conviction by a person that they intend to establish a new business venture and consciously plan to do so at some point in the future" (p. 676). The origins of the entrepreneurship intentions notion lie in the seminal cognitive psychology intentions models, specifically Ajzen and Fishbein's theory of reasoned action and Ajzen's theory of planned behaviour [3–5]. As the body of research on the concept grew over time, so did the number of variants of the entrepreneurship intention construct, as well as the cognate theories [6]. Entrepreneurial intentions are widely regarded as a reliable predictor of future entrepreneurial activity and have been widely used by various stakeholders around the world to forecast entrepreneurship propensity among young people [4].

Diverse entrepreneurship intention measurement instruments developed by scholars in universities and research institutes in developed countries are widely used by entrepreneurship scholars worldwide [2]. However, little attention has been paid to the consistency of these instruments' measurement properties across different contexts of use.

Thus, African entrepreneurship research, like any other field of primary research that uses psychological constructs, relies on measurement instruments developed in Western, educated, industrialised, rich, and democratic (WEIRD) societies to measure entrepreneurship intentions. This is done without regard for contextual differences or the possibility that the instrument's measurement properties will differ across cultural or demographic groups. The possible outcome is measurement inconsistency, which makes it challenging to compare, authenticate, synthesise, or add to earlier research outcomes [2]. Measurement errors can occur when measuring entrepreneurial intentions across contextual settings because of scalar non-equivalence. Scalar non-equivalence happens when scale scores vary across nations and the variation can be attributed to cultural or national differences [7]. When researchers use scales in surveys, they make the supposition that participants from various nations who have similar values for a specific variable would provide similar ratings on a scale [8]. Varying levels of knowledge of scaling styles, however, may lead to discrepancies.

Against this background, the purpose of this study was to evaluate the measurement invariance of the entrepreneurship goal intentions (EGI) and entrepreneurship implementation intentions (EII) scales (sub-dimensions of entrepreneurship intentions) when administered to male and female college students in Zimbabwe, an African country. The outcomes of the tests would either support or call into question the indiscriminate usage of such tools.

2. Materials and Methods

To accomplish the research goal, a positivist philosophy and quantitative research approach were used. In July 2019, data was collected from college students in Zimbabwe's Midlands province via a cross-sectional survey. A self-completion questionnaire, written in the English language, was used for the purpose. The mall-intercept approach was used to distribute the questionnaire to the respondents identified with the help of three trained research assistants. The respondents filled out the questionnaires and handed them back to the research assistant after completion. The respondents were chosen because they were college students and willing to engage in the study. Thus, participation in the study was entirely voluntary, and participants were assured of their right to confidentiality and privacy. The study aimed for a minimum of 200 participants, following Kline's sample size requirements for structural equation modelling [9]. To meet this expectation, 350 questionnaires were printed and distributed. Of those completed and returned to the researcher, only 262 had minimal cases of incomplete information and were therefore usable.

A six-item entrepreneurship goal intention scale was adapted from Liñán and Chen [10]. The respondents needed to indicate their level of agreement with each of the following items, which were based on a five-point Likert scale: "It is very likely that I will start a venture one day", "I am willing to make every effort to become an entrepreneur", "I have serious doubts whether I will ever start a venture", " I am determined to start a venture in the future", and "My professional objective is to be an entrepreneur". All scale points were labelled 1 (strongly disagree) to 5 (strongly agree).

The entrepreneurship implementation intention measure was adapted from [11] and used a three-item and five-point Likert scale with response categories ranging from 1 (Nothing at all) to 5 (I have it totally planned). The respondents needed to indicate how much they had thought about the following aspects in the creation of their business venture: "What specific steps I have to take to create my company", "When I will take each of the steps to create my company", and "Where I will carry out each of the steps to create my company".

The measurement invariance of the scales was ascertained using multi-group confirmatory factor analysis. Four levels of measurement invariance, namely configural, metric, scalar, and strict invariance were tested. Firstly, the configural invariance test was designed to ascertain whether the latent variables had the same pattern of free and fixed loadings. Secondly, metric invariance sought to test the equivalence of the item loadings on the latent variables, and the procedure entailed running a confirmatory factor analysis test with the item loadings on the two constructs constrained to be equivalent in males and females. Thirdly, scalar invariance, which implies that mean differences in the latent variables reflect all mean differences in the shared variance of the measuring items, was tested by restricting the item intercepts to be equal in the male and female groups and then running a confirmatory factor analysis of the model. Lastly, strict invariance which reflects the equivalence of item residuals of metric and scalar invariant items across the gender groups was evaluated by running a confirmatory factor analysis with the item residuals constrained to be equivalent in both males and females. Measurement invariance was supported if the overall model fitness was not significantly worse off at each stage of the test. The model-fit indices used in this study include the comparative fit index (CFI), goodness-of-fit index (GFI), and standardized root mean square residual (SRMR). CFI and GFI values greater than 0.90 imply that the model fitness is acceptable, while for SRMR, values less than 0.08 suggest an adequate model fit [12].

3. Results

Firstly, Figure 1 depicts the conceptual model tested, which comprised entrepreneurship goal intentions and entrepreneurship implementation intentions and their indicators. Secondly, Table 1 shows the demographic profile of the respondents, including their gender, age, marital status, field of study, highest qualification attained, and three life experience categories. Most of the respondents were males (52.29%, n = 137), aged between 21 and 30 years (71.76%, n = 188), were single (82.44%, n = 216), and had high school education as their highest qualification (79.39%, n = 208).

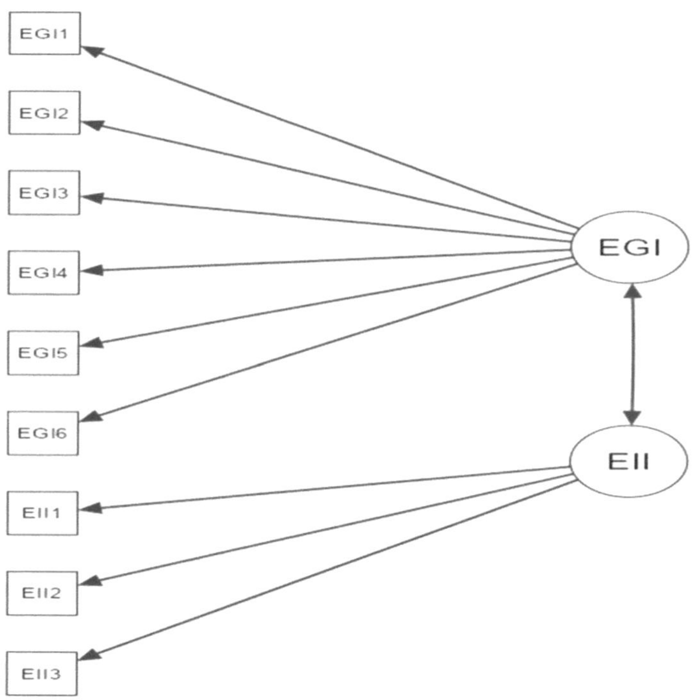

Figure 1. Conceptual model.

Table 1. Demographic profile of respondents.

Variable		Frequency	Percent
Gender	Male	125	47.710
	Female	137	52.290
		262	100
Age in years	Below 21	57	21.756
	21 to 30	188	71.756
	31 to 40	13	4.962
	41 to 50	1	0.382
	Missing values	3	1.145
		262	100
Marital status	Not married	216	82.443
	Married	46	17.557
		262	100
Qualification	High school	208	79.389
	Tertiary certificate	43	16.412
	Diploma/Degree	11	4.198
		262	100
Field of study	Applied Sciences	92	35.115
	Business/Commerce	44	16.794
	Engineering	126	48.092
		262	100

Note that EGI means entrepreneurship goal intentions and EII stands for entrepreneurship implementation intentions.

Thirdly, Table 2 summarises the results relating to the robustness of the measurement models, revealing the reliability and construct validity of the two scales across the different gender groups. For both latent variables, the findings suggest satisfactory levels of reliability and construct validity, as shown by the Cronbach alpha values of greater than 0.8 and the average variances extracted that were greater than 0.5 for males and females.

Table 2. Reliability and convergent validity indices.

Group	Variable	Number of Items	Cronbach Alpha (α)	Average Variance Extracted
Male	EGI	3	0.889	0.693
Male	EII	6	0.873	0.773
Female	EGI	3	0.840	0.592
Female	EII	6	0.844	0.711

Note that EGI means entrepreneurship goal intentions and EII stands for entrepreneurship implementation intentions.

Next, Table 3 shows whether the measurement properties of the scales differed between male and female respondents. The consistency of each measure was tested at four levels: configural, metric, scalar, and metric invariance. Finally, the results in Table 3 suggest that the conditions for the four levels of measurement invariance were satisfied given that most of the model-fit indices satisfied the minimum acceptable conditions expected.

Table 3. Measurement invariance results of the entrepreneurship goal and implementation intentions scale.

	χ^2	df	GFI	SRMR	CFI	Change in CFI
Configural	50.621	52	0.995	0.057	1	-
Metric	67.818	59	0.993	0.067	0.999	0.006
Scalar	79.380	84	0.992	0.061	1	0.007
Strict	79.380	84	0.992	0.061	1	0.007

(CFI Comparative fit index, GFI: Goodness-of-fit index, SRMR: standardized root mean square residual, df: degrees of freedom).

4. Conclusions

The study's goal was to establish the measurement invariance of the entrepreneurship goal intentions (EGI) and entrepreneurship implementation intentions (EII) scales when administered to male and female Zimbabwean college students. A multigroup confirmatory factor analysis test demonstrated that the scales of entrepreneurship goal intentions and entrepreneurship implementation intentions were invariant among the gender groups sampled. As a result, even though the two measurements were designed and verified in a developed-world setting, their measuring properties remained constant in a distinct cultural milieu. This discovery lends credence to the use of scales in various world areas. The results corroborate those of a study conducted in Greece by [13], which discovered that although there were variations in the country's levels of entrepreneurial intentions between men and women, these variations were not due to the scales' measurement characteristics. However, other studies conducted outside the context of Western culture [14,15] only succeeded in demonstrating the partial measurement invariance of entrepreneurial intentions measures.

The data is relevant to a wide range of players in Zimbabwe's economy. First, the data will be beneficial to entrepreneurship scholars since it gives information on the consistency of the psychometric features of an entrepreneurship intention testing instrument across different gender groups. Researchers interested in the study's topic can use the data in future replication studies. Second, the dataset will be beneficial to researchers, educators, business development assistance organisations, and policymakers who are looking for reliable tools to evaluate the level of entrepreneurial propensity among students to quantify the pool of future entrepreneurs. Third, authorities might utilise the data to create policies to enhance the interest of young people in entrepreneurship. Finally, causal links that can be used to generate entrepreneurship policy-related inferences can be tested by incorporating a new dataset on other variables that can either be antecedents or outcomes of entrepreneurial intent.

However, the generalizability of the study findings is limited due to the use of a convenient sample of respondents, as well as the small sample size, which may not accurately reflect all the qualities of the target population. Future research on the same topic should aim to use more representative samples.

Funding: This research did not receive any specific grant from funding agencies in the public, commercial, or not-for-profit sectors.

Institutional Review Board Statement: Ethical approval not required.

Informed Consent Statement: All respondents gave verbal informed consent and participated voluntarily. The author confirms that the research was carried out ethically. No ethical permission was sought.

Data Availability Statement: Underlying data: Mendeley Data: Measurement invariance of entrepreneurship intentions scales. https://doi.org/10.17632/74nhxtmrzx.1, accessed on 5 October 2022. This project contains the following underlying data: File 2.xlsx (data file). Extended data: File 1.doc (blank questionnaire file). Data are available under the terms of the Creative Commons Attribution 4.0 International license (CC-BY 4.0).

Acknowledgments: The author would like to thank all the respondents and research assistants who took part in the data-gathering process.

Conflicts of Interest: The author declares that he has no known competing financial interests or personal relationships that could appear to influence the work reported in this paper.

References

1. Fayolle, A.; Liñán, F. The future of research on entrepreneurial intentions. *J. Bus. Res.* **2014**, *67*, 663–666. [CrossRef]
2. Thompson, E.R. Entrepreneurial Intent: And Development Reliable Metric. *Entrep. Theory Pract.* **2009**, *33*, 669–694. [CrossRef]
3. Engle, R.L.; Dimitriadi, N.; Gavidia, J.V.; Schlaegel, C.; Delanoe, S.; Alvarado, I.; He, X.; Buame, S.; Wolff, B. Entrepreneurial intent: A twelve-country evaluation of Ajzen's model of planned behavior. *Int. J. Entrep. Behav. Res.* **2010**, *16*, 35–57. [CrossRef]
4. Fayolle, A.; Liñán, F.; Moriano, J.A. Beyond entrepreneurial intentions: Values and motivations in entrepreneurship. *Int. Entrep. Manag. J.* **2014**, *10*, 679–689. [CrossRef]

5. Tornikoski, E.; Maalaoui, A. Critical reflections—The Theory of Planned Behaviour: An interview with Icek Ajzen with implications for entrepreneurship research. *Int. Small Bus. J. Res. Entrep.* **2019**, *37*, 536–550. [CrossRef]
6. Liñán, F.; Fayolle, A. A systematic literature review on entrepreneurial intentions: Citation, thematic analyses, and research agenda. *Int. Entrep. Manag. J.* **2015**, *11*, 907–933. [CrossRef]
7. Bartram, D. Scalar equivalence of OPQ32: Big Five profiles of 31 countries. *J. Cross-Cult. Psychol.* **2013**, *44*, 61–83. [CrossRef]
8. Buil, I.; de Chernatony, L.; Martinez, E. Methodological issues in cross-cultural research: An overview and recommendations. *J. Target. Meas. Anal. Mark.* **2012**, *20*, 223–234. [CrossRef]
9. Kline, R.B. *Principles and Practice of Structural Equation Modeling*, 3rd ed.; Guilford Press: New York, NY, USA, 2011.
10. Liñán, F.; Chen, Y.W. Development and cross-cultural application of a specific instrument to measure entrepreneurial intentions. *Entrep. Theory Pract.* **2009**, *33*, 593–617. [CrossRef]
11. Liñán, F.; Fernández-Serrano, J. ELITE's Initial Questionnaire for Nascent Entrepreneurs. 2018. Available online: /https://www.researchgate.net/profile/Francisco-Linan/publication/323006169_Deliverable_E3-ELITE_Questionnaire_for_nascent_entrepreneurs/links/5a7c2689a6fdcce697d7ef7c/Deliverable-E3-ELITE-Questionnaire-for-nascent-entrepreneurs.pdf (accessed on 20 September 2022).
12. Hooper, D.; Coughlan, J.; Mullen, M. Structural equation modelling: Guidelines for determining model fit. *Electron. J. Bus. Res. Methods* **2008**, *6*, 53–60.
13. Zampetakis, L.A.; Bakatsaki, M.; Litos, C.; Kafetsios, K.G.; Moustakis, V. Gender-based Differential Item Functioning in the Application of the Theory of Planned Behavior for the Study of Entrepreneurial Intentions. *Front. Psychol.* **2017**, *8*, 451. [CrossRef] [PubMed]
14. Looi, K.H. Contextual motivations for undergraduates' entrepreneurial intentions in emerging Asian economies. *J. Entrep.* **2020**, *29*, 53–87. [CrossRef]
15. Moriano, J.A.; Gorgievski, M.; Laguna, M.; Stephan, U.; Zarafshani, K. A cross-cultural approach to understanding entrepreneurial intention. *J. Career Dev.* **2012**, *39*, 162–185. [CrossRef]

data

MDPI

Article

Density-Based Unsupervised Learning Algorithm to Categorize College Students into Dropout Risk Levels

Miguel Angel Valles-Coral *, Luis Salazar-Ramírez, Richard Injante, Edwin Augusto Hernandez-Torres, Juan Juárez-Díaz, Jorge Raul Navarro-Cabrera, Lloy Pinedo and Pierre Vidaurre-Rojas

Facultad de Ingeniería de Sistemas e Informática, Universidad Nacional de San Martín, Jr. Maynas, Tarapoto 22200, Peru
* Correspondence: mavalles@unsm.edu.pe

Abstract: Compliance with the basic conditions of quality in higher education implies the design of strategies to reduce student dropout, and Information and Communication Technologies (ICT) in the educational field have allowed directing, reinforcing, and consolidating the process of professional academic training. We propose an academic and emotional tracking model that uses data mining and machine learning to group university students according to their level of dropout risk. We worked with 670 students from a Peruvian public university, applied 5 valid and reliable psychological assessment questionnaires to them using a chatbot-based system, and then classified them using 3 density-based unsupervised learning algorithms, DBSCAN, K-Means, and HDBSCAN. The results showed that HDBSCAN was the most robust option, obtaining better validity levels in two of the three internal indices evaluated, where the performance of the Silhouette index was 0.6823, the performance of the Davies–Bouldin index was 0.6563, and the performance of the Calinski–Harabasz index was 369.6459. The best number of clusters produced by the internal indices was five. For the validation of external indices, with answers from mental health professionals, we obtained a high level of precision in the *F*-measure: 90.9%, purity: 94.5%, *V*-measure: 86.9%, and ARI: 86.5%, and this indicates the robustness of the proposed model that allows us to categorize university students into five levels according to the risk of dropping out.

Keywords: clustering; data mining; DBSCAN; K-Means; HDBSCAN

Citation: Valles-Coral, M.A.; Salazar-Ramírez, L.; Injante, R.; Hernandez-Torres, E.A.; Juárez-Díaz, J.; Navarro-Cabrera, J.R.; Pinedo, L.; Vidaurre-Rojas, P. Density-Based Unsupervised Learning Algorithm to Categorize College Students into Dropout Risk Levels. *Data* **2022**, *7*, 165. https://doi.org/10.3390/data7110165

Academic Editors: Antonio Sarasa Cabezuelo and Ramón González del Campo Rodríguez Barbero

Received: 18 October 2022
Accepted: 16 November 2022
Published: 18 November 2022

Publisher's Note: MDPI stays neutral with regard to jurisdictional claims in published maps and institutional affiliations.

1. Introduction

The university, within society, is the institution dedicated to teaching, research, and generation of new knowledge, where the student is the nucleus on which its main purposes and principles are governed [1]. Therefore, it is essential to have strategies and mechanisms that ensure the care and permanence of the student, as well as compliance with the basic conditions of educational quality that guarantee the adequate teaching–learning process [2].

During the life of the student, the university stage represents perhaps the most important challenge [3]; in general, this stage takes place at the end of secondary education, a moment when the student experiences big changes that range from the social to the emotional aspects [4]. Likewise, at that time, the student is exposed to new experiences and responsibilities that require high physical and mental effort [5].

This wears them down and makes them self-demanding, a combination that generates anxiety, a normal and involuntary response that produces physical and psychological symptoms [6]. Therefore, not meeting the psychological needs of the student generates problems that affect the learning process, their social development, and puts their health and life at risk [7]. In addition, the low or null attention to the problems of the students by those responsible, generates dissatisfaction and low motivation for the development of academic activities and increases the chances of abandoning studies partially or totally [8,9].

Thus, university tutoring has turned out to be the fundamental support mechanism for monitoring the student's training process [10,11]. In this sense, universities must respond to the challenges of providing comprehensive education, institutionalizing methods and procedures that guarantee the identification of students with academic risks and establishing preventive and corrective intervention programs to mitigate the probability of desertion [12]. Then, we can affirm that within the university, tutoring directs fundamental processes related to the attention and psychological follow-up of the student to promote professional development and projection [13].

Based on this, one of the topics of wide interest in university institutions is the need to have mechanisms and tools that help to face the phenomena related to the risk of student desertion [14]. Likewise, these tools must provide alert systems or strategies that allow us to intervene in the most vulnerable groups that have a higher level of risk of deserting [15]. Thus, we emphasize the use of ICT in the educational field [16–18], and automatic learning methods become the most viable option, since thanks to their characteristics they allow us to develop useful models capable of analyzing and discovering complex patterns in datasets. This allows us to model information for decision-making in the diagnosis and treatment of possible psychological interventions [19–21].

Consequently, our objective and main contribution is to propose an academic and emotional monitoring model that uses data mining and machine learning to group university students, according to their level of dropout risk. In this solution, we integrate ICT advances; specifically, chatbots for data collection and density-based unsupervised algorithms for student clustering, which serve as a basis for future projects and a precedent for other work and joint efforts between mental health professionals and ICT management.

The article is organized as follows: In Section 2, we describe the theoretical foundations of the variables under study, in Section 3 we present the experimental design, as well as the materials and methods we used, and in Section 4 we detail the analysis of the results found and the discussions. Finally, in Section 5, we presented conclusions and future implications of the research.

2. Theoretical Fundament

2.1. College Dropout

The phenomenon of university dropout is one of the main problems affecting educational systems [22], which is why it has been studied by different approaches, such as psychological, sociological, and economic [23]. Each approach has independently exposed the different perspectives and perceptions of the students regarding the main variables that motivated their attempt or action to dropout, encompassing them in two blocks, the academic aspect, and the individual (personal) aspect [24,25].

In the case of the academic aspect, the main variables related to desertion are the previous performance of the student (the knowledge that they have formed before entering the university, which can be used to develop their academic activities), emotional intelligence (ability to understand, use, and manage feelings and emotions appropriately), motivation, and individual learning objectives (the awareness of the knowledge acquired by the student, who puts it into practice in daily life).

For the individual aspect, the main variables are linked to the age and sex of the individual, their socioeconomic status, their social and interpersonal relationships, their mood, and their behavioral aspects [26–31].

These variables generate data and information of each student, and thanks to the advances of ICT in educational issues, mechanisms have been generated that apply data mining and machine learning, and data can be worked on and manipulated, regardless of whether they are static or dynamic. This facilitates analysis and control, and reflected in terms of precision, these mechanisms are very effective and require less effort for data processing, compared to the conventional methods used by mental health professionals [32–34].

Several studies use data mining techniques to find common and specific denominator patterns in student populations and group them to predict academic performance and the possibility of dropping out of academic activities, and based on this, generate immediate solutions that mitigate the cases found, minimizing the dropout rate of university students [35–37].

2.2. Density-Based Clustering

Clustering based on point cloud density is an unsupervised learning methodology whose function is to identify specific groups in the data, based on the fact that a cluster is a region within a contiguous data space of high density of elements, dissociated from other similar clusters by contiguous zones of low density [38,39]. For a better definition, there are different approaches to classify what characterizes different groups in the data.

- Procedurally, the various clustering methods attempt to partition the data into k clusters, such that we minimize within-cluster differences while we maximize between-group differences. We defined notions of dissimilarity within the cluster and dissimilarity between clusters using the distance function "d" [40,41].
- From a statistical point of view, the methods correspond to a parametric approach. We assume that the unknown density, $p(x)$, of the data is a mixture of k densities, $pi(x)$, each of which corresponds to one of the k groups in the data. We assume that $pi(x)$ comes from some parametric family (for example, Gaussian distributions) with unknown parameters, which we then estimate from the data [40,42].

Thus, density-based clustering takes a nonparametric approach, where the clusters in the data are high-density areas of density, $p(x)$. Density-based clustering methods do not require the number of clusters as input parameters, nor do they make any assumptions about the underlying density, $p(x)$, or the within-cluster variance that may exist in the data. Consequently, density-based clusters are not necessarily groups of points with high similarity within the cluster, as measured by the distance function d, but may have an "arbitrary shape" in feature space; sometimes, we also call them "natural pools" [40].

Likewise, we can evaluate the data density by analyzing the neighborhood of each data object. There are two possible ways to define the neighborhood of an object. First, when we express the neighborhood radius of an object as the Euclidean distance to the k-nearest neighbor, we define the neighborhood size dynamically depending on the data density. Object neighborhoods are relatively small in dense regions of the data space and considerably larger in less dense regions of the data space. The second option is to assume the same neighborhood radius for all data objects while pooling the data [43]. The density-based clustering algorithms that we used in the investigation are described in the following sections.

2.2.1. DBSCAN

Application density-based spatial clustering with noise (DBSCAN) is a density-based clustering algorithm proposed in [44], and we used it to evaluate the density of the data in a neighborhood of a predefined radius for each object and expressed it as the number of objects in that neighborhood. Therefore, we could identify three types of data objects in the DBSCAN pool: core objects, border objects, and peripheral objects [45].

- Core objects contains a predefined number of objects, k, in its neighborhood of radius r.
- We call the border objects if there are less than k objects in its neighborhood of radius r, but at least one of them is a core object.
- Peripheral objects is the object with less than k objects in its neighborhood of radius r, and none of them are a core object.

2.2.2. K-Means

K-Means clustering, or also known as the Lloyd–Forgy algorithm, is an unsupervised learning clustering algorithm first introduced in [46]. Its main objective is the classification of unlabeled data, based on characteristics; then, K-Means minimizes the intra-cluster variance and maximizes the inter-cluster variance, where each datum must be as close as possible to its group and as far as possible from another type of group [47,48]. Likewise, we must consider that for K-Means to find the optimal number of clusters, it is possible to apply certain techniques, the most popular being the elbow method [49,50].

2.2.3. HDBSCAN

The density-based clustering algorithm based on hierarchical density estimates (HDBSCAN) is the proposal of the authors of [51], who generated an advanced DBSCAN method, improving the theoretical and practical aspects of the algorithm. The execution of the algorithm in five stages according to [52] is:

- Space transformation (stage 1)

  We defined a new distance metric between points called "mutual reach distance" as:

$$d_{mreach-k}(a,\ b) = \max\{core_k(a),\ core_k(b),\ d(a,\ b)\} \tag{1}$$

Under this metric, dense points (with a low center distance) stay the same distance from each other, but sparser points move away to be at least their center distance from every other point.

- Construction of the minimum spanning tree (stage 2)

  We started by considering the data as a weighted graph with the data points as vertices and an edge between two points with a weight equal to the mutual reach distance of those points. We considered a threshold value, starting high and steadily going down. We released any weighted edges above that threshold. As we released the edges, the graph in connected components started to become disconnected. Eventually, we will obtain a hierarchy of connected components (from fully connected to fully disconnected) at different threshold levels.

- Construction of a cluster hierarchy (stage 3)

  We sorted the edges of the tree by distance (in increasing order) and then iterated, creating a new merge group for each edge.

- Condensation of the cluster hierarchy (stage 4)

  Via the HDBSCAN parameter 'min_samples', we obtained the value for the minimum cluster size, then we traversed the hierarchy and, at each split, noted if one of the new clusters created by the split had fewer points than the minimum cluster size.

  If it was the case that we obtained fewer points than the minimum size, we declared them as points that fall outside of a group. We resolved that the largest group retained the identity of the main group, marking which points fell out of the group and at what distance value that happened.

  If, on the other hand, the split was into two groups (each at least as large as the minimum size), then we considered that we were dealing with a true split of the group and kept that split persistent in the tree. After traversing the entire hierarchy and doing this, we obtained a much smaller tree with a small number of nodes, each of which had data on how the size of the cluster at that node decreased over distance.

- Extraction of stable clusters from the condensed tree (stage 5)

  Doing so involves calculating the stability of each previously formed group as follows:

$$\sum\nolimits_{p\ \in\ cluster}(\lambda_p - \lambda_{birth}) \tag{2}$$

where lambda $\lambda = \frac{1}{distance}$, where λ_p is the lambda value at which point p "got out of the group", which is a value somewhere between λ_{birth} (lambda value when the group broke up and became its own group) and λ_{death} (lambda value when the group was split into smaller groups).

We declared all leaf nodes as selected clusters. Therefore, we proceeded through the tree (in reverse topological sort order). If the sum of the stabilities of the secondary clusters was greater than the stability of the cluster, we set the stability of the cluster to be the sum of the secondary stabilities. If, on the other hand, the stability of the cluster was greater than the sum of its children, then we declared the cluster to be a selected cluster and deselected all its descendants. Once we reached the root node, we named it as current set of clusters.

2.3. Cluster Validation Techniques

Clustering methods have the objective of discovering characteristic groups present in a universe of data. In general, they tend to look for clusters whose members are close together (i.e., have a high degree of similarity) and are well-separated from other clusters [53]. Therefore, one of the most important problems in the field of cluster analysis is the validation of the results to find the number of groups or clusters best-suited to the data provided. For this, there are three approaches to verify the validity of the clusters: external, internal, and relative [54].

For the case of the study, given its nature and applied methodology, we used three internal validation indices:

- Silhouette coefficient

The Silhouette coefficient evaluates the validity of the clustering and selects the appropriate number of clusters. When the value of the coefficient is one or close to one, it indicates the good cohesion relationship between the elements of the cluster (internal) and the separability between the clusters (external). If the coefficient is zero or close to zero, it indicates that the clusters tend to overlap each other, and for values equal to or close to minus one, it indicates that the assignment to the cluster is incorrect, because the different clusters have greater similarity [55]:

$$s(i) = \frac{(b(i) - a(i))}{Max\{a(i), b(i)\}} \tag{3}$$

where $a(i)$ is the average distance within the cluster and $b(i)$ is the average distance of the nearest cluster for each sample [55].

- Calinski–Harabasz coefficient (*CH*)

The Calinski–Harabasz coefficient is the ratio of the sum of the inter-cluster spread to the within-cluster spread for all clusters (where we define the spread as the sum of the squared distances). For the case of this coefficient, we relate a higher score to a model with better-defined clusters [56]:

$$CH = \frac{trace(S_B)}{trace(S_w)} \cdot \frac{n_p - 1}{n_p - k} \tag{4}$$

where (S_B) is the intergroup dispersion matrix, (S_w) is the internal dispersion matrix, n_p is the number of grouped samples, and k is the number of clusters [57].

- Davies–Bouldin coefficient (*DB*)

This coefficient indicates the average "similarity" between clusters, where similarity is a measure that compares the distance between clusters to the size of the clusters themselves. Zero is the lowest possible score. Values closer to zero indicate better partitioning [58].

$$DB = \frac{1}{c} \sum_{i=1}^{c} Max_{i \neq j} \left\{ \frac{d(X_i) + d(X_j)}{d(c_i, c_j)} \right\} \tag{5}$$

where c denotes the number of clusters, i and j are labeled clusters, so $d(X_i)$ and $d(X_j)$ are all samples in clusters i and j to their respective cluster centroids, and $d(c_i, c_j)$ is the distance between these centroids [57].

We also used three external validation indices:

- *F*-measure

The F-measure combines the precision and recall concepts. Precision is the ratio of the number of true positives to the number of false positives and is intuitively the ability of the classifier not to label a sample that is negative as positive. Recall is the ratio of the number of true positives to the number of false negatives [59].

$$Recall(i,j) = \frac{n_{ij}}{n_i} \tag{6}$$

$$Precision(i,j) = \frac{n_{ij}}{n_j} \tag{7}$$

where n_{ij} is the number of elements of class i that are in cluster j, n_j is the number of elements in cluster j, and n_i is the number of elements in class i. We calculated the F-measure of cluster j and class i with [57]:

$$F(i,j) = \frac{2Recall(i,j)Precision(i,j)}{Precision(i,j) + Recall(i,j)} \tag{8}$$

The values of (8) are within the interval [0–1], and larger values indicate better quality of the grouping.

- *Purity*

Purity is the analysis of the clusters that yields the percentage value of the total number of elements that we correctly classified in the range of [0–1] [60]. For each cluster, the purity, $P_j = \frac{1}{n_j} Max_i(n_j^i)$, is the number of elements in j with class label i. Then, P_j represents a fraction of the total size of the cluster that the largest class of elements allocated. We obtained the total purity estimate from (9) [57]:

$$Purity = \sum_{j=1}^{m} \frac{n_j}{n} P_j \tag{9}$$

where n_j is the size of cluster j, m is the number of clusters, and n is the total number of elements.

- V-measure

We describe this measure as the harmonic mean between the measures of homogeneity and completeness [57]:

$$V = \frac{(1+\beta) * Homogeneity * Completness}{\beta * Homogeneity + Completness} \tag{10}$$

The result of this measurement varies in a range from 0 to 1, where 1 is the best value and 0 is the worst.

- Random Adjusted Rand Index

The Rand index computes a similarity measure between two clusters by considering all pairs of samples and counting the pairs that map to the same or different clusters in the predicted and true clusters [61]:

$$RI = \frac{a+b}{C_2^{n_{samples}}} \tag{11}$$

where:

- a: The number of times a pair of elements are in the same group for both the actual and predicted grouping.
- b: The number of times that a pair of elements are neither in the same group for the real grouping, nor in the predicted one.
- $C_2^{n_{samples}}$: Total number of possible pairs in the dataset.

We then "likelihood-adjusted" the raw "RI" score into the *ARI* score using the following scheme:

$$ARI = \frac{RI - Expected\_RI}{max(RI) - Expected\_RI} \quad (12)$$

The result provided by this coefficient varies in a range from −1 to 1, where −1 is the worst result, 0 is a random result, and 1 is a completely similar result.

3. Materials and Methods

3.1. Type, Level, and Design of the Investigation

We carried out applied descriptive-level research, where we carried out the collection of data from the observation and subsequent processing to obtain a solution. The design was non-experimental for technological development, we did not manipulate any variables and we only limited ourselves to the study and analysis of pre-existing data to develop a solution that could improve current techniques.

3.2. Population and Sample

The population was undergraduate students enrolled during the academic semester 2021-II of the National University of San Martín, Peru: 5575 individuals. We calculated the sample with the finite population formula, 95% confidence level, resulting in 670 students. To select the sample, we sent emails to the entire university community and selected the first 670 participants who provided their informed consent and completed the provided psychological evaluation questionnaires.

3.3. Proposed Model

We generated a model for grouping students, according to the level of dropout risk, based on their responses to psychological tests, in which we integrated data mining and machine learning to replace the conventional mechanisms dedicated to tutoring; in this way, we improved academic and emotional follow-up. Figure 1 illustrates the proposed model consisting of five stages.

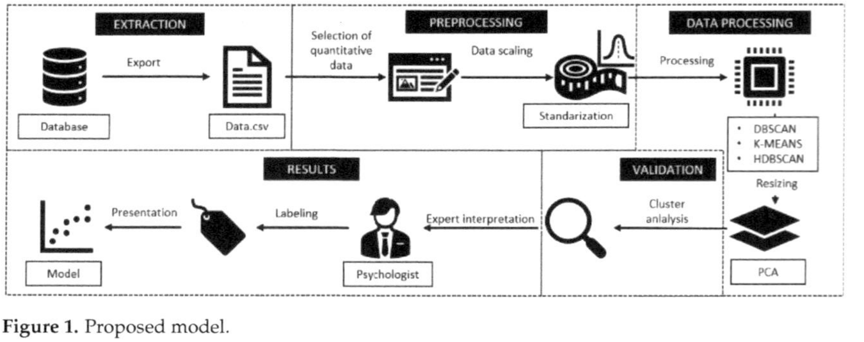

Figure 1. Proposed model.

3.4. Data Colection

We performed data collection through a web chatbot. The students responded to five psychological evaluation questionnaires validated and endorsed by previous studies, according to the following details:

- Study habits is a psychological questionnaire structured by 55 items. The questions evaluate the study habits and techniques used by students, which influence the learning process. The questionnaire is divided into five dimensions: how to organize to study, strategies used to solve tasks, methods used to prepare for an exam, the way you pay attention in class, and how do you study at home? The responses to the questionnaire are dichotomous (always/never). The main objective of the instrument is to categorize the academic performance of students [62,63].

- Adaptation to university life is a questionnaire focused on evaluating the academic, institutional, and social dimensions of the students, with 50 structured items. It has Likert-type assessment scale responses, from the most negative rating to the most positive rating (totally disagree/ sometimes disagree/sometimes agree/totally agree). Specifically, the questionnaire helps to determine the nature of the adaptive process of the university student [64].
- Zung's Self-Assessment Depression Scale (SDS) is a standardized questionnaire that can be self-administered, based on norms elaborated in percentiles, with 20 structured items. It evaluates the affective, cognitive, and somatic aspects of the patients, through questions with a Likert-type assessment scale (never/sometimes/most of the time/always). It has the aim of measuring the level of depression in a simple and specific way as a psychiatric disorder, allowing to categorize the depression level of an individual [65].
- The validated Spanish version of the Hamilton Anxiety Rating Scale (HARS) is a questionnaire and clinical assessment tool, structured in 14 items, with Likert-type responses (very disabling/severe/moderate/mild/none), which provide useful information about possible anxious-depressive symptoms to evaluate the symptomatology of an individual's level of anxiety [66].

For access to the chatbot web platform, we sent specific links to each student, periodically to their institutional email. We carried out this process during the academic semester 2021-II. Finally, we stored the data in a relational database.

3.5. Data Pre-Processing, Processing, and Visualization

During the investigation, we executed a set of data pre-processing, processing, and visualization techniques for further analysis. We stored the data in a digest schema database for simplicity, ease, and speed of processing. Then, we performed the data processing using an open-source integrated development environment (IDE) for scientific programming in Python, called Spyder V5.2.2.

With the purpose of executing the data processing through the DBSCAN, K-Means, and HDBSCAN unsupervised learning algorithms, we removed the data that did not contribute quantitative values to the model. To do this, we started by importing the data from the "Datos.csv" file to the "data" variable.

Once the data from the "data" variable file were imported, we eliminated the data from the "code" column, since they only had the function of identifying the student and did not provide relevant values for the model, as shown in Table 1.

Table 1. Data columns of the "data" variable.

Column	Type
code	string
study habits	int
adaptation and coexistence	int
depression	int
anxiety	int

Subsequently, we analyzed the descriptive statistics of the resulting set after the data selection and cleaning process, as shown in Table 2.

Table 2. Statistical data of the data columns in the "data" variable.

Index	Study Habits	Adaptation and Coexistence	Depression	Anxiety
count	670	670	670	670
mean	3.5731	0.6985	1.0940	0.2746
SD	0.9673	0.4784	0.3166	0.6975
min	1	0	0	0
25%	3	0	1	0
50%	4	1	1	0
75%	4	1	1	0
max	5	3	3	3

From this, we identified heterogeneity of the data ranges (maximum and minimum) of the different columns. This is because the scores of the instruments, due to the number of items, are different, as we can see in Table 3.

Table 3. Scale of possible values in the data columns of the "data" variable.

Column	Type	Labels
study habits	0–5	(very negative, negative, negative trend, positive trend, positive, very positive)
adaptation and coexistence	0–2	(low, medium, high)
depression	0–3	(normal, light, moderate, severe)
anxiety	0–3	(mild, moderate, serious, severe)

Next, we proceeded to scale the data through normalization methods to provide the unsupervised DBSCAN, K-Means, and HDBSCAN algorithms with data in the same format and scale.

In Figure 2, we show the process of "standardization", where we scaled the data based on a normal distribution, adjusting the mean to 0 and the variance to 1.

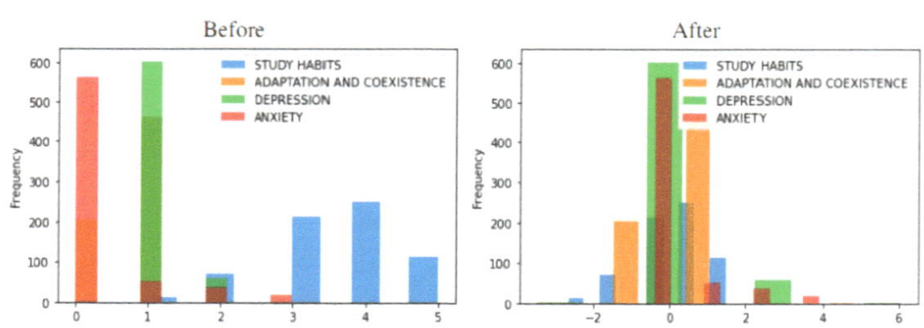

Figure 2. Contrast of data distributions before and after normalization.

- Processing with DBSCAN

We processed the data with the DBSCAN unsupervised learning algorithm. We presented the data in a multidimensional array, which we considered as a universal group. Considering the radius (*Eps*) for each of the points in a Euclidean space and through a minimum number of points (Min_pts), we defined the neighborhood of a point as:

$$N_{Eps}(p) = \{q \in D | dist(p, q) \leq Eps\} \quad (13)$$

Given the *Eps* and Min_pts parameters, DBSCAN randomly chooses a core point as a seed and retrieves all attainable density samples (within the *Eps* radius) from the seed to form a cluster, considering those points that do not belong to a cluster as noise.

To start the processing, it was crucial to have the parameters that the algorithm requires for its execution, *Eps* and MinPts. We calculated these parameters by iteratively executing the algorithm itself over a range of *Eps* and MinPts values to compile their results and contrast them with the project's objectives.

We based the method and criteria used for the selection of the algorithm parameters on the analysis of the coefficients: Silhouette coefficient, Calinski–Harabasz coefficient, and Davies–Bouldin coefficient.

We started by importing the necessary resources: numpy for numerical calculations, pandas for data manipulation in schemas called dataframes, and the "metrics" and "pre-processing" modules of the sci-kit learn library. We used product method of the itertools package to generate combinations based on the elements of two or more data lists. We executed the DBSCAN unsupervised learning algorithm based on the combinations of parameters generated by the product method.

Based on the objective of the investigation and with a range of *Eps* values from 0.2 to 2, we chose to assign an arbitrary Min_Pts range from 5 to 15. Then, with each of the proposed parameters, we generated the combinations based on the lists of generated data ranges and initialized the variables to store the data resulting from the iterative execution of the algorithm. We also applied a conditional filter to store results where the number of clusters was greater than 3, and less than 6.

From these results, we selected the parameters 1.7 and 6 for the *Eps* and Min_pts values, respectively, as shown in Table 4.

Table 4. Results obtained after DBSCAN execution.

Index	Number of Clusters	Silhouette	Calinski–Harabasz	Davies–Bouldin	*Eps*	MinPts	Noise
25	5	0.4972	190.7099	0.9571	1.7	6	9
35	4	0.4919	220.9307	1.1153	1.8	12	13
43	4	0.4919	220.9307	1.1153	1.9	12	13
51	4	0.4919	220.9307	1.1153	2	12	13

We established parameters according to Silhouette coefficient greater than 0, a high Calinski–Harabasz coefficient, a low Davies–Bouldin coefficient, a low number of noise-type points, and number of clusters greater than 3 and less than 6.

We applied DBSCAN to the obtained parameters. In the variable "labels", we stored the result of the computation of the algorithm on the dataset, resulting in a 670×1 list whose only column contains the labels of the clusters generated with its index in Y, the index corresponding to the student within the initial dataset.

From this result, we extracted the number of clusters = 5, the number of noise points = 9, the Silhouette coefficient = 0 4972, the Calinski–Harabasz coefficient =190.7099, and the Davies–Bouldin coefficient = 0.9571.

- Processing with K-Means

For data processing with the K-Means algorithm, we used the elbow method for the selection of the parameter "n_clusters" The elbow method consists of iteratively executing the clustering algorithm on a range of "n_clusters" that usually range from 1 to 10, and then, for each value of *k*, it calculates an average score for all the groups. We calculated the distortion score, which is the sum of the squared distances from each point to its assigned center.

When we plotted the values of these metrics, we could visually determine the best value for *k*. If the line graph looks like an arm, then the 'elbow' (the turning point in the curve) is the best value of k. The 'arm' can be up or down, but if there is a strong inflection point, it is a good indication that the underlying model is a better fit for that point.

After obtaining the optimal value for the "n_clusters" parameter, shown in Figure 3, we applied the K-Means algorithm to the data.

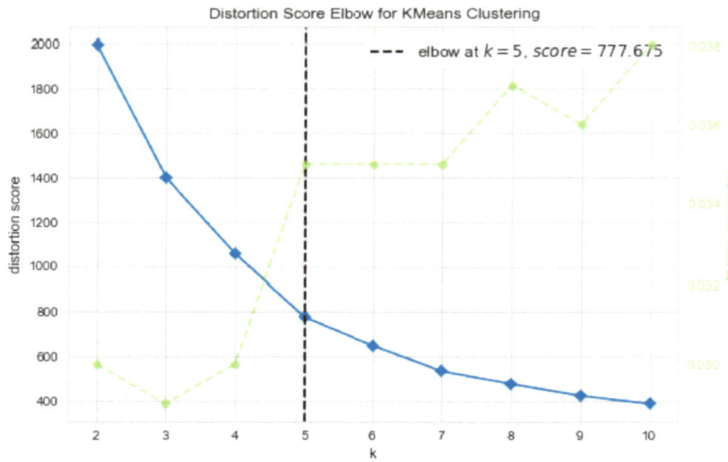

Figure 3. Result of the elbow method's application.

Based on the processing process, we obtained the following results: the number of clusters = 5, the Silhouette coefficient = 0.5586, the Calinski–Harabasz coefficient = 406.45089, and the Davies–Bouldin coefficient = 0.8001.

- Processing with HDBSCAN

In the case of HDBSCAN, it was possible to work with a single parameter: "min_cluster _size"; however, to have greater precision and control over the results of the coefficients, we decided to work with the second parameter "min_samples". Using a similar technique to the one we applied with DBSCAN, we based the selection method of the algorithm parameters on the analysis of the following coefficients: Silhouette coefficient, Calinski–Harabasz coefficient, and Davies–Bouldin coefficient.

We started by importing the necessary resources: numpy for number calculations, pandas for data manipulation in schemas called dataframes, the "metrics" and "preprocessing" modules of the sklearn library, and the product method of the itertools package to generate combinations based on the elements of two or more data lists. Then, we ran the HDBSCAN algorithm based on the combinations of parameters generated by the product method.

The range of values in the "min_cluster_size" parameter was from 15 to 80, and the range of values for the "min_samples" parameter was from 10 to 30. Next, with each of the proposed parameters, we generated the combinations based on to the lists of proposed data ranges and we initialized the variables to store the data resulting from the iterative execution of the algorithm. In addition, we applied a conditional filter to store results where the number of clusters was greater than 3, and less than 6, as well as a filter to avoid results that exceeded an amount of noise greater than 10% of the data.

Finally, after the execution of all cases of interest, we obtained 12 results. After this, we selected the parameters 55 and 19 as the values of "min_cluster_size" and "min_samples", respectively, as shown in Table 5.

As in the case of DBSCAN, we focused the parameter selection criteria on the analysis of internal validation indices: Silhouette coefficient with values greater than 0 and close to 1, the most ideal, a high Calinski–Harabasz coefficient, a low Davies–Bouldin coefficient, with values close to 0 as the ideal, a number of noise-like points, and number of clusters greater than 3 and less than 6.

Table 5. Results after executing HDBSCAN.

Index	Number of Clusters	Silhouette	Calinski–Harabasz	Davies–Bouldin	Minimum Cluster Size	Minimum Samples	Noise
8	5	0.6823	369.6459	0.6563	55	19	63
7	5	0.6704	349.5316	0.6677	55	18	59
2	5	0.6639	334.9714	0.6861	60	17	56
6	5	0.6639	334.9714	0.6861	60	17	56

We selected the number of clusters = 5, the number of noise points = 63, the Silhouette coefficient = 0.6823, the Calinski–Harabasz coefficient = 369.6459, and the Davies–Bouldin coefficient = 0.6563.

4. Analysis of Results and Discussion

When executing the algorithm on the proposed dataset, we obtained sets of results to which we applied two validation techniques (visual and internal) to verify the accuracy of the proposed model.

We performed the three-dimensional composition of the point cloud and its respective clusters for the graphic display of the results in each of the generated models; however, we generated the cloud in four dimensions, so it is impossible to represent it on a three-dimensional plane. For this reason, we resized the dataset using a data compression technique known as principal component analysis (PCA). For a graphical representation, we imported the PCA method from the "decomposition" module, which reduce the four-dimensional dataset to a three-dimensional dataset.

4.1. Visual Validation

We proceeded with the graphical representation of the resized dataset.

We could differentiate strongly defined structures after obtaining the graphical representation of the point cloud of the dataset and its respective clusters identified by colors in Figures 4–6. This three-dimensional representation is the result of resizing the four-dimensional dataset by applying PCA. However, despite not having the representation of the point cloud in its natural state, the feature reduction technique offers us a structure in which it is possible to visually recognize the clusters, being close to those elements of its own cluster and distant to those who do not belong or are alien to their group.

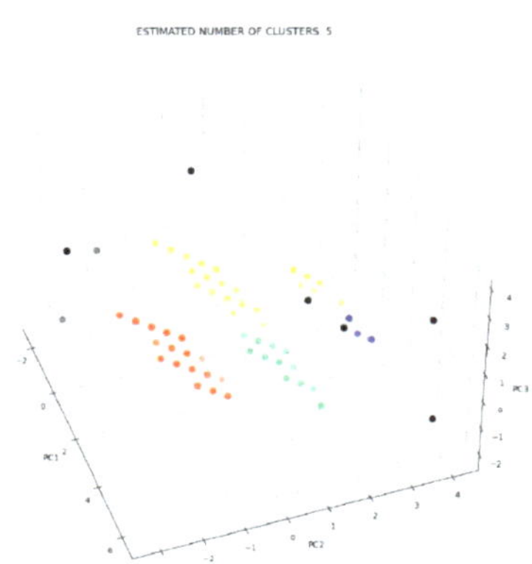

Figure 4. Visual representation of the three-dimensional clustered point cloud (DBSCAN).

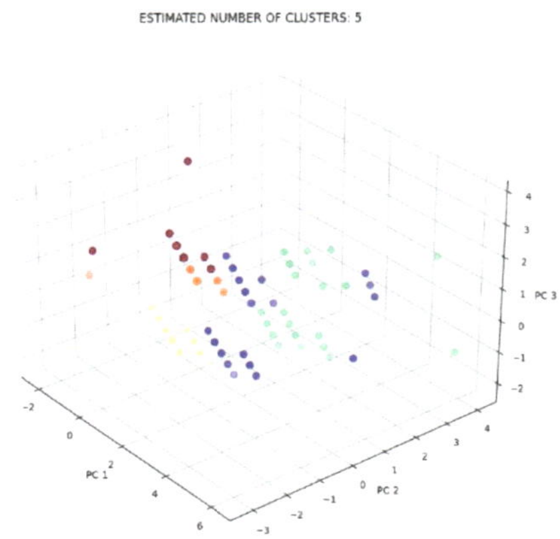

Figure 5. Visual representation of the clustered point cloud in three dimensions (K-Means).

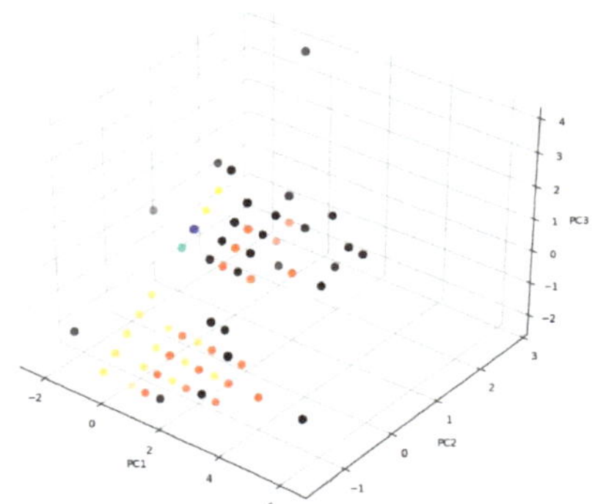

Figure 6. Visual representation of the three-dimensional clustered point cloud (HDBSCAN).

4.2. Internal Validation

Due to the lack of a previous classification of the sample used with which we can compare the external indices of the models generated by each algorithm, we considered it correct and prudent to make a comparison of the resulting internal indices: the Silhouette coefficient, the Calinski–Harabasz coefficient, and the Davies–Bouldin coefficient. Based on the results of the indices, we selected the best cases of each algorithm, as shown in Table 6.

Table 6. Comparative table of the best internal metrics resulting from the models.

Algorithm	Silhouette	Calinski–Harabasz	Davies–Bouldin	Number of Clusters	Noise
DBSCAN	0.4972	190.7099	0.9571	5	9
K-Means	0.5586	406.4509	0.8001	5	-
HDBSCAN	0.6823	369.6459	0.6563	5	63

From the results shown in Table 6, we found that the model generated with HDBSCAN was superior to the model generated with K-Means, and widely superior to the one generated with DBSCAN. This is because HDBSCAN had a better level of validity in two (Silhouette coefficient and Davies–Bouldin coefficient) of the three internal validation indices evaluated. However, for the third case, the Calinski–Harabasz coefficient generated with the K-Means model was higher. This is because to calculate the Calinski–Harabasz coefficient, we used the centroids of each cluster as parameters. Thus, we obtained a much higher score in convex-shaped clusters and greater affinity with the shape of the clusters formed by the K-Means algorithm, that tended to have an almost spherical convex shape [47,48].

Based on the validation results of the internal HDBSCAN indices, we decided to categorize university students into five clusters. according to their risk of dropping out.

4.3. Expert Validacion

In collaboration with a team of three mental health experts, we interpreted and identified the recognized patterns in each data cluster. In Figure 7, the patterns in the distribution of the results after the generation of the clusters are presented.

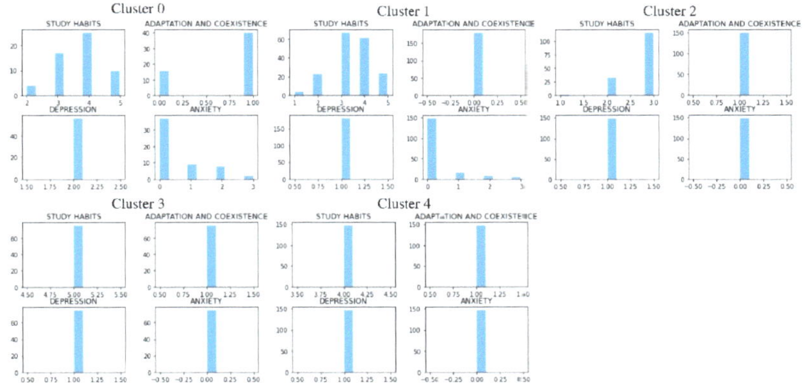

Figure 7. Distribution of results in each test with the members of the clusters 0, 1, 2, 3, and 4.

We proceeded with the labeling with the mean of the results in the different distributions and the criteria of the mental health professionals, based on the theories of the psychological questionnaires used [62–66].

From Table 7, the groupings of the sample of 670 students resulted in: very low dropout risk level = 75 students, low level = 147, medium level = 149, high level = 180, and very high level = 56, and 63 students belonged to the noise sector.

Table 7. Values of the mean of the data distribution according to clusters.

Cluster	Study Habits	Adaptation and Coexistence	Depression	Anxiety	Risk Level
Cluster 0	3.7321	0.71429	2	0.5536	5 = Very high
Cluster 1	3.4389	0	1	0.2944	4 = High
Cluster 2	2.7651	1	1	0	3 = Middle
Cluster 3	5	1	1	0	1 = Very low
Cluster 4	4	1	1	0	2 = Low

After the labeling process, we asked the team of experts to classify the students by evaluating the results of the applied psychological evaluation instruments. Then, we proceeded with the validation of external indices between the results of HDBSCAN and those of the experts to compare the accuracy of the model.

After we ran the validation of external indexes, we obtained Table 8 as a result.

Table 8. Values obtained after validation of external indexes.

Index	Score
F-measure	0.909
Purity	0.945
V-measure	0.869
Adjusted Rand Index	0.865

According to Table 8, the results obtained after comparing the classification performed through HDBSCAN with the classification provided by mental health professionals indicate a high level of precision in the *F*-measure (90.9%), which was consistent with the high similarity between the clusters calculated and the one predicted by the model with HDBSCAN (*Purity*: 94.5%, *V*-Measure: 86.9%, ARI: 86.5%).

The categorization of students provided a better picture of their dropout risk, which, associated with an early diagnosis, allows us to take corrective measures [34,35,67]. We must highlight that the use of data mining in conjunction with machine learning as tools allowed us to develop the main axis of the proposed model [15,18,21,31].

5. Conclusions

We developed a clustering model that integrates methodologies and data analysis and processing techniques, widely studied in the field of ICT, specifically in the field of unsupervised machine learning. This allowed us to obtain the successful categorization of undergraduate students from a Peruvian university into five levels based on the risk of desertion. HDBSCAN was the method that turned out to be the best option for data processing, as evidenced by the results of the internal validation indexes used to compare them with the K-Means and DBSCAN methods.

The resulting model serves as the basis of knowledge about the current view of university students. It can be replicated in other contexts, and it can be adjusted to other types of tests. For this, it would be necessary to standardize the input data types to generate values in less disperse ranges, to group them optimally. Likewise, it is scalable if we articulate joint efforts between mental health professionals and unsupervised learning techniques to generate a comprehensive solution that encompasses more dimensions of the psychological field. With this research, we contribute to the identification, prevention, and correction of various situations of psycho-emotional risk that university students may face.

Author Contributions: The manuscript was conceptualized by M.A.V.-C.; methodology, M.A.V.-C and J.J.-D.; software, L.S.-R. and R.I.; validation, J.R.N.-C. and L.P.; formal analysis, P.V.-R.; data curation, L.S.-R. and E.A.H.-T.; writing—proofreading and editing, M.A.V.-C. and J.R.N.-C. All authors have read and agreed to the published version of the manuscript.

Funding: Thanks to the Universidad Nacional de San Martín for the financing of the project "Caracterización del proceso de tutoría a estudiantes de la UNSM aplicando un modelo de atención virtual basado en chatbots", financed by Resolution No. 359-2021-UNSM/CU-R.

Institutional Review Board Statement: Not applicable.

Informed Consent Statement: We obtained informed consent from all subjects involved in the study.

Data Availability Statement: The data of the survey carried out on students are available upon request at: The Research Unit of the Universidad Nacional de San Martín.

Conflicts of Interest: The authors declare no conflict of interest.

References

1. Díaz-Méndez, M.; Paredes, M.R.; Saren, M. Improving Society by Improving Education through Service-Dominant Logic: Reframing the Role of Students in Higher Education. *Sustainability* **2019**, *11*, 5292. [CrossRef]
2. Zarouk, M.Y.; Olivera, E.; Peres, P.; Khaldi, M. The Impact of Flipped Project-Based Learning on Self-Regulation in Higher Education. *Int. J. Emerg. Technol. Learn.* **2020**, *15*, 127–147. [CrossRef]
3. Chacón-Cuberos, R.; Martínez-Martínez, A.; Puertas-Molero, P.; Viciana-Garófano, V.; González-Valero, G.; Zurita-Ortega, F. Bienestar Social En La Etapa Universitaria Según Factores Sociodemográficos En Estudiantes de Educación. *Rev. Electrónica Investig. Educ.* **2020**, *22*, e03. [CrossRef]
4. Mejía-Navarrete, J. El Proceso de La Educación Superior En El Perú. La Descolonialidad Del Saber Universitario. *Cinta de Moebio* **2018**, *61*, 56–71. [CrossRef]
5. Barreto-Osma, D.; Salazar-Blanco, H.A. Agotamiento Emocional En Estudiantes Universitarios Del Área de La Salud. *Univ. y Salud* **2021**, *23*, 30–39. [CrossRef]
6. Vargas, M.; Talledo-Ulfe, L.; Heredia, P.; Quispe-Colquepisco, S.; Mejia, C.R. Influencia de Los Hábitos En La Depresión Del Estudiante de Medicina Peruano: Estudio En Siete Departamentos. *Rev. Colomb. Psiquiatr.* **2018**, *47*, 32–36. [CrossRef]
7. Castillo Riquelme, V.; Cabezas Maureira, N.; Vera Navarro, C.; Toledo Puente, C. Ansiedad Al Aprendizaje En Línea: Relación Con Actitud, Género, Entorno y Salud Mental En Universitarios. *Rev. Digit. Investig. Docencia Univ.* **2021**, *15*, e1284. [CrossRef]
8. Zulu, W.V.; Mutereko, S. Exploring the Causes of Student Attrition in South African TVET Colleges: A Case of One KwaZulu-Natal Technical and Vocational Education and Training College. *Interchange* **2020**, *51*, 385–407. [CrossRef]
9. Aina, C.; Baici, E.; Casalone, G.; Pastore, F. The determinants of university dropout: A review of the socio-economic literature. *Socio-Econ. Plan. Sci.* **2021**, *79*, 101102. [CrossRef]
10. Aguilera García, J.L. La Tutoría Universitaria Como Práctica Docente: Fundamentos y Métodos Para El Desarrollo de Planes de Acción Tutorial En La Universidad. *Pro-Posições* **2019**, *30*, e20170038. [CrossRef]
11. Buring, S.M.; Williams, A.; Cavanaugh, T. The life raft to keep students afloat: Early detection, supplemental instruction, tutoring, and self-directed remediation. *Curr. Pharm. Teach. Learn.* **2022**, *14*, 1060–1067. [CrossRef] [PubMed]
12. Sánchez Cabezas, P.d.P.; Luna Álvarez, H.E.; López Rodríguez del Rey, M.M. La Tutoría En La Educación Superior y Su Integración En La Actividad Pedagógica Del Docente Universitario. *Conrado* **2019**, *15*, 300–305.
13. Alonso-García, S.; Rodríguez-García, A.M.; Cáceres-Reche, M.P. Analysis of the Tutorial Action and Its Impact on the Overall Development of the Students. The Case of the University of Castilla La Mancha, Spain. *Form. Univ.* **2018**, *11*, 63–72. [CrossRef]
14. Mi, H.; Gao, Z.; Zhang, Q.; Zheng, Y. Research on Constructing Online Learning Performance Prediction Model Combining Feature Selection and Neural Network. *Int. J. Emerg. Technol. Learn.* **2022**, *17*, 94–111. [CrossRef]
15. Guzmán-Castillo, S.; Körner, F.; Pantoja-García, J.I.; Nieto-Ramos, L.; Gómez-Charris, Y.; Castro-Sarmiento, A.; Romero-Conrado, A.R. Implementation of a Predictive Information System for University Dropout Prevention. *Procedia Comput. Sci.* **2022**, *198*, 566–571. [CrossRef]
16. Chen, M.; Yan, Z.; Meng, C.; Huang, M. The Supporting Environment Evaluation Model of ICT in Chinese University Teaching. In Proceedings of the 2018 International Symposium on Educational Technology (ISET), Osaka, Japan, 31 July 2018–2 August 2018; IEEE: Piscataway, NJ, USA, 2018; pp. 99–103. [CrossRef]
17. Delerna Rios, G.E.; Levano Rodriguez, D. Importancia de Las Tecnologías de Información En El Fortalecimiento de Competencias Pedagógicas En Tiempos de Pandemia. *Rev. Científica Sist. Inf.* **2021**, *1*, 69–78. [CrossRef]
18. Ghareeb, S.; Hussain, A.J.; Al-Jumeily, D.; Khan, W.; Al-Jumeily, R.; Baker, T.; Al Shammaa, A.; Khalaf, M. Evaluating Student Levelling Based on Machine Learning Model's Performance. *Discov. Internet Things* **2022**, *2*, 1–25. [CrossRef]
19. Gonzalez Salas Duhne, P.; Delgadillo, J.; Lutz, W. Predicting Early Dropout in Online versus Face-to-Face Guided Self-Help: A Machine Learning Approach (Authors Masked for Peer Review). *Behav. Res. Ther.* **2022**, *159*, 104200. [CrossRef]
20. Narayanasamy, S.K.; Elçi, A. An Effective Prediction Model for Online Course Dropout Rate. *Int. J. Distance Educ. Technol.* **2020**, *18*, 94–110. [CrossRef]
21. Mduma, N.; Kalegele, K.; Machuve, D. A Survey of Machine Learning Approaches and Techniques for Student Dropout Prediction. *Data Sci. J.* **2019**, *18*, 1–10. [CrossRef]
22. Castro-Lopez, A.; Silva Almeida, L.; Fernández Rivas, S.; Guzmán, A.; Barragán, S.; Cala-Vitery, F. Comparative Analysis of Dropout and Student Permanence in Rural Higher Education. *Sustainability* **2022**, *14*, 8871. [CrossRef]
23. Guzmán, A.; Barragán, S.; Cala Vitery, F. Dropout in Rural Higher Education: A Systematic Review. *Front. Educ.* **2021**, *6*, 351. [CrossRef]
24. Yi, S.; Dianatinasab, M.; Faria De Moura Villela, E.; Khanal, P.; Lin, Y.; Maluenda-Albornoz, J.; Infante-Villagrán, V.; Galve-González, C.; Flores-Oyarzo, G.; Berríos-Riquelme, J. Early and Dynamic Socio-Academic Variables Related to Dropout Intention: A Predictive Model Made during the Pandemic. *Sustainability* **2022**, *14*, 831. [CrossRef]
25. Bernardo, A.B.; Galve-González, C.; Núñez, J.C.; Almeida, L.S. Settings Open AccessFeature PaperArticle A Path Model of University Dropout Predictors: The Role of Satisfaction, the Use of Self-Regulation Learning Strategies and Students' Engagement. *Sustainability* **2022**, *14*, 1057. [CrossRef]
26. Kanetaki, Z.; Stergiou, C.; Bekas, G.; Troussas, C.; Sgouropoulou, C. Analysis of Engineering Student Data in Online Higher Education During the COVID-19 Pandemic. *Int. J. Eng. Pedagog.* **2021**, *11*, 27–49. [CrossRef]

27. Tayebi, A.; Gomez, J.; Delgado, C. Analysis on the Lack of Motivation and Dropout in Engineering Students in Spain. *IEEE Access* **2021**, *9*, 66253–66265. [CrossRef]
28. Pavelea, A.M.; Moldovan, O. Why Some Fail and Others Succeed? Explaining the Academic Performance of PA Undergraduate Students. *NISPAcee J. Public Adm. Policy* **2020**, *13*, 109–132. [CrossRef]
29. Zapata-Lamana, R.; Sanhueza-Campos, C.; Stuardo-Álvarez, M.; Ibarra-Mora, J.; Mardones-Contreras, M.; Reyes-Molina, D.; Vásquez-Gómez, J.; Lasserre-Laso, N.; Poblete-Valderrama, F.; Petermann-Rocha, F.; et al. Anxiety, Low Self-Esteem and a Low Happiness Index Are Associated with Poor School Performance in Chilean Adolescents: A Cross-Sectional Analysis. *Int. J. Environ. Res. Public Health* **2021**, *18*, 11685. [CrossRef]
30. Mena, M.; Godoy, W.; Tisalema, S. Analysis of Causes of Early Dropout of Students Higher Education. *Minerva* **2021**, *2*, 79–89. [CrossRef]
31. Núñez-Naranjo, A.F.; Ayala-Chauvin, M.; Riba-Sanmartí, G. Prediction of University Dropout Using Machine Learning. In Proceedings of the International Conference on Information Technology & Systems, Libertad, Ecuador, 4–6 February 2021; Springer: Cham, Switzerland, 2021; pp. 396–406. [CrossRef]
32. Dalipi, F.; Imran, A.S.; Kastrati, Z. MOOC Dropout Prediction Using Machine Learning Techniques: Review and Research Challenges. In Proceedings of the 2018 IEEE Global Engineering Education Conference (EDUCON), Santa Cruz de Tenerife, Spain, 17–20 April 2018; IEEE Computer Society: Piscataway, NJ, USA, 2018; pp. 1007–1014. [CrossRef]
33. Albreiki, B.; Zaki, N.; Alashwal, H. A Systematic Literature Review of Student' Performance Prediction Using Machine Learning Techniques. *Educ. Sci.* **2021**, *11*, 552. [CrossRef]
34. Mohamed Nafuri, A.F.; Sani, N.S.; Zainudin, N.F.A.; Rahman, A.H.A.; Aliff, M. Clustering Analysis for Classifying Student Academic Performance in Higher Education. *Appl. Sci.* **2022**, *12*, 9467. [CrossRef]
35. Freitas, F.A.d.S.; Vasconcelos, F.F.X.; Peixoto, S.A.; Hassan, M.M.; Ali Akber Dewan, M.; de Albuquerque, V.H.C.; Rebouças Filho, P.P. IoT System for School Dropout Prediction Using Machine Learning Techniques Based on Socioeconomic Data. *Electronics* **2020**, *9*, 1613. [CrossRef]
36. Rovira, S.; Puertas, E.; Igual, L. Data-Driven System to Predict Academic Grades and Dropout. *PLoS ONE* **2017**, *12*, e0171207. [CrossRef]
37. Sansone, D. Beyond Early Warning Indicators: High School Dropout and Machine Learning. *Oxf. Bull. Econ. Stat.* **2019**, *81*, 456–485. [CrossRef]
38. Duque Hernández, J.I.; Rodríguez-Chávez, M.H.; Polanco-Martagón, S. Caracterización Del Aprendizaje de Algoritmos Mediante Minería de Datos En El Nivel Superior. *Dilemas Contemp. Educ. Política y Valores* **2021**, *9*, 1–18. [CrossRef]
39. Zuo, W.; Hou, X. An Improved Probability Propagation Algorithm for Density Peak Clustering Based on Natural Nearest Neighborhood. *Array* **2022**, *15*, 100232. [CrossRef]
40. Webb, G.I.; Fürnkranz, J.; Fürnkranz, J.; Fürnkranz, J.; Hinton, G.; Sammut, C.; Sander, J.; Vlachos, M.; Teh, Y.W.; Yang, Y.; et al. Density-Based Clustering. In *Encyclopedia of Machine Learning*; Springer: Boston, MA, USA, 2011; pp. 270–273. [CrossRef]
41. Tavakkol, B.; Choi, J.; Jeong, M.K.; Albin, S.L. Object-Based Cluster Validation with Densities. *Pattern Recognit.* **2022**, *121*, 108223. [CrossRef]
42. Xie, H.; Li, P. A Density-Based Evolutionary Clustering Algorithm for Intelligent Development. *Eng. Appl. Artif. Intell.* **2021**, *104*, 104396. [CrossRef]
43. Daszykowski, M.; Walczak, B. Density-Based Clustering Methods. *Compr. Chemom.* **2009**, *2*, 635–654. [CrossRef]
44. Ester, M.; Kriegel, H.-P.; Sander, J.; Xu, X. A Density-Based Algorithm for Discovering Clusters in Large Spatial Databases with Noise. In Proceedings of the Second International Conference on Knowledge Discovery and Data Mining (KDD-96), Portland, OR, USA, 2–4 August 1996; AAAI Press: Washington, DC, USA, 1996; pp. 226–231.
45. Li, M.; Bi, X.; Wang, L.; Han, X. A Method of Two-Stage Clustering Learning Based on Improved DBSCAN and Density Peak Algorithm. *Comput. Commun.* **2021**, *167*, 75–84. [CrossRef]
46. Lloyd, S.P. Least Squares Quantization in PCM. *IEEE Trans. Inf. Theory* **1982**, *28*, 129–137. [CrossRef]
47. Meng, Y.; Liang, J.; Cao, F.; He, Y. A New Distance with Derivative Information for Functional K-Means Clustering Algorithm. *Inf. Sci.* **2018**, *463–464*, 166–185. [CrossRef]
48. Wang, F.; Wang, Q.; Nie, F.; Li, Z.; Yu, W.; Ren, F. A Linear Multivariate Binary Decision Tree Classifier Based on K-Means Splitting. *Pattern Recognit.* **2020**, *107*, 107521. [CrossRef]
49. Yuan, C.; Yang, H. Research on K-Value Selection Method of K-Means Clustering Algorithm. *J. Multidiscip. Sci. J.* **2019**, *2*, 226–235. [CrossRef]
50. Liu, F.; Deng, Y. Determine the Number of Unknown Targets in Open World Based on Elbow Method. *IEEE Trans. Fuzzy Syst.* **2021**, *29*, 986–995. [CrossRef]
51. Campello, R.J.G.B.; Moulavi, D.; Sander, J. Density-Based Clustering Based on Hierarchical Density Estimates. In *Lecture Notes in Computer Science (Including Subseries Lecture Notes in Artificial Intelligence and Lecture Notes in Bioinformatics)*; Springer: Berlin/Heidelberg, Germany, 2013; Volume 7819, pp. 160–172. [CrossRef]
52. McInnes, L.; Healy, J.; Astels, S. Hdbscan: Hierarchical Density Based Clustering. *J. Open Source Softw.* **2017**, *2*, 205. [CrossRef]
53. Draszawka, K.; Szymański, J. External Validation Measures for Nested Clustering of Text Documents. *Stud. Comput. Intell.* **2011**, *369*, 207–225. [CrossRef]

54. Haouas, F.; Ben Dhiaf, Z.; Hammouda, A.; Solaiman, B. A New Efficient Fuzzy Cluster Validity Index: Application to Images Clustering. In Proceedings of the 2017 IEEE International Conference on Fuzzy Systems (FUZZ-IEEE), Naples, Italy, 9–12 July 2017; pp. 1–6. [CrossRef]
55. Rousseeuw, P.J. Silhouettes: A Graphical Aid to the Interpretation and Validation of Cluster Analysis. *J. Comput. Appl. Math.* **1987**, *20*, 53–65. [CrossRef]
56. Caliñski, T.; Harabasz, J. A Dendrite Method Foe Cluster Analysis. *Commun. Stat.* **1974**, *3*, 1–27. [CrossRef]
57. Rendón, E.; Abundez, I.; Arizmendi, A.; Quiroz, E.M. Internal versus External Cluster Validation Indexes. *Int. J. Comput. Commun.* **2011**, *5*, 27–34.
58. Davies, D.L.; Bouldin, D.W. A Cluster Separation Measure. *IEEE Trans. Pattern Anal. Mach. Intell.* **1979**, *PAMI-1*, 224–227. [CrossRef]
59. Zhang, E.; Zhang, Y. F-Measure. In *Encyclopedia of Database Systems*; Springer: New York, NY, USA, 2018; pp. 1492–1493. [CrossRef]
60. Bagunaid, W.; Chilamkurti, N.; Veeraraghavan, P. AISAR: Artificial Intelligence-Based Student Assessment and Recommendation System for E-Learning in Big Data. *Sustainability* **2022**, *14*, 10551. [CrossRef]
61. Rovetta, S.; Masulli, F.; Cabri, A. The "Probabilistic Rand Index": A Look from Some Different Perspectives. In *Smart Innovation, Systems and Technologies*; Springer Science and Business Media Deutschland GmbH: Singapore, 2020; Volume 151, pp. 95–105. [CrossRef]
62. Benitez Molina, A.; Caballero Badillo, M.C. Psychometric Study of the Depression, Anxiety and Family Dysfunction Scales in Students at Universidad Industrial de Santander. *Acta Colomb. Psicol.* **2017**, *20*, 221–231. [CrossRef]
63. De la Parra Paz, E. *Herencia de Vida Para Tus Hijos: Crecimiento Integral Con Técnicas PNL*; Grijalbo Mondadori: Barcelona, Spain, 2004.
64. Almeida, L.S.; Soares, A.P.C.; Ferreira, J.A. Questionário de Vivências Académicas (QVA-r): Avaliação Do Ajustamento Dos Estudantes Universitários. *Avaliação Psicológica* **2002**, *1*, 81–93.
65. Hamilton, M. The Assessment of Anxiety States by Rating. *Br. J. Med. Psychol* **1959**, *32*, 50–55. [CrossRef] [PubMed]
66. Lobo, A.; Chamorro, L.; Luque, A.; Dal-Ré, R.; Badia, X.; Baró, E. Validación de Las Versiones En Español de La Montgomery-Asberg Depression Rating Scale y La Hamilton Anxiety Rating Scale Para La Evaluación de La Depresión y de La Ansiedad. *Med. Clin. (Barc.)* **2002**, *118*, 493–499. [CrossRef]
67. Evangelista, E.D. A Hybrid Machine Learning Framework for Predicting Students' Performance in Virtual Learning Environment. *Int. J. Emerg. Technol. Learn.* **2021**, *16*, 255–272. [CrossRef]

data

MDPI

Data Descriptor

Predicting Student Dropout and Academic Success

Valentim Realinho [1,2,*], Jorge Machado [2], Luís Baptista [2] and Mónica V. Martins [2]

[1] VALORIZA—Research Center for Endogenous Resource Valorization, Instituto Politécnico de Portalegre, 7300-555 Portalegre, Portugal
[2] Escola Superior de Tecnologia e Gestão, Instituto Politécnico de Portalegre, 7300-555 Portalegre, Portugal
* Correspondence: vrealinho@ipportalegre.pt

Abstract: Higher education institutions record a significant amount of data about their students, representing a considerable potential to generate information, knowledge, and monitoring. Both school dropout and educational failure in higher education are an obstacle to economic growth, employment, competitiveness, and productivity, directly impacting the lives of students and their families, higher education institutions, and society as a whole. The dataset described here results from the aggregation of information from different disjointed data sources and includes demographic, socioeconomic, macroeconomic, and academic data on enrollment and academic performance at the end of the first and second semesters. The dataset is used to build machine learning models for predicting academic performance and dropout, which is part of a Learning Analytic tool developed at the Polytechnic Institute of Portalegre that provides information to the tutoring team with an estimate of the risk of dropout and failure. The dataset is useful for researchers who want to conduct comparative studies on student academic performance and also for training in the machine learning area.

Dataset: https://doi.org/10.5281/zenodo.5777339.

Dataset License: CC BY 4.0

Keywords: academic performance; machine learning in education; imbalanced classes; multi-class classification; educational data mining; learning management system; prediction

Citation: Realinho, V.; Machado, J.; Baptista, L.; Martins, M.V. Predicting Student Dropout and Academic Success. *Data* **2022**, *7*, 146. https://doi.org/10.3390/data7110146

Academic Editors: Antonio Sarasa Cabezuelo and Ramón González del Campo Rodríguez Barbero

Received: 11 October 2022
Accepted: 25 October 2022
Published: 28 October 2022

1. Introduction

Academic success in higher education is vital for jobs, social justice, and economic growth. Dropout represents the most problematic issue that higher education institutions must address to improve their success. There is no universally accepted definition of dropout. The proportion of students who dropout varies between different studies depending on how dropout is defined, the data source, and the calculation methods [1]. Frequently, dropout is analyzed in the research literature based on the timing of the dropout (early vs. late) [2]. Due to differences in reporting, it is not possible to compare dropout rates across institutions [3]. In this work, we define dropouts from a micro-perspective, where field and institution changes are considered dropouts independently of the timing these occur. This approach leads to much higher dropout rates than the macro-perspective, which considers only students who leave the higher education system without a degree.

According to the independent report for the European Commission, too many students drop out before the end of their higher education courses [4]. Even in the most successful country (Denmark), only around 80% of students complete their studies, while in Italy, this rate is only 46%. This report highlights key factors that lead students to drop out, with the major cause being socioeconomic conditions.

Namoun and Alshanqiti [5] performed an exhaustive search that found 62 papers published in peer-reviewed journals between 2010 and 2020, which present intelligent models to predict student performance. Additionally, in recent years, early prediction of student outcomes has attracted increasing research interest [6–9]. However, despite the research interest and the considerable amount of data that the universities generate, there is a need to collect more and better administrative data, including dropout and transfer reasons [2].

This descriptor presents a dataset created from a higher education institution (acquired from several disjoint databases) related to students enrolled in different undergraduate degrees, such as agronomy, design, education, nursing, journalism, management, social service, and technologies. The dataset includes information known at the time of student enrollment (academic path, demographics, and macroeconomics and socioeconomic factors) and the students' academic performance at the end of the first and second semesters. The data are used to build classification models to predict student dropout and academic success. The problem is formulated as a three-category classification task (dropout, enrolled, and graduate) at the end of the normal duration of the course. These classification models are part of a Learning Analytic tool that includes predictive analyses which provide information to the tutoring team at our higher education institution with an estimate of the risk of dropout and failure. With this information, the tutoring team provides more accurate help to students.

The dataset contained 4424 records with 35 attributes, where each record represents an individual student and can be used for benchmarking the performance of different algorithms for solving the same type of problem and for training in the machine learning area.

In addition to this introduction section, the rest of the descriptor is organized as follows. Section 2 provides the details of the dataset. Section 3 presents the methodology that was followed for the development of this dataset and also presents a brief exploratory data analysis. Section 4 presents the conclusions, which are followed by references.

2. Data Description

The dataset includes demographic data, socioeconomic and macroeconomic data, data at the time of student enrollment, and data at the end of the first and second semesters. The data sources used consist of internal and external data from the institution and include data from (i) the Academic Management System (AMS) of the institution, (ii) the Support System for the Teaching Activity of the institution (developed internally and called PAE), (iii) the annual data from the General Directorate of Higher Education (DGES) regarding admission through the National Competition for Access to Higher Education (CNAES), and (iv) the Contemporary Portugal Database (PORDATA) regarding macroeconomic data.

The data refer to records of students enrolled between the academic years 2008/2009 (after the application of the Bologna Process to higher education in Europe) to 2018/2019. These include data from 17 undergraduate degrees from different fields of knowledge, such as agronomy, design, education, nursing, journalism, management, social service, and technologies. The final dataset is available as a comma-separated values (CSV) file encoded as UTF8 and consists of 4424 records with 35 attributes and contains no missing values.

Table 1 describes each attribute used in the dataset grouped by class: demographic, socioeconomic, macroeconomic, academic data at enrollment, and academic data at the end of the first and second semesters. Appendix A contains the descriptions of possible values for the attributes, and the URL referenced in the Supplementary Material contains more detailed information.

Table 1. Attributes used grouped by class of attribute.

Class of Attribute	Attribute	Type
Demographic data	Marital status	Numeric/discrete
	Nationality	Numeric/discrete
	Displaced	Numeric/binary
	Gender	Numeric/binary
	Age at enrollment	Numeric/discrete
	International	Numeric/binary
Socioeconomic data	Mother's qualification	Numeric/discrete
	Father's qualification	Numeric/discrete
	Mother's occupation	Numeric/discrete
	Father's occupation	Numeric/discrete
	Educational special needs	Numeric/binary
	Debtor	Numeric/binary
	Tuition fees up to date	Numeric/binary
	Scholarship holder	Numeric/binary
Macroeconomic data	Unemployment rate	Numeric/continuous
	Inflation rate	Numeric/continuous
	GDP	Numeric/continuous
Academic data at enrollment	Application mode	Numeric/discrete
	Application order	Numeric/ordinal
	Course	Numeric/discrete
	Daytime/evening attendance	Numeric/binary
	Previous qualification	Numeric/discrete
Academic data at the end of 1st semester	Curricular units 1st sem (credited)	Numeric/discrete
	Curricular units 1st sem (enrolled)	Numeric/discrete
	Curricular units 1st sem (evaluations)	Numeric/discrete
	Curricular units 1st sem (approved)	Numeric/discrete
	Curricular units 1st sem (grade)	Numeric/continuous
	Curricular units 1st sem (without evaluations)	Numeric/discrete
Academic data at the end of 2nd semester	Curricular units 2nd sem (credited)	Numeric/discrete
	Curricular units 2nd sem (enrolled)	Numeric/discrete
	Curricular units 2nd sem (evaluations)	Numeric/discrete
	Curricular units 2nd sem (approved)	Numeric/discrete
	Curricular units 2nd sem (grade)	Numeric/continuous
	Curricular units 2nd sem (without evaluations)	Numeric/discrete
Target	Target	Categorical

3. Materials and Methods

This section describes the process that was followed for building the dataset and also presents a brief exploratory data analysis highlighting some relevant issues that may help other researchers quickly get their hands on the dataset and work with it, such as the imbalanced nature of data, the multicollinearity found in the features, and the results of permutation feature importance using the most used algorithms in similar problems shown in the literature.

3.1. Data Preprocessing

The data are collected in three different formats: (i) as Microsoft Access databases from CNAES; (ii) as comma-separated values (CSV) files from the AMS; and (iii) as manual data collected from the site of PORDATA concerning macroeconomics data.

Apart from the data received from CNAES, which are processed through a Visual Basic for Applications (VBA) program in a Microsoft Windows system, all the other code (in Python) runs on the Ubuntu operating system on an NVIDIA DGX Station computer with 2 CPU Intel Xeon E5-2698V4 with 20 core 2.2 GHz, 256 GB of memory, and 4 NVIDIA Tesla V100 GPU. This same computer was also used for training the machine learning models and to predict students' performance, which is part of the Learning Analytics tool developed.

Figure 1 shows the workflow designed to create the dataset, which contains four steps that are described next.

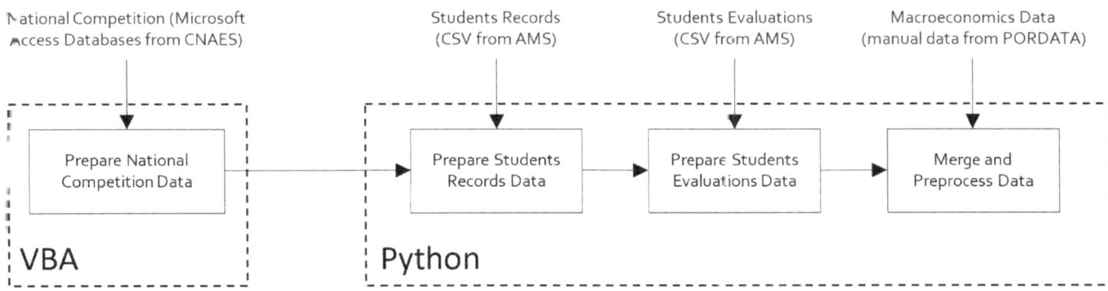

Figure 1. Workflow designed to create the dataset.

1. Prepare National Competition Data. The data relating to the National Competition for Access to Higher Education (CNAES) are received, every year, after the results of the competition, as a Microsoft Access database. We developed a Visual Basic for Applications (VBA) program that collects, from the different Microsoft Access databases (one for each year), the information needed and exports a CSV file (competition.csv) that contains one row for each student with fields related to the group "Data at Enrollment" described in Table 1.
2. Prepare Student Records Data. In this step, the CSV received from the AMS with students' records is prepared to be processed in the next steps. This file contains 13,992 rows and 398 columns, with a significant number of rows and columns that are duplicated or irrelevant to our study. To resume, this step comprises the deletion of students' records enrolled in old courses that do not currently accept enrollments, the deletion of students' records with irrelevant ways of enrollment such as Erasmus, the selection and renaming of relevant columns, and the elimination of duplicated rows. At the end of this step, all data related to the groups "Demographics Data" and "Socioeconomics Data" (see Table 1) are gathered to be used in the next steps.
3. Prepare Student Evaluations Data. In this step, the CSV file with all the information related to the evaluation attempts of students is processed. For each student that results from the processing in the previous step, the attributes related to the groups "Academic data at the end of 1st semester" and "Academic data are calculated at the end of 2nd semester" (see Table 1).
4. Merge and Preprocessing Data. All data gathered in the previous steps are merged into one single dataset in which are added the attributes related to "Macroeconomics Data". Then, we performed rigorous data preprocessing to handle anomalies, unexplainable outliers, and missing values. Finally, each student is classified as a dropout, enrolled, or graduate depending on their situation at the end of the normal duration of the course (3 years, except Nursing which has 4 years). The result is the final dataset, available at https://doi.org/10.5281/zenodo.5777339 (accessed on 10 October 2022).

3.2. Data Analysis

We performed a brief exploratory data analysis in Python 3 using the Pandas library version 1.4.3, the Scikit-learn library version 1.1.1, and the Bokeh library version 2.4.3 for visualizations.

3.2.1. Descriptive Analysis

Tables 2–8 contain basic statistics about all the attributes. These tables include a histogram of attribute values, the central tendency of each attribute value (mode for categorical attributes and mean for numeric attributes), the median of each attribute value, the dispersion of the attribute values (the entropy of the value distribution for categorical attributes and coefficient of variation for numeric attributes), and the minimum and maximum value for numerical attributes only.

Table 2. Basic statistics information about demographic data.

Attribute	Distrib.	Mean	Median	Dispersion	Min.	Max.
Marital status		1.180	1	0.510	1	6
Nationality		1.250	1	1.390	1	21
Displaced		0.548	1	0.907	0	1
Gender		0.352	0	1.358	0	1
Age at enrollment		23.130	20	0.320	17	70
International		0.025	0	6.262	0	1

Table 3. Basic statistics information about socioeconomics data.

Attribute	Distrib.	Mean	Median	Dispersion	Min.	Max.
Father's qualification		16.460	14	0.670	1	34
Mother's qualification		12.320	13	0.730	1	29
Father's occupation		7.820	8	0.620	1	46
Mother's occupation		7.320	6	0.550	1	32
Educational special needs		0.012	0	9.260	0	1
Debtor		0.114	0	2.792	0	1
Tuition fees up to date		0.881	1	0.368	0	1
Scholarship holder		0.248	0	1.739	0	1

Table 4. Basic statistics information about macroeconomics data.

Attribute	Distrib.	Mean	Median	Dispersion	Min.	Max.
Unemployment rate		11.566	11.100	0.230	7.600	16.200
Inflation rate		1.228	1.400	1.126	−0.800	3.700
GDP		0.002	0.320	1152.820	−4.100	3.500

Table 5. Basic statistics information about academic data at enrollment.

Attribute	Distrib.	Mean	Median	Dispersion	Min.	Max.
Application mode		6.890	8	0.770	1	18
Application order		1.730	1	0.760	1	9
Course		9.900	10	0.440	1	17
Daytime/evening attendance		0.891	1	0.350	0	1
Previous qualification		2.530	1	1.570	1	17

Table 6. Basic statistics information about academic data at end of the first semester.

Attribute	Distrib.	Mean	Median	Dispersion	Min.	Max.
Curricular units 1st sem (credited)		0.710	0	3.320	0	20
Curricular units 1st sem (enrolled)		6.270	6	0.400	0	26
Curricular units 1st sem (evaluations)		8.300	8	0.500	0	45
Curricular units 1st sem (approved)		4.710	5	0.660	0	26
Curricular units 1st sem (grade)		10.641	12.286	0.455	0.000	18.875
Curricular units 1st sem (without evaluations)		0.140	0	5.020	0	12

Table 7. Basic statistics information about academic data at end of the second semester.

Attribute	Distrib.	Mean	Median	Dispersion	Min.	Max.
Curricular units 2nd sem (credited)		0.540	0	3.540	0	19
Curricular units 2nd sem (enrolled)		6.230	6	0.350	0	23
Curricular units 2nd sem (evaluations)		8.060	8	0.490	0	33
Curricular units 2nd sem (approved)		4.440	5	0.680	0	20
Curricular units 2nd sem (grade)		10.230	12.200	0.509	0.000	18.571
Curricular units 2nd sem (without evaluations)		0.150	0	5.010	0	12

Table 8. Basic statistics information about Target.

Attribute	Distrib.	Center	Median	Dispersion	Min.	Max.
Target			Graduate	1.02		

3.2.2. Imbalanced Data

The problem was formulated as a three-category classification task, in which there is a strong imbalance towards one of the classes (Figure 2). The majority class, Graduate, represents 50% of the records (2209 of 4424) and Dropout represents 32% of total records (1421 of 4424), while the minority class, Enrolled, represents 18% of total records (794 of 4424). This might result in a high prediction accuracy driven by the majority class at the expense of a poor performance of the minority class. Therefore, anyone using this dataset should pay attention to this problem and address it with a data-level approach or with an algorithm-level approach. At the data-level approach, a sampling technique such as the Synthetic Minority Over Sampling Technique (SMOTE) [10] or the Adaptive Synthetic Sampling Approach (ADASYN) [11] or any variant thereof can be applied. At the algorithm-level approach, a machine learning algorithm that already incorporates balancing steps must be used, such as Balanced Random Forest [12] or Easy Ensemble [13], or bagging classifiers with additional balancing, such as Exactly Balanced Bagging [14], Roughly Balanced Bagging [15], Over-Bagging [14], or SMOTE-Bagging [16].

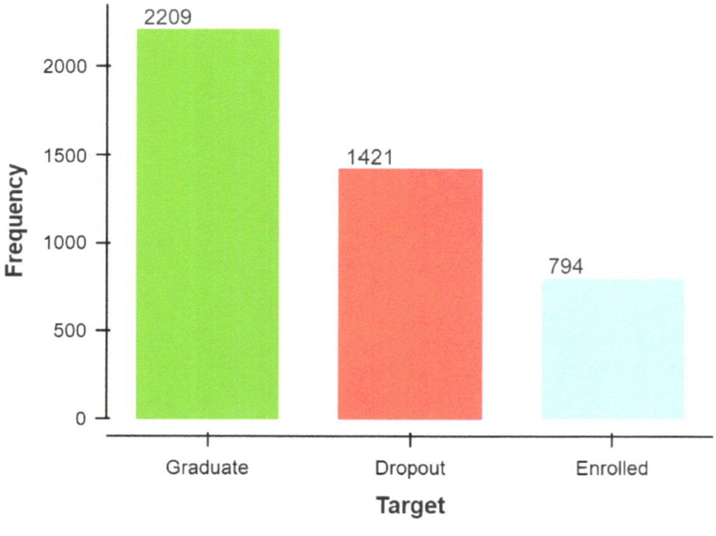

Figure 2. Distribution of student records among the three categories considered for academic success.

Figure 3 shows the same imbalanced nature of data when grouping the student outcomes by course, gender, student displaced, tuition fees up to date, scholarship holder, and evening/daytime attendance. Figure 3a shows that the most successful courses are Nursing and Social Service, with 72% and 70% of the students, respectively, receiving their degree within the normal duration of the course. On the opposite side, the technologies field with the courses of Biofuel Production Technologies and Informatics Engineering presents the most unsuccessful results, with only 8% of the students receiving their degree within the normal duration of the course. Dropout is also higher in these two courses (67% and 54%, respectively), along with the Equiniculture course with 55% dropout. Figure 3b shows that females are most successful, as well as the students that hold a scholarship and have their tuition fees up to date. Regarding the attendance regime (daytime or evening), the results show that students with daytime attendance finish the course earlier than evening students, as well as the students that are displaced from their homes.

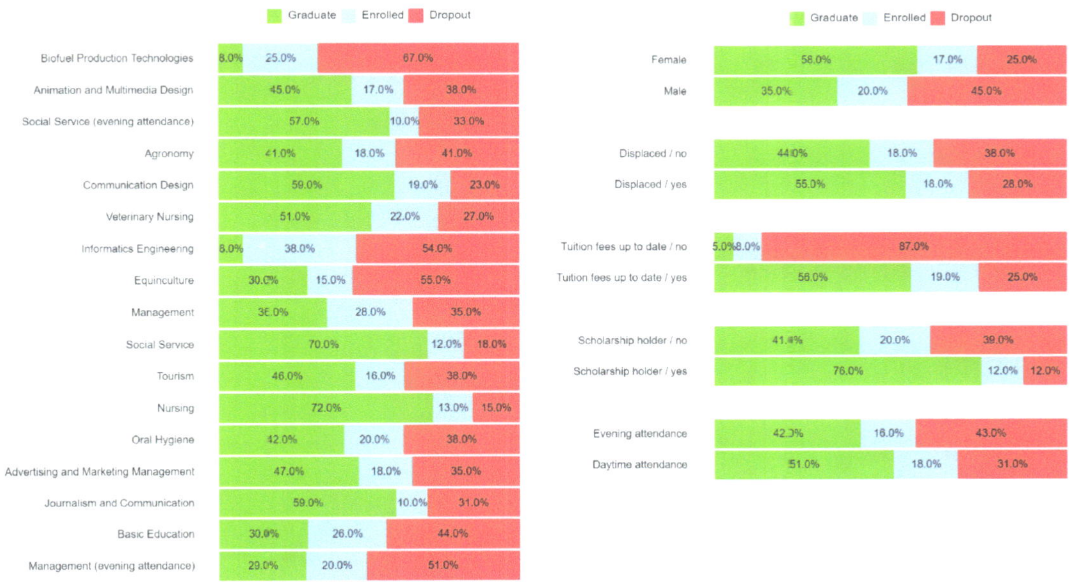

(a) (b)

Figure 3. Student outcomes grouped by: (**a**) course; (**b**) gender, student displaced, tuition fees up to date, scholarship holder, and evening/daytime attendance.

3.2.3. Multi-collinearity

Collinearity (or multi-collinearity) may be an issue that must be considered in some types of problems. The analysis of the heatmap (Figure 4), using the Pearson correlation coefficient, shows that there are some pairs of features having high correlation coefficients, which increases multi-collinearity in the dataset. In Figure 4, the blues represent the heatmap between demographics features, the oranges between socioeconomics features, the greens between macroeconomics features, the reds between academics features at enrollment time, the purples between academics features at the end of the first semester, the browns at the end of the second semester and, the grays represent collinearity between groups of features.

The collinearity is strongest within the same group of features, but we can also find higher values of correlation between groups. Table 9 shows a Pearson correlation coefficient greater than 0.7, which shows that the correlation is the strongest in features in the same groups, such as "Nationality" and "International" or "Mother's occupation" and "Father's occupation", but also between the groups related with the performance at the end of the first semester and the second semester, such as "Curricular units 1st sem (approved)" and "Curricular units 2nd sem (approved)".

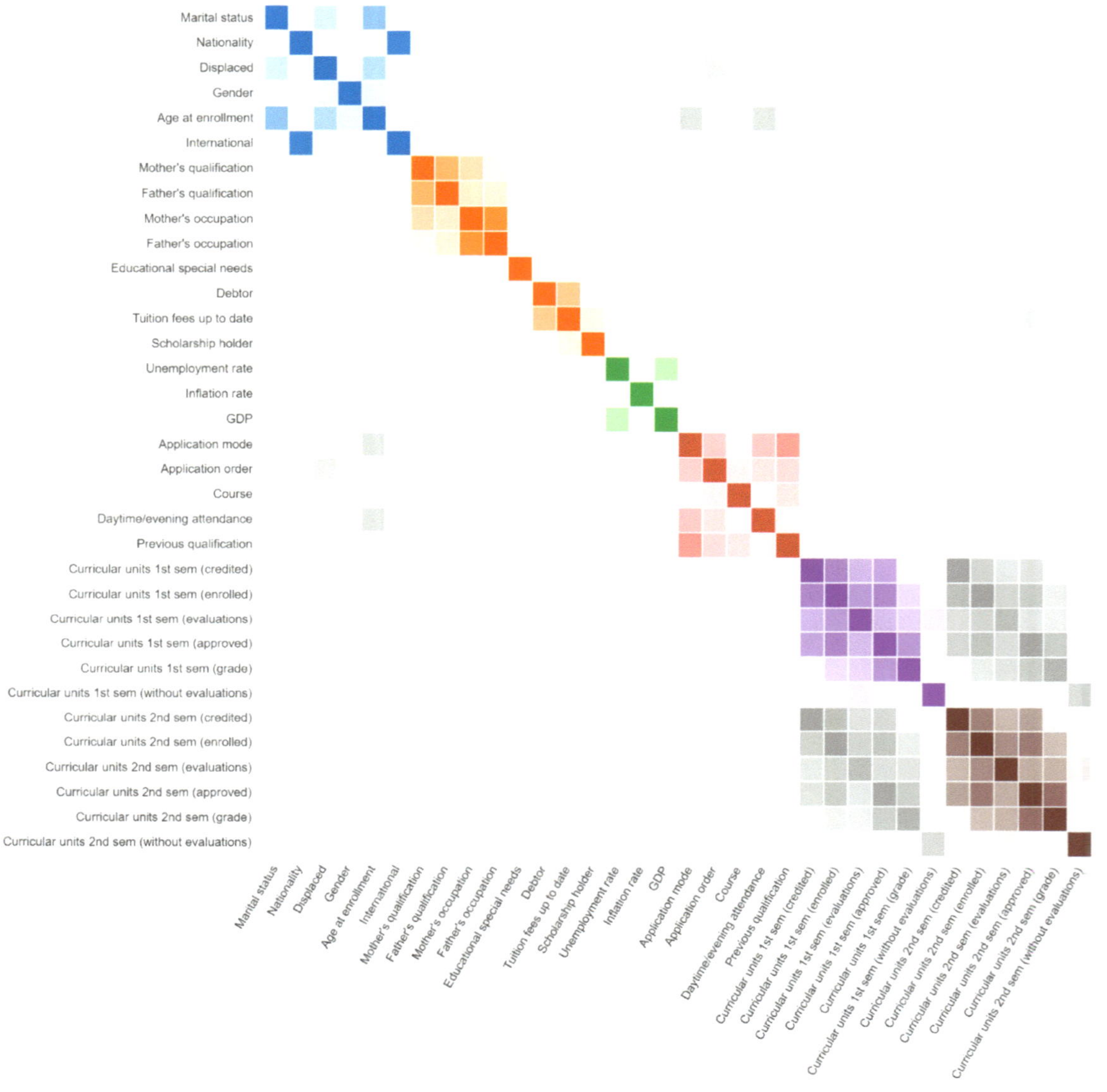

Figure 4. Heatmap with Pearson correlation.

3.2.4. Feature Importance

Feature importance plays an important role in understanding the data and also in the improvement and interpretation of the machine learning models. On the other hand, useless data results in bias that messes up the final results of a machine learning problem, so feature importance is frequently used to reduce de number of features used. The most important features differ depending on the technique used to calculate the importance of each feature and also the machine learning algorithm used [17]. One of the simplest and most used techniques to measure feature importance is Permutation Feature Importance. In this technique, feature importance is calculated by noticing the increase or decrease in error when we permute the values of a feature. If permuting the values causes a huge change in the error, it means the feature is important for our model.

Table 9. Collinearity between features with Pearson correlation coefficient greater than 0.7.

Feature	Collinearity with	Pearson
Curricular units 1st sem (credited)	Curricular units 2nd sem (credited) Curricular units 1st sem (enrolled)	0.9448 0.7743
Curricular units 1st sem (enrolled)	Curricular units 2nd sem (enrolled) Curricular units 1st sem (approved) Curricular units 2nd sem (credited)	0.9426 0.7691 0.7537
Nationality	International	0.9117
Curricular units 1st sem (approved)	Curricular units 2nd sem (approved) Curricular units 2nd sem (enrolled)	0.9040 0.7338
Curricular units 1st sem (grade)	Curricular units 2nd sem (grade)	0.8372
Curricular units 1st sem (evaluations)	Curricular units 2nd sem (evaluations)	0.7789
Curricular units 2nd sem (approved)	Curricular units 2nd sem (grade)	0.7608
Mother's occupation	Father's occupation	0.7240
Curricular units 2nd sem (enrolled)	Curricular units 2nd sem (approved)	0.7033

We performed a test to determine the most important features considering the Permutation Feature Importance, using F1 as the error metric, which is a metric more adequate for imbalanced data, taking into account the trade-off between precision and recall. The Permutation Feature Importance was applied to some of the most interesting results reported in the literature for multiclass imbalanced classification [18,19]. We used the ensemble method Random Forest (RF) [20] and three general boosting methods: Extreme Gradient Boosting (XGBOOST) [21], Light Gradient Boosting Machine (LIGHTGBM) [22], and CatBoost (CATBOOST) [23]. Figure 5 shows the 10 biggest changes in the F1-score metric using the Permutation Feature Importance technic for each machine learning algorithm considered. The analysis of these results shows that five features are considered important in all algorithms: "Curricular units 2nd sem (approved)", "Curricular units 1st sem (approved)", "Curricular units 2nd sem (grade)", "Course", and "Tuition fees up to date". The features "Curricular units 1st sem (enrolled)", "Curricular units 1st sem (evaluations)", "Curricular units 2nd sem (enrolled)", and "Curricular units 2nd sem (evaluations)" are important in three of the algorithms.

3.3. Compliances

All data are anonymized, and compliance with the Privacy and Personal Data Processing Policy of the institution is ensured according to the General Data Protection Regulation (GDPR). This dataset is also compliant with the FAIR (Findability, Accessibility, Interoperability, and Reusability) principles for scientific data management [24].

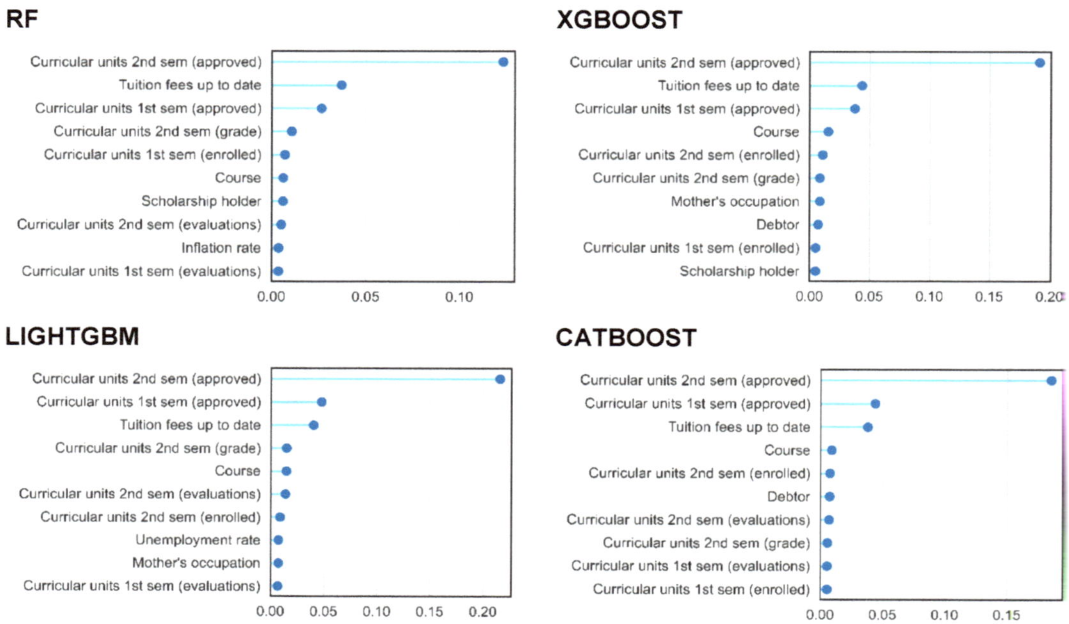

Figure 5. Plot of top 10 Permutation Feature Importance for each machine learning algorithm considered.

4. Conclusions

This descriptor presents a dataset created from the Polytechnic Institute of Portalegre (acquired from several disjoint databases) related to students enrolled in different undergraduate degrees, such as agronomy, design, education, nursing, journalism, management, social service, and technologies. It contains 4424 records with 35 attributes that include information known at the time of student enrollment, demographics, socioeconomics, macroeconomics data, and students' academic performance at the end of the first and second semesters.

The dataset is useful for researchers who want to conduct comparative studies on student academic performance and also for training in the machine learning area.

Supplementary Materials: The document with detailed features information can be consulted at: http://valoriza.ipportalegre.pt/piaes/features-info-stats.html (accessed on 10 October 2022).

Author Contributions: Conceptualization, V.R., J.M., L.B. and M.V.M.; methodology, M.V.M., J.M. and V.R.; software, V.R.; validation, V.R. and M.V.M.; resources, V.R.; data curation, V.R. and M.V.M.; writing—original draft preparation, V.R.; writing—review and editing, L.B. and M.V.M.; visualization, V.R.; project administration, V.R.; funding acquisition, V.R. All authors have read and agreed to the published version of the manuscript.

Funding: This research was funded by the program SATDAP—Capacitação da Administração Pública under grant number POCI-05-5762-FSE-000191.

Institutional Review Board Statement: Privacy issues related to the use and publication of the dataset were validated by the Data Protection Officer (DPO) of the Polytechnic Institute of Portalegre according to the General Data Protection Regulation (GDPR) directives.

Informed Consent Statement: Not applicable.

Data Availability Statement: The data presented in this study are publicly available at https://doi.org/10.5281/zenodo.5777339 (accessed on 10 October 2022).

Acknowledgments: The authors would like to thank the Polytechnic Institute of Portalegre for providing support for this project, particularly to the Academic Services Department for providing the data and explaining the attributes used.

Conflicts of Interest: The authors declare no conflict of interest.

Abbreviations

The following abbreviations are used in this manuscript:

AMS	Academic Management System
CATBOOST	CatBoost
CSV	Comma-separated values
DGES	Direção Geral do Ensino Superior
DPO	Data Protection Officer
GDPR	General Data Protection Regulation
LIGHTGBM	Light Gradient Boosting Machine
PAE	Enterprise Application Platform
RF	Random Forest
XGBOOST	Extreme Gradient Boost

Appendix A

Table A1. Marital status values.

Attribute	Values
Marital status	1—Single 2—Married 3—Widower 4—Divorced 5—Facto union 6—Legally separated

Table A2. Nationality values.

Attribute	Values
Nationality	1—Portuguese 2—German 3—Spanish 4—Italian 5—Dutch 6—English 7—Lithuanian 8—Angolan 9—Cape Verdean 10—Guinean 11—Mozambican 12—Santomean 13—Turkish 14—Brazilian 15—Romanian 16—Moldova (Republic of) 17—Mexican 18—Ukrainian 19—Russian 20—Cuban 21—Colombian

Table A3. Application mode values.

Attribute	Values
Application mode	1—1st phase—general contingent 2—Ordinance No. 612/93 3—1st phase—special contingent (Azores Island) 4—Holders of other higher courses 5—Ordinance No. 854-B/99 6—International student (bachelor) 7—1st phase—special contingent (Madeira Island) 8—2nd phase—general contingent 9—3rd phase—general contingent 10—Ordinance No. 533-A/99, item b2) (Different Plan) 11—Ordinance No. 533-A/99, item b3 (Other Institution) 12—Over 23 years old 13—Transfer 14—Change in course 15—Technological specialization diploma holders 16—Change in institution/course 17—Short cycle diploma holders 18—Change in institution/course (International)

Table A4. Course values.

Attribute	Values
Course	1—Biofuel Production Technologies 2—Animation and Multimedia Design 3—Social Service (evening attendance) 4—Agronomy 5—Communication Design 6—Veterinary Nursing 7—Informatics Engineering 8—Equiniculture 9—Management 10—Social Service 11—Tourism 12—Nursing 13—Oral Hygiene 14—Advertising and Marketing Management 15—Journalism and Communication 16—Basic Education 17—Management (evening attendance)

Table A5. Previous qualification values.

Attribute	Values
Previous qualification	1—Secondary education 2—Higher education—bachelor's degree 3—Higher education—degree 4—Higher education—master's degree 5—Higher education—doctorate 6—Frequency of higher education 7—12th year of schooling—not completed 8—11th year of schooling—not completed

Table A5. *Cont.*

Attribute	Values
	9—Other—11th year of schooling 10—10th year of schooling 11—10th year of schooling—not completed 12—Basic education 3rd cycle (9th/10th/11th year) or equivalent 13—Basic education 2nd cycle (6th/7th/8th year) or equivalent 14—Technological specialization course 15—Higher education—degree (1st cycle) 16—Professional higher technical course 17—Higher education—master's degree (2nd cycle)

Table A6. Mother's and Father's values.

Attribute	Values
Mother's qualification Father's qualification	1—Secondary Education—12th Year of Schooling or Equivalent 2—Higher Education—bachelor's degree 3—Higher Education—degree 4—Higher Education—master's degree 5—Higher Education—doctorate 6—Frequency of Higher Education 7—12th Year of Schooling—not completed 8—11th Year of Schooling—not completed 9—7th Year (Old) 10—Other—11th Year of Schooling 11—2nd year complementary high school course 12—10th Year of Schooling 13—General commerce course 14—Basic Education 3rd Cycle (9th/10th/11th Year) or Equivalent 15—Complementary High School Course 16—Technical-professional course 17—Complementary High School Course—not concluded 18—7th year of schooling 19—2nd cycle of the general high school course 20—9th Year of Schooling—not completed 21—8th year of schooling 22—General Course of Administration and Commerce 23—Supplementary Accounting and Administration 24—Unknown 25—Cannot read or write 26—Can read without having a 4th year of schooling 27—Basic education 1st cycle (4th/5th year) or equivalent 28—Basic Education 2nd Cycle (6th/7th/8th Year) or equivalent 29—Technological specialization course 30—Higher education—degree (1st cycle) 31—Specialized higher studies course 32—Professional higher technical course 33—Higher Education—master's degree (2nd cycle) 34—Higher Education—doctorate (3rd cycle)

Table A7. Mother's and Father's occupation.

Attribute	Values
Mother's occupation Father's occupation	1—Student 2—Representatives of the Legislative Power and Executive Bodies, Directors, Directors and Executive Managers 3—Specialists in Intellectual and Scientific Activities 4—Intermediate Level Technicians and Professions 5—Administrative staff 6—Personal Services, Security and Safety Workers, and Sellers 7—Farmers and Skilled Workers in Agriculture, Fisheries, and Forestry 8—Skilled Workers in Industry, Construction, and Craftsmen 9—Installation and Machine Operators and Assembly Workers 10—Unskilled Workers 11—Armed Forces Professions 12—Other Situation; 13—(blank) 14—Armed Forces Officers 15—Armed Forces Sergeants 16—Other Armed Forces personnel 17—Directors of administrative and commercial services 18—Hotel, catering, trade, and other services directors 19—Specialists in the physical sciences, mathematics, engineering, and related techniques 20—Health professionals 21—Teachers 22—Specialists in finance, accounting, administrative organization, and public and commercial relations 23—Intermediate level science and engineering technicians and professions 24—Technicians and professionals of intermediate level of health 25—Intermediate level technicians from legal, social, sports, cultural, and similar services 26—Information and communication technology technicians 27—Office workers, secretaries in general, and data processing operators 28—Data, accounting, statistical, financial services, and registry-related operators 29—Other administrative support staff 30—Personal service workers 31—Sellers 32—Personal care workers and the like 33—Protection and security services personnel 34—Market-oriented farmers and skilled agricultural and animal production workers 35—Farmers, livestock keepers, fishermen, hunters and gatherers, and subsistence 36—Skilled construction workers and the like, except electricians 37—Skilled workers in metallurgy, metalworking, and similar 38—Skilled workers in electricity and electronics 39—Workers in food processing, woodworking, and clothing and other industries and crafts 40—Fixed plant and machine operators 41—Assembly workers 42—Vehicle drivers and mobile equipment operators 43—Unskilled workers in agriculture, animal production, and fisheries and forestry 44—Unskilled workers in extractive industry, construction, manufacturing, and transport 45—Meal preparation assistants 46—Street vendors (except food) and street service providers

Table A8. Gender values.

Attribute	Values
Gender	1—male 0—female

Table A9. Attendance regime values.

Attribute	Values
Daytime/evening attendance	1—daytime 0—evening

Table A10. Yes/No attributes.

Attribute	Values
Displaced Educational special needs Debtor Tuition fees up to date Scholarship holder International	1—yes 0—no

References

1. Behr, A.; Giese, M.; Teguim Kamdjou, H.D.; Theune, K. Motives for Dropping out from Higher Education—An Analysis of Bachelor's Degree Students in Germany. *Eur. J. Educ.* **2021**, *56*, 325–343. [CrossRef]
2. Kehm, B.M.; Larsen, M.R.; Sommersel, H.B. Student Dropout from Universities in Europe: A Review of Empirical Literature. *Hungarian Educ. Res. J.* **2020**, *9*, 147–164. [CrossRef]
3. Atchley, W.; Wingenbach, G.; Akers, C. Comparison of Course Completion and Student Performance through Online and Traditional Courses. *Int. Rev. Res. Open Distance Learn.* **2013**, *14*, 104–116. [CrossRef]
4. Quinn, J. *Dropout and Completion in Higher Education in Europe among Students from Under-Represented Groups*; An Independent report authored for the NESET network of experts; European Commission: Brussels, Belgium, 2013.
5. Namoun, A.; Alshanqiti, A. Predicting Student Performance Using Data Mining and Learning Analytics Techniques: A Systematic Literature Review. *Appl. Sci.* **2020**, *11*, 237. [CrossRef]
6. Saa, A.A.; Al-Emran, M.; Shaalan, K. Mining Student Information System Records to Predict Students' Academic Performance. *Adv. Intell. Syst. Comput.* **2020**, *921*, 229–239. [CrossRef]
7. Akçapınar, G.; Altun, A.; Aşkar, P. Using Learning Analytics to Develop Early-Warning System for at-Risk Students. *Int. J. Educ. Technol. High. Educ.* **2019**, *16*, 40. [CrossRef]
8. Daud, A.; Lytras, M.D.; Aljohani, N.R.; Abbas, F.; Abbasi, R.A.; Alowibdi, J.S. Predicting Student Performance Using Advanced Learning Analytics. In Proceedings of the 26th International World Wide Web Conference 2017, WWW 2017 Companion, Perth, Australia, 3–7 April 2017; pp. 415–421. [CrossRef]
9. Martins, M.V.; Tolledo, D.; Machado, J.; Baptista, L.M.T.; Realinho, V. Early Prediction of Student's Performance in Higher Education: A Case Study. *Adv. Intell. Syst. Comput.* **2021**, *1365*, 166–175. [CrossRef]
10. Chawla, N.V.; Bowyer, K.W.; Hall, L.O.; Kegelmeyer, W.P. SMOTE: Synthetic Minority Over-Sampling Technique. *J. Artif. Intell. Res.* **2002**, *16*, 321–357. [CrossRef]
11. He, H.; Bai, Y.; Garcia, E.A.; Li, S. ADASYN: Adaptive Synthetic Sampling Approach for Imbalanced Learning. In Proceedings of the International Joint Conference on Neural Networks, Hong Kong, China, 1–8 June 2008; pp. 1322–1328. [CrossRef]
12. Chen, C.; Liaw, A.; Breiman, L. Using Random Forest to Learn Imbalanced Data. *Univ. Calif. Berkeley* **2004**, *110*, 1–12.
13. Liu, X.Y.; Wu, J.; Zhou, Z.H. Exploratory Undersampling for Class-Imbalance Learning. *IEEE Trans. Syst. Man Cybern. Part B Cybern.* **2009**, *39*, 539–550. [CrossRef]
14. Maclin, R.; Opitz, D. An Empirical Evaluation of Bagging and Boosting. In Proceedings of the National Conference on Artificial Intelligence, Providence, RI, USA; 1997; pp. 546–551.
15. Hido, S.; Kashima, H.; Takahashi, Y. Roughly Balanced Bagging for Imbalanced Data. *Stat. Anal. Data Min.* **2009**, *2*, 412–426. [CrossRef]
16. Wang, S.; Yao, X. Diversity Analysis on Imbalanced Data Sets by Using Ensemble Models. In Proceedings of the 2009 IEEE Symposium on Computational Intelligence and Data Mining, Nashville, TN, USA, 30 March–2 April 2009; pp. 324–331. [CrossRef]
17. Saarela, M.; Jauhiainen, S. Comparison of Feature Importance Measures as Explanations for Classification Models. *SN Appl. Sci.* **2021**, *3*, 272. [CrossRef]

18. Spelmen, V.S.; Porkodi, R. A Review on Handling Imbalanced Data. In Proceedings of the 2018 International Conference on Current Trends towards Converging Technologies (ICCTCT), Coimbatore, India, 1–3 March 2018. [CrossRef]
19. Ali, H.; Salleh, M.N.M.; Saedudin, R.; Hussain, K.; Mushtaq, M.F. Imbalance Class Problems in Data Mining: A Review. *Indones. J. Electr. Eng. Comput. Sci.* **2019**, *14*, 1552–1563. [CrossRef]
20. Ho, T.K. Random Decision Forests. In Proceedings of the 3rd International Conference on Document Analysis and Recognition, Montreal, QC, Canada, 14–16 August 1995; Volume 1, pp. 278–282. [CrossRef]
21. Chen, T.; Guestrin, C. XGBoost: A Scalable Tree Boosting System. In Proceedings of the 22nd ACM SIGKDD International Conference, San Francisco, CA, USA, 13–17 August 2016. [CrossRef]
22. Ke, G.; Meng, Q.; Finley, T.; Wang, T.; Chen, W.; Ma, W.; Ye, Q.; Liu, T.Y. LightGBM: A Highly Efficient Gradient Boosting Decision Tree. *Adv. Neural Inf. Process. Syst.* **2017**, *30*, 3147–3155. [CrossRef]
23. Prokhorenkova, L.; Gusev, G.; Vorobev, A.; Dorogush, A.V.; Gulin, A. CatBoost: Unbiased Boosting with Categorical Features. *arXiv* **2017**, arXiv:1706.09516v5. [CrossRef]
24. Wilkinson, M.D.; Dumontier, M.; Aalbersberg, I.J.; Appleton, G.; Axton, M.; Baak, A.; Blomberg, N.; Boiten, J.W.; da Silva Santos, L.B.; Bourne, P.E.; et al. The FAIR Guiding Principles for Scientific Data Management and Stewardship. *Sci. Data* **2016**, *3*, 160018. [CrossRef] [PubMed]

data

MDPI

Article

Thematic Analysis of Indonesian Physics Education Research Literature Using Machine Learning

Purwoko Haryadi Santoso [1,2,*], Edi Istiyono [1,3], Haryanto [1] and Wahyu Hidayatulloh [3]

1 Graduate School of Educational Research and Evaluation, Universitas Negeri Yogyakarta, Sleman 55281, Indonesia
2 Department of Physics Education, Universitas Sulawesi Barat, Majene 91412, Indonesia
3 Department of Physics Education, Universitas Negeri Yogyakarta, Sleman 55281, Indonesia
* Correspondence: purwokoharyadi.2021@student.uny.ac.id or purwokoharyadisantoso@unsulbar.ac.id

Abstract: Abundant physics education research (PER) literature has been disseminated through academic publications. Over the years, the growing body of literature challenges Indonesian PER scholars to understand how the research community has progressed and possible future work that should be encouraged. Nevertheless, the previous traditional method of thematic analysis possesses limitations when the amount of PER literature exponentially increases. In order to deal with this plethora of publications, one of the machine learning (ML) algorithms from natural language processing (NLP) studies was employed in this paper to automate a thematic analysis of Indonesian PER literature that still needs to be explored within the community. One of the well-known NLP algorithms, latent Dirichlet allocation (LDA), was used in this study to extract Indonesian PER topics and their evolution between 2014 and 2021. A total of 852 papers (~4 to 8 pages each) were collectively downloaded from five international conference proceedings organized, peer reviewed, and published by Indonesian PER researchers. Before their topics were modeled through the LDA algorithm, our data corpus was preprocessed through several common procedures of established NLP studies. The findings revealed that LDA had thematically quantified Indonesian PER topics and described their distinct development over a certain period. The identified topics from this study recommended that the Indonesian PER community establish robust development in eight distinct topics to the present. Here, we commenced with an initial interest focusing on research on physics laboratories and followed the research-based instruction in late 2015. For the past few years, the Indonesian PER scholars have mostly studied 21st century skills which have given way to a focus on developing relevant educational technologies and promoting the interdisciplinary aspects of physics education. We suggest an open room for Indonesian PER scholars to address the qualitative aspects of physics teaching and learning that is still scant within the literature.

Keywords: thematic analysis; Indonesia; physics education research; machine learning

Citation: Santoso, P.H.; Istiyono, E.; Haryanto; Hidayatulloh, W. Thematic Analysis of Indonesian Physics Education Research Literature Using Machine Learning. *Data* **2022**, *7*, 147. https://doi.org/10.3390/data7110147

Academic Editors: Antonio Sarasa Cabezuelo and Ramón González del Campo Rodríguez Barbero

Received: 1 August 2022
Accepted: 19 September 2022
Published: 28 October 2022

1. Introduction

Several decades of physics education research (PER) have established an enormous body of literature related to physics teaching and learning. Outside the context of the Indonesian PER community, many thousands of PER articles have been published in several high impact journals, such as *The Physics Teacher* (TPT), *The American Journal of Physics* (AJP), and *Physical Review Physics Education Research* (PRPER) (previously announced as *Physical Review Special Topics Physics Education Research*) since 1933, 1963, and 2005, respectively. We term it as "outside" since representatives of Indonesian PER scholars within these journals are still scant. It must be considered that unique findings from the Indonesian environment are still missing based on these references.

Rare representation of Indonesian PER scholars covered in these journals cannot be translated as the absence of PER development within the Indonesian community. Since 2014 to date, several international conferences in the area of PER have been annually organized by several Indonesian teacher education institutions (TEIs). The five oldest conferences on science, technology, engineering, and mathematics (STEM) education have included the topic of physics education research (PER) for publication. They comprise the International Conference on Research, Implementation, & Education of Mathematics and Science (ICRIEMS, since 2014) [1] and the International Seminar on Science Education (ISSE, since 2015) organized by Universitas Negeri Yogyakarta (UNY) [2], the International Conference on Mathematics & Science Education (ICMSE, since 2014) organized by Universitas Negeri Semarang (UNNES) [3], the International Conference on Mathematics and Science Education (ICMScE, since 2016) organized by Universitas Pendidikan Indonesia (UPI) [4], and the International Conference on Mathematics and Science Education (ICoMSE, since 2017) organized by Universitas Negeri Malang (UM) [5]. These selected international conferences have substantially contributed to our research insights into the Indonesian PER field. Otherwise, peer-reviewed journals were only published nationally during the same timeframe and a smaller number of publications than the aforementioned conferences. Furthermore, they have attracted PER scholars of various backgrounds from novice researchers (graduate students) to PER experts (senior scholars and professors) funded through research grants from the Indonesian government. Mostly, the authors have been affiliated with several Indonesian institutions and a few with neighboring countries, particularly from Southeast Asia region.

Essentially, this volume of publications provides a convincing challenge for PER scholars to understand how the research community has progressed and possible future work that should be emphasized. Nevertheless, it can be troublesome to synthesize whole articles published within a large number of publications. Most researchers tend to review only the most relevant research articles for their work. There is always a possibility that they have neglected some academic resources within the collection of literature. We believe that it is imperative to have insight into PER researchers to further their understanding of PER. These cases are more complicated for novice researchers, who should exhaustively review the extensive development of the field [6]. Consequently, they are usually more dependent on the given suggestions either provided by communities, research groups, or indexing databases like Google Scholar [7].

On the other hand, the number of works could inevitably be perceived as the Indonesian PER field having currently developed to a phase of maintaining its research merit of theoretical and methodological practice through their continued existence for a certain time. Hence, this body of literature is valuable in explaining the characteristics of the Indonesian PER field and its development of topics over time. To synthesize a comprehensive story of PER topics outside the Indonesian PER field, one must consult the previously ambitious work that has been disseminated by McDermott & Redish [8], Docktor and Mestre [9], Meltzer and Otero [10], Odden et al. [11], and Yun [12]. These great works admittedly have guided the PER community in several parts of the world, including the Indonesian PER scholars. Nevertheless, as clearly mentioned before, the representation of Indonesian scholars covered by these disseminations is still limited to best capture the Indonesian PER findings. It might be less appropriate to understand the characteristics and development of the Indonesian PER topics if we merely considered those resources without sufficient involvement of Indonesian PER scholars. Therefore, our current paper extends the intention of previous works to analyze the Indonesian PER field through the methodology of thematic analysis. We believe that addressing this issue should be considered a potential contribution to enrich the merit of previous references. In this paper, we studied 852 proceeding papers organized, peer reviewed, and published by the Indonesian PER community that are unknown from previous works. To the best of our knowledge, Indonesian PER researchers have not yet performed work to analyze their research literature using the similar method performed by our study. Instead, a recent study by Hartono, et al. [13] (Indonesian author) investigated a data corpus outside the context of Indonesia.

Although Indonesian PER research is still scant with regard to performing a thematic analysis, we must admit that other aims related to Indonesian PER have made several efforts in this area, particularly through the conventional method of content analysis on science education [14], scientific literacy [15], teacher education [16], and learning media [17]. However, one may argue that conducting a thematic analysis through traditionally reading and summarizing the vast amount of literature is inefficient. For instance, a recent study on science education research reported by Faisal et al. [14] even argued that performing this sort of analysis on a large number of articles was "tricky", as mentioned in their introduction of a paper about mapping the research trends in Indonesian science education research. Hence, they considered that a content analysis approach on the keywords of proposed titles of research grants was more doable to simplify their study. In their conclusion, Faisal et al. [14] conceded that the selection of this method of keyword-based analysis was problematic in representing the final state of research dissemination. The initial title of the research grant was more likely to be improved after the work had been finished, and either theoretical or methodological considerations may have made it possible for some improvements to occur. Publication of their work might have slightly evolved from the proposed title of the initial announcement of the research grant.

Furthermore, the traditional method of content analysis fails to satisfy the principle of the distributional hypothesis of topics established by the linguistic field [18]. The nature of research topics should demonstrate a mixture of words instead of a single keyword [19]. Consequently, the principle of thematic analysis needs several words to represent a literature topic. Therefore, the mixed membership idea and the distributional hypothesis of topics should be consulted to shed more light on the analysis of literature topics. For this reason, a new more efficient and significant method of thematic analysis should be approached to complete our understanding about the literature topics.

Over the past few years, machine learning (ML) has rapidly become a powerful tool to respond to the growing size of data emerging in the digital era. Textual data is one of the data structures studied within this field. Natural language processing (NLP) is one of the ML studies concerned with sets of texts. NLP proposes a method of thematic analysis to extract our understanding of textual data based on a large collection of literature. Recent studies by Odden et al. [20] have performed this sort of analysis towards *Physics Education Research Conference* (PERC) proceedings [11] and Yun [12] towards the *American Journal of Physics* (AJP) and the *Physical Review Physics Education Research* (PRPER). In this paper, we extend these efforts to analyze Indonesian PER literature using the NLP algorithm. We have performed one of the popular NLP algorithms, latent Dirichlet allocation (LDA) [21,22], to automate a thematic analysis of Indonesian PER literature selected from the five longest running international conference proceedings organized, peer reviewed, and published by the Indonesian PER community between 2014 and 2021. Throughout the LDA topic modeling, we have extracted eight characteristics of Indonesian PER topics and how those topics have been developed within the field over a certain period.

Our contribution to this paper is intended to demonstrate the LDA algorithm in Indonesian PER literature. It has the potential ability to help PER scholars extract valuable information from the vast number of Indonesian PER literature. It inevitably could extract the discovered Indonesian PER topics based on the nature of topics and their associated rise and fall within the field over a certain publication time frame. This study then will be guided by the following two research questions:

RQ1. Using LDA topic modeling towards the five Indonesian PER publications, what are the topic characteristics studied between 2014 and 2021?

RQ2. How has the development of these topics occurred between 2014 and 2021?

The extracted Indonesian PER topics from this study are dedicated to enriching our knowledge about research activities that have been attempted and suggesting areas of further investigation. The demonstration of the promising analytic approach would be our trigger to the wider academic publications within the Indonesian PER community.

2. Theoretical Review

Thematic analysis is one type of literature research methodology used in collecting, reviewing, summarizing, and synthesizing previous studies about specific domains [23]. Naturally, thematic analysis is established in the climate of qualitative inquiry. It is constructed, and has similarities, with other systematic procedures of qualitative analysis as demonstrated by grounded theory, narrative analysis, interpretative phenomenological analysis, and content analysis in analyzing personal experience about phenomena [24,25]. The early research practices of a literature review using thematic analysis is undertaken through the constructivist paradigm that the researcher is the main actor in the data collection and analysis [25,26]. Therefore, human-based analysis plays a vital role to conduct the time-consuming literature review using traditional thematic analysis [27]. As briefly discussed in the introduction above, this way encounters serious disadvantages when the number of pieces of literature significantly increases [14]. It also has the potential to make unstable findings, particularly those that are undertaken by novice researchers [28]. Snyder [23] even argues that traditional thematic analysis often produces a lack of thoroughness and rigor-specific methodology. Therefore, several researchers recommend the enhancement of this conventional way to strengthen its robustness for literature reviews. They propose automation technology [29], computational toolkit [30], as well as using machine learning (ML) technology, as demonstrated by the current paper.

Natural language processing (NLP) is the subfield of ML studies that performs topic modeling or text analysis from a set of documents. Broadly speaking, there are two types of ML models, namely supervised and unsupervised algorithms. The supervised ML model specifies a predetermined set of labels in fitting, predicting, or classifying the trained subset of data. Conversely, unsupervised ML models do not specify the desired labels in advance. Accordingly, in an unsupervised NLP model, we do not have a predetermined set of results before processing the text analysis. They rather intend to extract latent entities from a set of documents without knowing the desired results previously. Thereafter, this technique naturally may be troublesome for the interpretation of extracted topics due to the absence of predetermined labels. However, this disadvantage simultaneously often occurred in common text analysis studies [31]. Therefore, NLP researchers must evaluate their interpretations of the extracted topics through several procedures of evaluation metrics explained in the subsequent methodological section of this paper.

Latent Dirichlet allocation (LDA) is a popular unsupervised NLP algorithm that has been commonly used to extract the essence of diverse literature. Even though this text analysis technique has been disadvantaged with some simplifications as explained above, several fields have employed this method persuasively. Since Blei et al. [21] published their LDA algorithm in 2003, LDA has been employed for several purposes such as analyzing customers' opinions in agricultural companies [32], commercial reviews [33], political issues [34], and topics in online news portals [35]. Additionally, LDA also has been implemented in the educational environment to analyze informatics engineering studies [36], project reports [37], undergraduate theses [38], scientific papers [39], and online educational resources [40,41]. Therefore, these numerous LDA implementations offer a promising tool in many fields, including physics education research (PER). Recently, the LDA method has been implemented for the subject of PER [11] in the analysis of large numbers of individual papers from physics education research conference proceedings (PERC) [11]. However, this previous attempt was intended to cover outside the Indonesian context. Thus, it can be less representative for grasping the full knowledge about the development of Indonesian PER studies. To enrich the insight into Indonesian PER development, we believe that analyzing the Indonesian PER literature using the LDA algorithm could be the potential contribution of our paper. Thus, it should be worthwhile since there is little known about how our Indonesian PER community has been established and where we are going further to develop our community.

Broadly speaking, LDA is a generative probabilistic model to analyze the latent topics from a set of documents or the data corpus. Using topic modeling, the document is presented as a collection of latent topics and each topic is a collection of representative words. The LDA algorithm can be used to identify the latent topics from a set of documents by counting the word co-occurrence within the document. It then should conclude the number of distinctive topics (K) based on a coherence measure, which is defined as how well these topics "hang together" to represent the extracted latent topics [42]. After the most representative model has been trained through the iterative findings of the optimum setting of several parameters (discussed in the methods section), the LDA result will extract the most representative words in each topic and the distribution of those topics within the document. Eventually, we can interpret these distinct groups of words to understand the properties of topics (RQ1). According to this LDA result, and by carefully reading the content of representative documents, the term for each topic can be defined.

Mathematically, there are two matrices as the input and output of the LDA algorithm. The entry of a matrix row represents the distribution of word co-occurrence, as illustrated in Figure 1. The input matrix corresponds to the documents row (D) and the words column (N) across the entire dataset (dimension $D \times N$, D is the number of documents and N is the count of words), termed as "document–word matrix". Each entry of a document–word matrix represents the count of words co-occurred in each document. This input matrix will be modeled by the LDA algorithm to create two output matrices. They are a document–topic matrix (θ_D) and a topic–word matrix (β_K) (Figure 1) that distribute the previous former document–word matrix using throughout a set of topics ($T_{1:K}$). The document–topic matrix (θ_D) corresponds to the document rows (D) and the topic columns (K) (size $D \times K$, D is the number of documents and K is the number of topics). The entries of a θ_D matrix represent the co-occurrence of each topic within a single document. The topic–word matrix (β_K) corresponds to the topic rows (K) and the word columns (N) (size $K \times N$, K is the number of topics and N is the number of words). The entries of a β_K matrix demonstrate the count of representative words in each latent topic. The interpretation of the LDA algorithm through this point of view is known as probabilistic matrix factorization, introduced by Hoffman et al. [43].

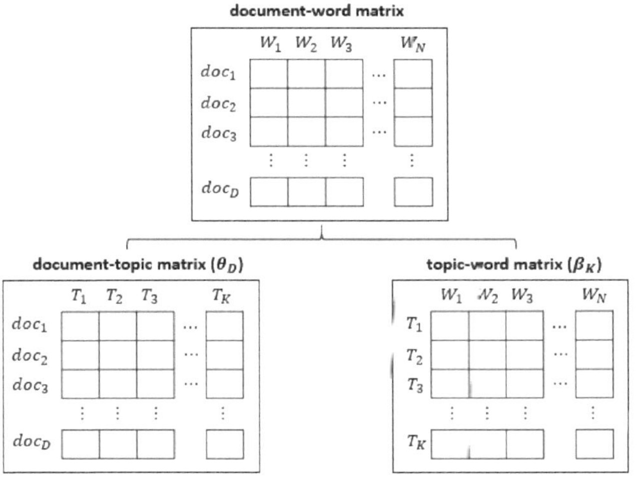

Figure 1. LDA interpretation through the concept of probabilistic matrix factorization (Adapted version from Odden, et al. [11]).

By the probabilistic matrix factorization, the LDA algorithm lies on three assumptions that must be taken into consideration by the user. The first assumption is that LDA does not consider the order of words in the analysis. Thus, it specifically disregards the nuance of language for text analysis. Indeed, it merely considers the number of words within the document. Despite the existence of this major assumption, this is commonly assumed in topic modeling studies. As proposed by Grimmer & Stewart [19], the principle of text analysis is that "all quantitative models of language are wrong, but some are useful"

The second assumption of LDA is that all documents should contain a mixture of several topics rather than a single topic. Specifically, LDA believes in a mixed membership model of a topic, rather than a single model of topic contained in the document [44]. Fortunately, we argue that this second assumption should lead to the impactful merit of the LDA model in performing automated text analysis from the interdisciplinary nature including PER studies. We typically investigate specific research problems in PER. We often bring, share, and combine insights, theories, or methods from another related field. For instance, research-based physics instructions are evaluated through the administration of assessment tools validated in advance. In the interdisciplinary context, the PER community should consult several resources from curriculum and instruction studies and the field of educational measurement to support assessment validity.

The third assumption of LDA assumes that the representative words of a distinct topic will be more likely to be mentioned than another word within the data corpus. Then, this greater probability of a word in a topic means that that distinct word will tend to co-occur more frequently in each topic. This assumption is known as the distributional hypothesis of linguistics [18]. For instance, if the current topic of a document is "culinary recipes", the words belong to "food", "ingredient", "taste", or "cook" will be more frequently co-occurred rather than the less relevant words, i.e., "representation", "mechanics", "item", or even "conceptual understanding".

3. Method

Our study involved three common steps of LDA topic modeling, as demonstrated in Figure 2. In this section, we will explain the details of these stages consecutively.

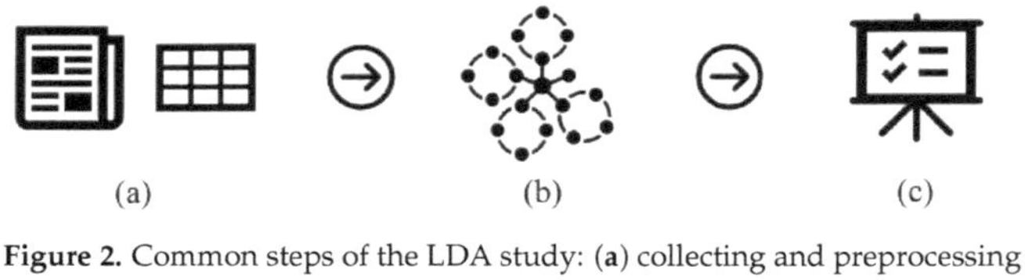

Figure 2. Common steps of the LDA study: (**a**) collecting and preprocessing the data; (**b**) modeling and evaluating the results; and (**c**) discussing the topical results to answer the research questions.

3.1. Collecting and Preprocessing the Data

In this step, we collected the PDFs by manually downloading the open access-based articles from five international conference proceedings between 2014 and 2021. Collectively, our dataset of Indonesian PER literature was sourced from 852 documents (~4 to 8 pages each). They were organized, peer-reviewed, and published by the Indonesian PER community. We involved the most five leading academic meetings within the Indonesian PER community including *International Conference on Research, Implementation, & Education of Mathematics and Science* (ICRIEMS) (n = 152) [45–54], *International Seminar on Science Education* (ISSE) (n = 220) [55–61], *International Conference on Mathematics & Science Education* (ICMSE) (n = 125) [62–69], *International Conference on Mathematics and Science Education* (ICMScE) (n = 291) [70–76], and *International Conference on Mathematics and Science Education* (ICoMSE) (n = 64) [77–80] to best capture the landscape of the Indonesian PER characteristics (RQ1) and their immediate development (RQ2). All those papers were published in the Scopus indexed proceedings (Journal of Physics: Conference Series by Institute of Physics (IOP) Publishing, Conference Proceedings by American Institute of Physics (AIP) Publishing, Advances in Social Science, Education and Humanities Research by Atlantis Press), and web-based repository of each conference hosted by the universities. Those conferences inevitably had multidisciplinary topics with other STEM education research.

Thus, we should ensure the downloaded file must be relevant to the PER aims only. In each conference, there was a clear section in which to choose the PER cluster.

We decided to analyze those conference proceedings since they are the oldest international conference organizers among the Indonesian Teacher Education Institutions (TEIs) and even within the Indonesian PER community. Furthermore, most of the authors were affiliated with several Indonesian TEIs and had various research experiences (graduate students to PER experts), and somehow attracted a few authors from neighboring Southeast Asian countries. The nature of "international" conferences inevitably had to involve non-Indonesian authors even if the conferences were organized by Indonesians. One can argue that these led to the misinterpretation that the currently selected papers failed to represent the Indonesian PER landscape. Nonetheless, this perception should be invalid if we remember that they are organized, peer-reviewed, and published by Indonesian PER scholars or even discussed and presented during a parallel session in the seminar. Moreover, the representation of authors affiliated as Indonesian was still the largest group from the data corpus. The contribution of authors from neighboring countries cannot be avoided since they could implicitly influence the development of the established Indonesian PER literature. Hence, there would be a likelihood that these overseas authors could inspire us and they are cited by the Indonesian PER scholars in their papers.

Furthermore, the authors of those publications came from outside of the organizing committees and from several regions of Indonesia hence it could represent a wider snapshot of Indonesian diversity. Additionally, those articles had also been peer-reviewed throughout using robust processes until the accepted decision was endorsed by the committee of publication. This criterion applied to our dataset should satisfy the eligibility standards for publications within the Indonesian PER community. We must admit that the selected proceeding papers analyzed in this paper could be arguable among other potential papers in Indonesian PER literature, i.e., other conferences or even academic journals. We see, however, the promising area of these other Indonesian PER literature that can be engaged in future thematic analysis studies.

After the articles had been gathered, we extracted the PDFs as a collection of words in each document using the "pdfminer" library within the python programming language. Then, we followed the common steps of data cleaning processes using the "nltk" library [81] which were admittedly time-consuming processes in the text analysis study [82]. First, we checked the downloaded files to ensure that they were in a good condition to be scraped as plain texts. Second, we removed the section headers ('Abstract', 'Keywords', 'Figure', 'Introduction', 'Table', 'Method', 'Conclusion'), authors' names, affiliations, references, and acknowledgment sections (if any) from the individual PDFs. Third, we deleted the numbers, symbols, punctuations, and stopwords based on the English vocabulary using the "nltk" library. Finally, the preprocessed texts were tokenized into a list of single words in each document as our document-word matrix (see Figure 1).

After that, we employed the "gensim" library [83] for lemmatizing and finding the bigrams. Lemmatization is the procedure to find the stem of some words in favor of the same meaning. For instance, "student" and "students" in the previous tokenized results should be lemmatized as "student". We then looked for the frequently mentioned pairs of words within the dataset, bigrams. For instance, "conceptual understanding", "problem solving" "scientific approach", "critical thinking", and so on (see more examples in Table 1). Bigrams should be combined by an underscore connecting the tokens. Finally, we had a "bag of words" containing 199,578 raw words and bigrams with 10,109 unique words. The tenth most frequent words in this current unfiltered data corpus are illustrated in Figure 3 below with their word frequency and fraction in each document (division between frequency and total of documents). The top five words that often co-occurred through our data corpus are "student", "learning", "physic", "skill", and "concept". These representative words demonstrate the scope of PER literature has been satisfied in our dataset. Nevertheless, these frequent words should be filtered to make for more efficient computing time and to make the extracted PER topics more distinct.

Table 1. Characteristics of the Indonesian PER topics based on their most representative words.

Topic Number	Top 10 Representative Word	Weight	Topic Name
1	critical_thinking st_century ability creative_thinking information technology data communication creativity need	0.053 0.025 0.020 0.016 0.014 0.012 0.011 0.011 0.010 0.008	21st-century skill
2	test assessment instrument item level question ability measure development analysis	0.053 0.036 0.032 0.019 0.017 0.014 0.013 0.012 0.009 0.008	Assessment
3	scienc eeducation scientific_literacy thinking_skill thinking ability school knowledge scientific level	0.034 0.019 0.015 0.013 0.012 0.012 0.012 0.012 0.010 0.009	Interdisciplinary aspect of physics education
4	misconception understanding representation conception conceptual_understanding scientific level phenomenon difficulty science	0.031 0.030 0.017 0.010 0.010 0.010 0.009 0.009 0.009 0.008	Conceptual understanding
5	model activity science_process inquiry achievement class science learning_outcome scientific knowledge	0.032 0.021 0.018 0.011 0.011 0.010 0.010 0.010 0.009 0.008	Research based instruction

Table 1. *Cont.*

Topic Number	Top 10 Representative Word	Weight	Topic Name
6	problem	0.035	Problem solving
	problem_solving	0.028	
	ability	0.023	
	knowledge	0.012	
	solve_problem	0.011	
	improve	0.010	
	understanding	0.010	
	problemsolving_skill	0.009	
	approach	0.009	
	model	0.009	
7	medium	0.037	Educational technology
	development	0.022	
	material	0.021	
	technology	0.017	
	use	0.016	
	education	0.010	
	online	0.009	
	school	0.008	
	teaching_material	0.008	
	module	0.008	
8	experiment	0.020	Physics laboratory
	course	0.013	
	laboratory	0.012	
	motion	0.010	
	method	0.010	
	experimental	0.009	
	tool	0.009	
	practicum	0.008	
	understanding	0.007	
	activity	0.007	

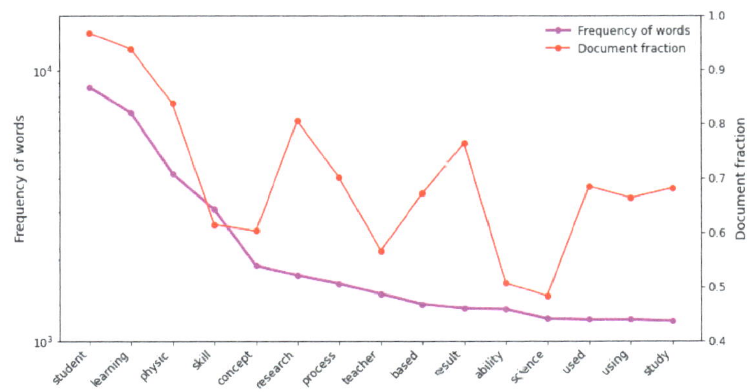

Figure 3. Distribution of the unfiltered word co-occurrence. The left axis represents the count of words and the right axis visualizes the document fraction within the unfiltered dataset.

Before we model the bag of words using the LDA algorithm, the next filtering processes for the most frequent and the rarest words should be followed. These words make our topical results difficult to identify. We want to discover unique terms to distinguish the research topics. Thus, the following step of data filtering was removing the most frequent words, and the rarest words that co-occurred within the bag of words. This removal action should be substantially noticed because the most-mentioned words might obscure the character of the topic studied in the literature. Our extracted topics should be concerned with the most specific words rather than the most frequent words. Thereafter, removing the rarest co-occurring words would also make our dataset more efficient. The larger size of the data corpus with many noises (typos, names, locations, specific terms) would extend the running time of the LDA algorithm, hence the process will become inefficient. Several selections of the filtering parameters should be evaluated to achieve the optimum coherence value (described below). This process should be exhaustively repeated to ensure the most representative topics with the optimum coherence measure. A detailed description of the coherence measure will be explained in the next subsection.

In this paper, we elected to exclude the most frequent words whose frequency was greater than 55% within the dataset. Furthermore, we also excluded the rarest words whose frequency was less than eight times within the data corpus. They were selected based on several evaluation processes to obtain the most optimum coherence measure. Admittedly, this selection was also inspired by the previous practices of thematic analysis by Odden, et al [11]. Obviously, it eliminated a substantial number of unique words and bigrams, approximately 7724 words. Then, we had the cleaned data as many as 2385 total words and bigrams for the next LDA analysis. This was actually a huge number of removals, but they did not contribute towards distinguishing the specific description of a topic [82]. As explained above, this filtered dataset would make the modeling time of the LDA algorithm more efficient since it would mathematically reduce the dimension of the LDA matrices (see Figure 1). These filtering processes decreased the size of our dataset from 10,109 to 2385 unique words and bigrams (see Figure 4). These filtered versions of the dataset determined the final LDA model of the Indonesian PER topics which were evaluated by multiple iterative modeling processes based on the mixtures of the number of topics (K), hyperparameter α, and random initialization (seed number) to obtain the most coherent topics within the literature. Furthermore, these topics must be qualitatively evaluated by PER experts to strengthen the solid topical description based on their experiences.

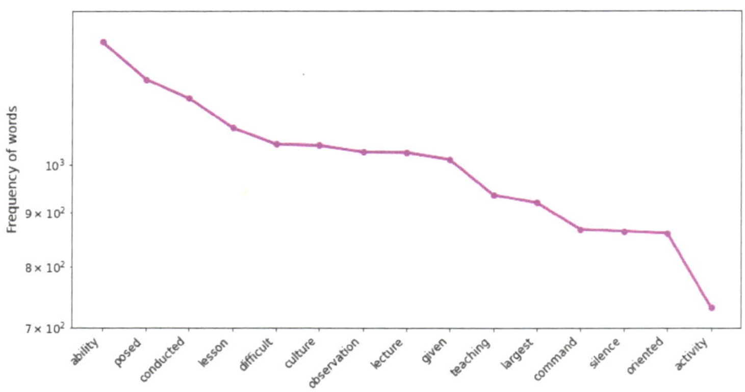

Figure 4. Distribution of words co-occurrence after the filtering process to the most frequent and the rarest words.

3.2. Modeling the Indonesian PER Topics through the LDA and Evaluating the Results

After the cleaned dataset had been served, we maintained it as a "pickle" file. Therefore, it could be imported directly without running the former code of data preprocessing and filtering processes. In this step, we conducted the iterative LDA modeling of the data corpus. The unsupervised nature of LDA requires us to manage several procedures of the evaluation process to find the final and the most representative LDA results. We must guarantee that their results make sense and do not deviate significantly based on the actual story of the research practice within the Indonesian PER field. In practice, users often implement one or multiple methods of evaluation to examine the LDA results [19,31,84]. Several pieces of literature have described some possible methods of evaluation. Accordingly, this study considered two choices of evaluation methods from the literature i.e., coherence score and face validity. In this subsection, the iterative processes of tuning the final LDA model are described through these two evaluation processes.

3.2.1. Coherence of Descriptors in Identified Topics

Essentially, the coherence value is defined as an external evaluation metric of how mixed the descriptors (the most representative words) are in each topic. In other words, this measure quantifies whether these descriptors in each topic have supported each other to represent the topics. Basically, this is recommended by the distributional hypothesis of linguistics which believes that there must be some central words in a certain topic. The set of words in a single topic will occur differently in another topic [18,31]. Hence, this will measure how we can distinguish the extracted topics from the diverse set of words within the data corpus. Coherence values will be normalized between 0 and 1. The LDA results can be concluded as "more coherent" when it raises a higher value and is near unity [42]. The best value of coherence will determine the final set of filtering processes above and several hyperparameters that will be tuned in training the best LDA result.

Several hyperparameters that should be tuned during the iterative process of LDA modeling are the alpha (α), random seed number, and the number of topics (K) [42]. Alpha is a hyperparameter that determines the relative "mixedness" of topics extracted by LDA. Moreover, the previous study has considered the potential issue during the training of LDA model, namely the random initialization seed [11]. It could cause a significantly different set of topics extracted from a single LDA model. Therefore, the LDA results are recommended to be interpreted from multiple random seed numbers. To find the most optimum model based on the coherence measure, we should train a high amount of LDA model in considering the mixtures of different numbers of topics (K), alpha (α), and random seed number. In this study, we selected a mixture of eleven numbers of topics (4 to 14), five alpha values (1, 5, 7.5, 10, 12.5), and ten selected different seed numbers. The different seed numbers were inspired by the method of repeated measurement in the physics laboratory [85]. The calculation of coherence values is represented by the moving dots in Figure 5 around the average coherence value (red dot). From these combinations, we trained 550 LDA models represented by the spread of coherence values in Figure 5.

Figure 5. Coherence score (α) within the mixture of number of topics (K), alpha (α), and random seed number.

Using an elbow plot, Figure 5 is provided to summarize the behavior of our LDA model within these combinations. The spread of white dots in this figure are the varying coherence values within a single LDA model of a certain number of topics. Our obtained coherence values are between 0.31–0.42 as an acceptable measure for the results of the LDA model reported by the previous studies [11,12,20]. The red marker visualizes the average value from the variation of each K-value and their respective standard deviations. To determine the best selected parameters for the final LDA model, we employed the "elbow" method as suggested by the previous literature [31]. The best model would be diagnosed by the flat pattern from the elbow plot in Figure 5. We can see that coherence values are greater with the increased number of topics and there is a leveling off pattern between six to 10 topics. This pattern can be an indication of diminishing returns. Based on these results, we choose the center of this range, $K = 8$, as our selection of the number of topics (K) for our final LDA model. This selection should be accompanied by the subsequent face validity from the PER experts to empower its representativeness within the literature.

3.2.2. Evaluating the Face Validity to the PER Experts

Face validity is a procedure to qualitatively evaluate the LDA results from the PER experts that are experienced with the established PER publications within the community. This will make sure the representativeness of our results based on their expertise and experience [86]. More technically, face validity requires experts who are familiar with the publication of the Indonesian PER field to judge how coherent the LDA results are based on their expertise, knowledge, and experience [31]. The second author of this paper is a professor in the Indonesian PER field with more than 20 years of research experience, particularly in the assessment and evaluation of physics problem solving and higher order thinking skills (HOTS). The third author of this paper is an associate professor of electrical engineering that has more than 20 years of research and teaching experience in programming language and artificial intelligence (AI) studies. These two authors confirmed the extracted topics that have been analyzed using the LDA algorithm. The second author presents to interpret the PER aspect and the third author contributes to guaranteeing our LDA algorithm in extracting the PER topics reported by this paper.

3.3. *Answering the Research Questions Based on the Final Trained LDA Model*

After the final LDA model has been trained to the most optimum coherence value, it will show the topical results derived from the data corpus. The aim of our study is to answer the two proposed research questions based on the most representative LDA model. This topic modeling results (see the next section) are then interpreted either to answer the proposed research question of the study or to re-evaluate the optimum model during the LDA training. The final model was trained from multiple phases of trial and evaluation toward different tuning of parameters described above. These processes should be exhaustively iterated in accordance with the most coherent results. After we discovered the coherence has been optimum, the final tuning of the LDA model would be selected.

In RQ1, the interpretation of the LDA model was explained in two ways. First, LDA results were understood by carefully examining the most representative words in each topic. In Table 1, we provide the top ten words of each topic. Our interpretation of these would be confirmed when these words have made sense based on the face validity. Accordingly, we can enumerate these results as eight research themes. Second, once the name of distinctive Indonesian PER topics had been determined, we then performed the subsequent strengthening interpretations to explore the most influential papers in each topic (see Table 2). In this table, we merely provide the five best representative papers of each topic to maintain the readability of this paper. In fact, we considered fifty representative papers in each topic to further study the characteristics of eight Indonesian PER topics. This analysis was necessary to obtain the next face validity to the extracted topics as well as to define the clear definition of the topic. Thereafter, the final terminology of each topic was decided according to these two steps of consideration. In RQ2, the evolution of each topic between 2014 and 2021 was measured by the "prevalence" parameter. In this study, the prevalence was defined as the percentage of each topic in each year within the collection of the annual documents [11]. A highly prevalent topic may be greatly studied in certain years but less focused on in other years. Eventually, it will illustrate the clear evolution of Indonesian PER studies for seven years that have been attempted. These results visualized what has been worked on by the Indonesian PER community and the potential room for future studies that could be addressed in the further journey.

4. Results

4.1. Characteristics of the Indonesian PER Topics between 2014–2021 (RQ1)

A final trained LDA model was employed to describe the characteristics of eight distinct Indonesian PER topics. They are reported in Table 1 with their representative set of words and in Table 2 with their representative set of papers in each topic as our baseline to interpret the LDA results and to understand how the Indonesian PER community has attempted the academic works. In this section, we will describe them in a consecutive way with supplemental interesting visualization in Figure 6 below.

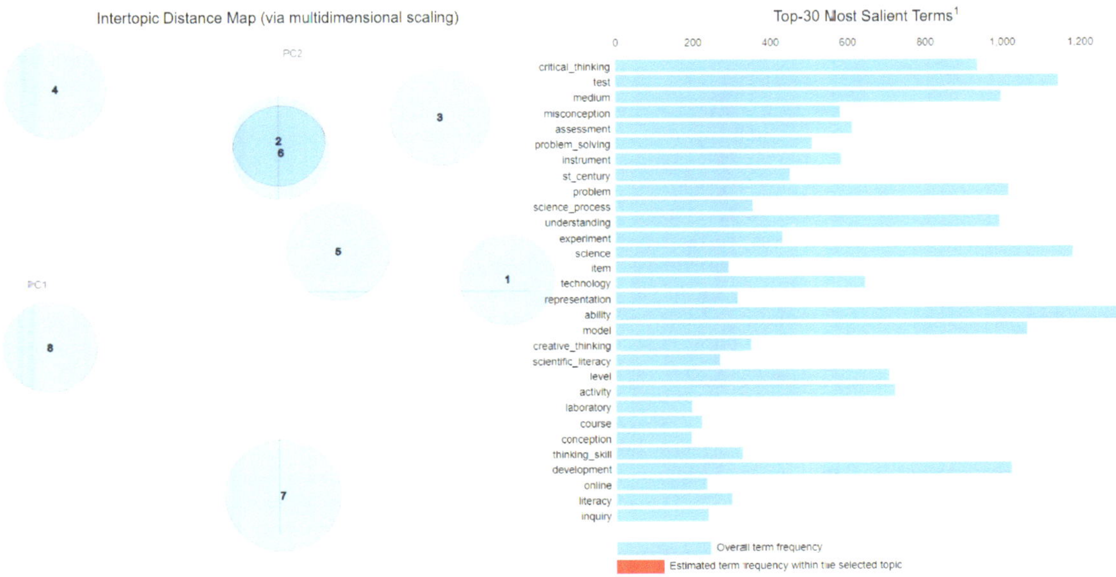

Figure 6. This figure is designed interactively thus if we select one of these thirty most salient words, we will obtain certain influential topics that are highly constituted by this word. This is the implementation of the distributional hypothesis of linguistics performed by the LDA algorithm. For instance, if we select "critical_thinking", this figure will make the zoomed bubble in the largest circle of topic 1 (21st century skill), smaller one of topic 3 (an interdisciplinary aspect of PER), and several tiny dots in other topics. It can illustrate that these zoomed topics (topic 1 and topic 3) have closely connected to each other and small dots at other topics have little connection to these topics. In this example of "critical_thinking", Indonesian PER researchers approached this skill as influential as 21st century skill and other interdisciplinary factors. ([1] Salience measure is calculated based on Chuang, et al. [87]).

As the first procedure of interpretation, we should initially notice the most representative words and weights of each topic number in Table 1. Essentially, the LDA results have no results about the research themes extracted from the literature. In practice, we situate Table 1 as being read from the left column to produce our interpretation of the topical name in the right column. It is implied that the right column of Table 1 is produced by the left part of the results. Our topic weights were probabilistic representations of each word in each topic which will become more relevant once the value is greater than a certain topic. Our findings report the spectrum of topic weights between 0.8% and 5.3%, which was also reported as acceptable measures by previous studies [11,20]. The order of the topic number is arranged based on the greater weight that represents how mixed the topic is within the literature.

Using the "pyLDAvis" library provided by the python programming language, the relationship among the Indonesian PER topics can be determined in Figure 6. In this study, we characterized eight distinct Indonesian PER topics studied between 2014 and 2021. The size of the displayed bubble in Figure 6 represents the most influential research theme within the Indonesian PER literature, namely "educational technology". The distance between the bubbles articulates the relative relationship of the topical results within a set of documents. We relatively found clear differences among the eight Indonesian PER topics produced in Figure 6. Even though "educational technology" has attracted the greatest focus of Indonesian PER scholars, it must be noted that the inter-topic distance map is constructed based on the multidimensional scaling of principal components (PC) emerging in the corpus. It is often assumed that it can be projected as a two-dimensional figure, as presented in Figure 6. Through this simplified visualization, we are assisted in illustrating the inter-topic relation that could be present among the emergent topics. This can be translated as the interdisciplinary nature of PER studies, as explained above.

Obviously, this will lead us to understand the disciplinary network that emerged within the PER community. Topic 2 (assessment) is closely correlated with topic 6 (problem solving). We suspect that this pattern is produced because the Indonesian PER scholars tended to develop and administer measurement tools to promote one form of students' performance, namely problem solving. Topic 2 (assessment) is also closely located with topic 5 (research-based instruction). It can be understood as the necessary evaluation metric after the implementation of several transformed physics learning within the PER community. Assessment must be required to measure the extent to which our physics learning reforms have effectively improved the students' learning process. To complement these aims, several students' performances from the national call of Indonesian curriculum are presented around these topics, including topic 1 (21st century skill) and topic 3 (an interdisciplinary aspect of physics education). In the next description, we will understand why this 21st century skill is connected to topic 6 (problem solving). This skill is one factor that should influence the critical and creative thinking of students as well as scientific literacy promoted by an interdisciplinary aspect of physics education. The advancement of technological development recently encourages students to contribute more to perform more sophisticated modern learning in 21st century society. These five topics can be clustered in quadrant I (positive x axes, and positive y axes) with their shorter relative distance from each other rather than the remaining topics, i.e., topic 4 (conceptual understanding), topic 7 (educational technology), and topic 8 (a physics laboratory). The separated relative distance from the quadrant I topics can be understood as the uniqueness of these topics within the analyzed literature.

The greater weight of the most representative words in Table 1 represents the more mixed the topics should be within the literature. Nevertheless, instead of Table 1, we recommended that one must interpret based on the most representative papers in each topic as further presented in Table 2. We admit that Table 1 can be troublesome since there are likely disconnected words of a topic, particularly in the case of small weights and making the interpretation trickier. Therefore, we supplement it by qualitatively cross-checking the content of the most representative papers on each topic in Table 2. This manner of literature reading is different from a traditional content analysis that was approached by the previous Indonesian author in [13–17]. Instead, we were aided by the topical results from Table 1, thus we merely explored the characteristics of each topic based on our clustered understanding in Table 1. In Table 2, we provide the prevalence, which is a quantitative measure of how mixed the paper is within a certain topic. For example, the 0.875 prevalence of Supahar's paper [88] in Table 2 articulates that it is composed of 87.5% of the assessment topic and the remaining values are lasid on the other mixture across all other topics. After the presentation of these tables, we detail the distinctive ways to differentiate the Indonesian PER topics that consider our results in Tables 1 and 2. This will justify the reason for which we interpret LDA results towards eight Indonesian PER topics.

Table 2. Representative articles, author, year, respective conference, and prevalence in each Indonesian PER topic.

Topic	Article	Author	Year	Conference	Prevalence
21st century skill	Profile of students' critical thinking ability in project-based learning integrated science technology engineering and mathematics	Eja, Ramalis, & Suwarma [89]	2019	ICMScE	0.812
	Gender differences in digital literacy among prospective physics teachers	Rizal, et al. [90]	2020	ICMScE	0.799
	Profile of senior high school in-service physics teachers' technological pedagogical and content knowledge (TPACK)	Masrifah, et al. [91]	2018	ICRIEMS	0.776
	Developing creative thinking skills of STKIP weetebula students through physics crossword puzzle learning media using eclipse crossword app	Anggraeni & Sole [92]	2019	ICMScE	0.771
	Evaluation of critical thinking skills of class x high school students on the material of Newton's laws	Febriana & Sinaga [93]	2020	ICMScE	0.759
Assessment	Applying content validity ratios (CVR) to the quantitative content validity of physics learning achievement tests	Supahar [88]	2015	ICRIEMS	0.875
	An eight-category partial credit model as very appropriate for four-tier diagnostic test scoring in physics learning	Istiyono, et al. [94]	2021	ISSE	0.873
	Developing of Bloomian HOTS Physics Test: Content and Construct Validation of The PhysTeBloHOTS	Istiyono Dwandaru, Muthmainah [95]	2019	ICRIEMS	0.866
	Instrument test physics-based computer adaptive test to meet the Islam economic community literature review	Ermansah, et al. [96]	2016	ISSE	0.861
	Implementation of Item Response Theory at Final Exam Test in Physics Learning: Rasch Model Study	Asriadi & Hadi [97]	2020	ISSE	0.858
Interdisciplinary aspects of physics education	Mapping of professional, pedagogical, social, and personal competence of senior high school physics teachers in Yogyakarta special region	Jumadi, Prasetyo, & Wilujeng [98]	2014	ICRIEMS	0.772
	Analysis of Scientific Literacy Through PISA 2015 Framework	Arsyad, Sopandi, & Chandra [99]	2016	ICMScE	0.766
	Shifting attitude from receiving to characterization as an interdisciplinary learning toward ecological phenomena	Napitupulu, et al. [100]	2017	ISSE	0.735
	Promoting metacognition and students' care attitude towards the environment through learning physics with STEM	Rahzianta & Purnama [101]	2016	ISSE	0.708
	Analysis of senior high school students' higher order thinking skills in physics learning	Maulita, Sukarmin, & Marzuki [102]	2018	ICRIEMS	0.690
Conceptual understanding	Alternative conception of high school students related to the concepts in the simple electric circuit subject matter	Wardiyah, Suhandi, & Samsudin [103]	2018	ICMScE	0.879
	Identification of student misconception about static fluid	Setiawan, Saputra, & Rusdiana [104]	2018	ICMScE	0.874
	External representation to overcome misconception in physics	Handhika, et al. [105]	2015	ICMSE	0.870
	Teachers, pre-service teachers, and students understanding about the heat conduction	Anam, Widodo, & Sopandi [106]	2018	ICMScE	0.869
	Identify students' conception and level of representations using five-tier test on wave concepts	Wiyantara, Widodo, & Prima [107]	2020	ICMScE	0.849

Table 2. *Cont.*

Topic	Article	Author	Year	Conference	Prevalence
Research based instruction	The effectiveness of local culture-based physics model of teaching in developing physics competence and national character	Suastra [108]	2015	ICRIEMS	0.846
	Cooperative learning model design based on collaborative game-based learning approach as a soft scaffolding strategy: preliminary research	Nurulsari, Suyatna, Abdurrahman [109]	2016	ICMScE	0.783
	Effect of free inquiry models to learning achievement and character of student class XI	Kaleka [110]	2018	ICRIEMS	0.773
	Training students' science process skills through didactic design on work and energy	Ramayanti, Utari, & Saepuzaman [111]	2017	ICMScE	0.769
	The effects of cooperative learning model think pair share assisted by animation media on learning outcomes of physics in high school	Astra, Susanti, & Sakinah [112]	2019	ICMScE	0.765
Problem solving	The effect of e-learning based worksheet to improve problem solving ability of senior high school students	Septiyono, Prasetyo, & Ihwan [113]	2020	ISSE	0.812
	The analysis of students' problem-solving ability in the 5e learning cycle with formative e-assessment	Yuliana, et al. [114]	2019	ICoMSE	0.797
	The development of physics e-book based on contextual teaching and learning to increase student problem-solving skill	Fitriadi, Latumalukita, & Warsono [115]	2021	ISSE	0.791
	Improving students' problem-solving skills through quick on the draw model assisted by the optical learning book integrated the Pancasila	Himawan & Wilujeng [116]	2019	ISSE	0.785
	Profile of problem-solving ability of Islamic senior high school students on momentum and impuls	Sakti, Wilujeng, & Alfianti [117]	2021	ISSE	0.766
Educational technology	Developing whiteboard animation video through local wisdom on work and energy materials as physics learning solutions during the covid-19 pandemic	Anggraini, et al. [118]	2020	ISSE	0.874
	Android-based carrom game comics integrated with discovery learning for physics teaching	Rahayu, Kuswanto, & Pranowo [119]	2020	ICRIEMS	0.864
	Development of physics mobile learning media in optical instruments for senior high school student using android studio	Aji, et al. [120]	2019	ISSE	0.843
	Smartphone-based learning media on microscope topic for high school students	Nadhiroh, et al. [121]	2020	ISSE	0.831
	Android for the 21st century learning media and its impact on students	Adi, et al. [122]	2016	ISSE	0.825
Physics laboratory	Simple vertical upward motion experiment using smartphone based phyphox app for physics learning	Janah, Ishafit, & Dwandaru [123]	2021	ISSE	0.865
	The Atwood machine experiment assisted by smartphone acceleration sensor for enhancing classical mechanics experiments	Listiaji, Darmawan, & Dahnuss [124]	2020	ICMSE	0.853
	Development of sound wave experimentation tools influenced by wind velocity	Maisyaroh, et al. [125]	2019	ISSE	0.840
	Analysis of simple harmonic spring motion using tracker software	Mu'iz, et al. [126]	2017	ICMScE	0.827
	Real laboratory-based learning using video tracker on terminal velocity	Ristanto, Novita, & Saptaningrum [127]	2016	ISSE	0.824

4.1.1. Topic 1: 21st Century Skills

This topic is the most mixed cluster based on the descending order of the weight measures of topical results. Promoting 21st century skills is discovered as the main concern from papers published within the Indonesian PER community. Keywords including "critical_thinking", "creative_thinking", and "communication" are several components of students' performances in 21st century learning. Students are expected (refers to "need") to grasp the well-known four components of 21st century learning skills (4Cs) [128]. Additionally, the abundance of digital technology in the past few decades encourages our physics educators to approach their physics learning with digital platforms represented by the terms of "information", "data", and "technology". It is undoubtedly also connected with the focus of the seventh topic below (educational technology). The vast development of the digital age during this century motivates physics educators to be concerned in this area. Therefore, this topic could be stated as the most influential party and increasingly takes much attention within the Indonesian PER literature for the past few years.

The research questions studied under the 21st century skill topic are predominantly made up of several categories: technological developments for physics learning and laboratory reforms in promoting 21st century skill [92,129–140], small- to large-scale survey in evaluating physics learner performance on this skill [89,90,141,142], correlational study toward another form of students' performance [143–145], and designing measurement tools to probe this skill on physics learning and instruction [146–150]. One could consider that this vast amount of literature is closely connected with other topics discussed below. For instance, technological development in this topic overlaps with the seventh topic (educational technology), and the emergence of physics laboratories in this topic is closely connected with the eighth topic (physics laboratory), and obviously with the second topic (assessment). Nevertheless, we argue that the uniqueness of the current topic is underlined by the focused aims to address the modern idea of 21st century learning. It promotes 21st learner skills including creative thinking [92,137–139,143,144,146,150], critical thinking [89,131,134,142,143,145–148,150,151], collaborative problem solving [130], data literacy [132,133,135,136], and digital literacy [90,152]. Moreover, Indonesian PER scholars are also attracted to approaches beyond high school physics instruction. Several studies have attempted to support pedagogical competence for professional physics teachers [153] or even prospective physics teachers [90,154–157]. These efforts can be made to ensure the physics educator as a mastermind of the physics classroom has to collectively support the intention of 21st century physics learning. Thus, they are expected to engage with this vision in physics learning responsively.

4.1.2. Topic 2: Assessment

This topic focuses on developing, validating, and disseminating measurement tools that are needed in performing assessments throughout the physics learning process and evaluating research-based instructions within the PER community. It is composed of several representative words for which we designed and developed measurement tools including "test", "instrument", "item", "question", and "measure". These tools are disseminated to define the quantitative measure of "ability" within physics learning or students' performance in physics classrooms. Moreover, several modern measurement theories including item response theory and Rasch modeling are mainly discussed by the Indonesian PER members within this topic. The emergent "level" keyword can be related to the other topics below, particularly with the third and fourth topics of our topical results. It could articulate several assessment concerns to factors that were mainly highlighted on students' performance within the Indonesian PER community.

In this second topic, several measurement tools have been developed and disseminated within the Indonesian PER community. They are comprised of performance tests and diagnostic tests. Performance tests are designed to measure diverse forms of students' performance on physics learning, including cognitive test [97], higher order thinking skills (HOTS) [95,158,159], critical thinking skill [160], representation [161–166], data literacy [167], digital literacy [168], science process skills [169,170], problem solving skills [171,172], inductive thinking [173], visual literacy [174], communication skills [175], analytical thinking skills [176], and scientific literacy [177]. Moreover, several diagnostic tests are also established by the Indonesian PER authors to detect potential students' misconceptions [178–182], lack of representation ability [183–185], lack of higher order thinking skills (HOTS) [186], lack of critical thinking skills [160], lack of problem-solving skills [184], lack of data literacy [187], as well as the lack of understanding throughout astronomy class [188]. On the other hand, one can argue that this topic seems to be similar to the other extracted topics currently discussed. For instance, in this topic, we discover that several research-based assessments (RBAs) are addressed to measure 21st century skills. They are critical thinking, data literacy, digital literacy, and problem solving. Additionally, the same set of physics learning skills emerged as discussed further in the third topic (an interdisciplinary aspect of PER) and the fifth topic (problem solving). We argue that this second topic can be distinguished from other topics in its focus on the dissemination of the robust methodology to design, examine, and evaluate the developed measurement tools for physics education. Several validity studies have been introduced including content validity [88], factor analysis [175], Rasch model [97,188–191], and engaging modern measurement theory of dichotomous and polytomous response model [94,192] from item response theory (IRT). Additionally, our RBAs are designed through several mediums including computer aided tests [160,193], computerized adaptive tests [194], two- to six-tiered tests [160,180–182,188], and other forms of the test let [195].

4.1.3. Topic 3: Interdisciplinary Aspect of Physics Education

The topic of 21st century skill guides the Indonesian PER scholars to a focus on the interdisciplinary aspect of physics learning. Physics can be studied as an integral part of science, engineering, technology, and mathematics (STEM) education. Physics should be taught to understand complex understanding about contextual phenomena. The phase of the 2013 Indonesian curriculum oriented the physics teachers to engage the philosophy of "scientific approach" in their learning [196]. Due to our dataset being drawn from 2014 to 2021 literature, it is reasonable when this topic can be situated to address the implementation of this ongoing curriculum. We enumerate this topic as an interdisciplinary aspect since the nature of physics education during this timeframe should involve an "integrated" understanding of science. Physics is closely connected with other STEM subjects such as mathematics, biology, and chemistry. The terms "science" and "education" can emerge within this topic due to most of the Indonesian PER studies believing that their physics learning should be adjusted to solve contextual phenomena using physical knowledge supplementing with another scientific knowledge. For instance, Napitupulu, et al. [100] engage ecological phenomena assumed as crucial factors to which physics education should address. Moreover, physics education can be transformed to harness moral values about the environmental aspects of the ecological issue. Using a metacognitive framework, Rahzianta and Pratama [101] support the previous idea of Napitupulu, et al. [100] that physics education can foster the value of awareness toward environmental attitude. Through physics instruction, students were also expected to be critically aware of the challenge about the integrated issue of science education.

Essentially, the 21st century skill topic above inevitably correlates to this movement in preparing physics students to face the future complex challenge of their modern real world. Students are expected to acquire several skills that they learn through physics learning in terms of scientific literacy (refers to keywords "scientific_literacy", "knowledge", "scientific"), higher order thinking skills (HOTS) (refers to "thinking_skill" and "higher_order") [197–199], and another form of "thinking" processes [200–206]. Research movements on scientific literacy in this topic can be driven by the international announcements of *Programme for International Student Assessment* (PISA) assessment for Indonesian secondary students [207–217]. PER members are one of discipline-based education research (DBER) on STEM education (refers to keywords "science", "education", "school") that is responsible for this duty call in improving students' performance on PISA results. In addition to the focus of this topic, the keyword of "thinking_skill" is particularly relevant to "higher_order" in the eleventh rank of representative words in this topic, nevertheless, it could not be shown in Table 1. Higher order thinking skills (HOTS) are also considered as part of other students' performance that are associated with other factors including scientific literacy and the first topic (21st century learning skills) [218,219]. Furthermore, unique findings from the Indonesian PER literature are discovered in promoting character values through physics education [220–223].

4.1.4. Topic 4: Conceptual Understanding

This topic is relevant to the previous results in Docktor and Mestre's [9] synthesis results of international PER literature for several decades. The earliest movement of PER literature underlined conceptual understanding as fundamental for physics learning. Docktor and Mestre [9] place this topic as the first theme of their thematic results. Our findings can be different from the results reported by Docktor and Mestre [9] since our conceptual understanding is discovered as the fourth topic. As previously described, the Indonesian PER community is encouraged mostly to the first topic (21st century skills) due to the national call for a scientific approach curriculum (2013 curriculum). Nevertheless, conceptual understanding could not be ignored from the Indonesian PER development. Indeed, we must admit that this topic is still imperative for physics learning among the other students' thinking skills and problem solving skills formerly mentioned. The name of conceptual understanding could be concluded in this topic because there are several representative keywords in Table 1 including "misconception", "understanding", "conception", and obviously the bigram of "conceptual_understanding". Using the LDA topic modeling, Yun [12] also recognized this current topic as an "introductory physics" theme in their results toward data corpus from *The American Journal of Physics* (AJP) and *Physical Review Physics Education Research* (PRPER). The keyword "conceptual" in Yun's results emerged in the first theme extracted from the AJP dataset.

Furthermore, "representation" of students' understanding is considered as a specific form of conceptual physics understanding [9,11]. Odden et al. [11] even discovered "representation" as their first topical results extracted from the same methodology of LDA algorithm. The term "difficulty" in conceptual understanding is also studied in our result. Likewise, other interdisciplinary aspects of physics understanding, such as "scientific", "phenomenon", and "science" emerge because of our movement to the third topic above. As discussed earlier, conceptual understanding of topics obviously influences other topics within the data corpus. The term "level" interestingly occurred in this topic as mentioned in the third topic (assessment).

One of the research questions explored in this topic is identifying conceptual knowledge about physics performed by Indonesian students [106,107,224–227] or physics teachers [106,228–230]. They investigated conceptual physics understanding on mechanics [107,227,231,232], electricity [224,228,230], magnetism [226], fluid [104,229], work and energy [225,233,234], thermodynamics [106], and modern physics [235]. Within the context of the Indonesian PER literature, we propagate conceptual understanding in another form of multiple representations [225,236,237], including external representation [105], mental model [238], drawing ability on free-body diagram [239,240], and mathematical representation [232].

Furthermore, diverse difficulties also have been discovered within the literature [227,234]. Various terminologies have emerged from Indonesian PER literatures to define the students' lack of understanding about conceptual physics, namely alternative conception [103,241], misconception [104,231,233,235,236,242–250], and misunderstanding [251]. To address this limitation on students' conceptual understanding, the Indonesian PER scholars have designed and examined vast learning reforms or interventions, i.e., conceptual construction-reconstruction oriented instruction (CCROI) [252], remedial programs [253], authentic learning [254], cognitive conflict instruction (CCI) [243], electronic conceptual development conceptual change text (E-CDCCText) [255], conceptual change-oriented text (CCO-Text) [235,248], and conceptual change laboratory (CC-Lab) [256]. Their purpose is to address students' misconceptions thus students can be supported to follow the conceptual progression [252,255], learning progression [253], or conceptual change [235,254,256].

Eventually, studying conceptual understanding through correlational inquiry has also been worthwhile to conduct [241].

4.1.5. Topic 5: Research Based Instruction

In improving the students' performance (refers to "achievement", "knowledge", "learning_outcome") on physics learning, several learning transformations and curricular developments (refer to "model", "activity", "class") have been attempted by the Indonesian PER members in this topic. As briefly discussed above, due to the national call of the 2013 curriculum, Indonesian physics education during this timeframe was encouraged to approach science process skills as five cycles of learning paces in physics learning. The paces include observing, asking questions, experimenting, explaining (or reasoning), and presenting (or reporting) abbreviated as "5M" in the Indonesian language [257]. They can be translated as inquiry-based learning in practice. Our fifth topic makes sense if the Indonesian PER literature mentioned keywords including "science", "science_process", "inquiry", and "scientific" in this topic. The term "activity" in the LDA results also implied that the "scientific" approach recommended by the 2013 Indonesian curriculum was engaged in students' activities within physics learning. Admittedly, they are also closely connected with the interdisciplinary topic on the third of our LDA results above.

We consider that this topic is one of the most diverse groups within our LDA results. Nonetheless, most of them are essentially designed based on the philosophical lens of constructivist learning. Indonesian physics education has a long history of adopting the student-centered learning approach since the establishment of the 1968 Indonesian curriculum [258]. We have probed several students' performance on the physics learning approaches above. Research-based instruction is generally designed and implemented to promote them through constructivist learning. On the other hand, we discover distinct aspects derived from the Indonesian PER literature that cover studies to approach the indigenous, cultural, or local context of Indonesian physics learning. Several learning reforms were inspired by culturally relevant aspects of Indonesian diversity, as reported by Suastra [108]. This learning tradition makes different colors emerge in Indonesian physics education besides the five scientific cycles-oriented learning approaches in the implementation of the 2013 national curriculum. They are reported by diverse papers, particularly in addressing inquiry-based learning [259–262], project-based learning [263–265], and problem-based learning [266–268]. Moreover, Indonesian PER scholars are motivated to adapt physics learning through the lens of a cooperative framework (social learning theory), i.e., collaborative game-based learning [109], think pair share (TPS) [112], time token [269], and social learning cycle [270].

4.1.6. Topic 6: Problem Solving

Relevant to the fourth topic above, this topic is also precisely reported by Docktor and Mestre's [9] synthesis analysis. They discuss this topic as the second position of their thematic result. Currently, our LDA model discovers several terms in this topic related to problem solving definition, including "problem", "problem_solving", "solve_problem", and "problemsolvingskill". In supporting students' success in physics learning, apart from the conceptual understanding discussed above, problem solving skills (several termed as ability) is also a fundamental factor to be a successful physics learner. Content knowledge of physics is primarily discovered through critical problem-solving steps to explore and understand how our physical circumstance works. Moreover, several terms including "improve", "approach", and "model" represent that the Indonesian PER scholars propagate it as a learning strategy to endorse this imperative topic as recently discussed in the fifth topic. They cover particularly the implementation of a problem-based learning model. Eventually, physics education could contribute to improving problem solving skills that inevitably correlate with 21st century skills for students' future.

As described in other studies focused on students' learning, this topic mainly commenced with the profiling of students' performance in solving physics problems [117,271–275]. These reports can be cited as a basis for Indonesian PER scholars to develop physics learning reforms [116,276–280], curricular developments [113,281–284], and computer-aided instruction [285,286] to improve the Indonesian students' performance in physics problem solving. Contextual issues within Indonesian society were on several occasions engaged with by the Indonesian PER authors, including cultural context [287] and disaster mitigation awareness [285,288–290]. The immediate movement of this contextual learning is grounded on physics as an interplay within STEM education. Therefore, physics educators have great expectations that students can learn complex thing from physics and make concrete efforts within their social communities.

4.1.7. Topic 7: Educational Technology

Admittedly, the first topic of our LDA results has been tremendously influenced by the emergence of this seventh topic within the Indonesian PER literature. The keyword "medium" in this topic is lemmatized from "media" during the preprocessing step of the LDA modeling. Physics instruction is motivated to follow the disruptive effect of the digital age in the 21st century era. The existence of digital technology makes our learning transform in response to these circumstances. We discover that this topic is frequently mentioned in several papers with regard to developing learning material (refers to keywords "material", "teaching material", "module") through technology-enhanced learning (refers to "technology", "online") implemented in physics classrooms. Broadly speaking, technology can be flourished from the manifestation of our understanding of science. Digital technologies, i.e., computers and mobile devices, have tremendously encouraged Indonesian PER scholars to be involved in physics learning and instruction. Complex applications within education makes this topic definitely diverse and broad. The demand for 21st century learning, the national call of the 2013 curriculum, and the rapid development of the digital age have been impactful for Indonesian PER scholars in the development of a vast number of technical assistances within physics learning, including audio-visual media [118,291–298], web-based applications [299,300], android applications [119–122,301–308], augmented reality [309], and distance learning [310–312]. The cultural context of Indonesian society is presented through the delivery of educational technology [306,313–319]. The former interdisciplinary aspect of physics education and the demands of 21st century learning drives an intention during the design and implementation of educational technology on physics [304,320,321].

4.1.8. Topic 8: Physics Laboratory

In Table 1, we discover several keywords bringing us to the definition of this topic as our learning scheme within the physics curriculum. Experimental physics is considered as one vital path through which physics knowledge might be taught to all physics people. We name this topic as a physics laboratory since "experimental" physics learning typically occurs in the laboratory setting. This topic focuses on how physics learning or "course" can be delivered through real [125,322–326] or virtual "laboratory" [327–331] in conducting the physics experiment (refers to "activity" and "practicum"). Several papers also have developed their own physical measurement "tool" and data acquisition using microcontrollers, trackers, or smartphones [124,332–337] that could be employed to enhance students' experience within physics laboratories. Eventually, through this channel, PER studies also consider addressing their learning transformation to improve "understanding" of physics [338,339]. The appearance of the keyword "motion" in this topic represents that a physics topic is mostly addressed on Newtonian mechanics as also reported by Yun's results based on *The American Journal of Physics* (AJP) journal [12].

4.2. Development of the Indonesian PER Topics between 2014 and 2021 (RQ2)

In the second research question, we investigate the development of the extracted Indonesian PER topics between 2014 and 2021 through the measure of topic prevalence. We adopt the definition of prevalence that has been approached by a previous study by

Odden, et al. [11]. Prevalence of a particular topic is defined as the sum of documents that are categorized on that topic within the amount of literature published in a certain year. This measure is represented as a percentage that could be aggregated both cumulatively (Figure 7) and averaged (Figure 8) by year. For instance, a 10% prevalence of topic 1 in a certain year has a two-fold meaning. First, it represents the average prevalence of topic 1 for that year as many as 10%. Then, the cumulative prevalence of topic 1 for that year is its multiplication with n, in which n is the number of documents published in that year. If the annual cumulative prevalence of all topics is summed up, then it would correspond to the total of documents published in that year.

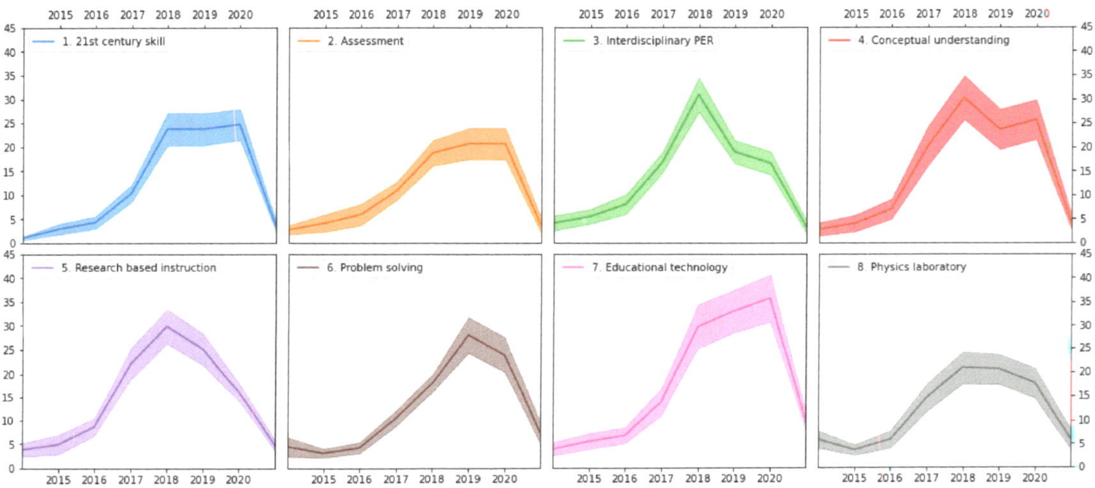

Figure 7. Cumulative prevalence of Indonesian PER topics development between 2014 and 2021.

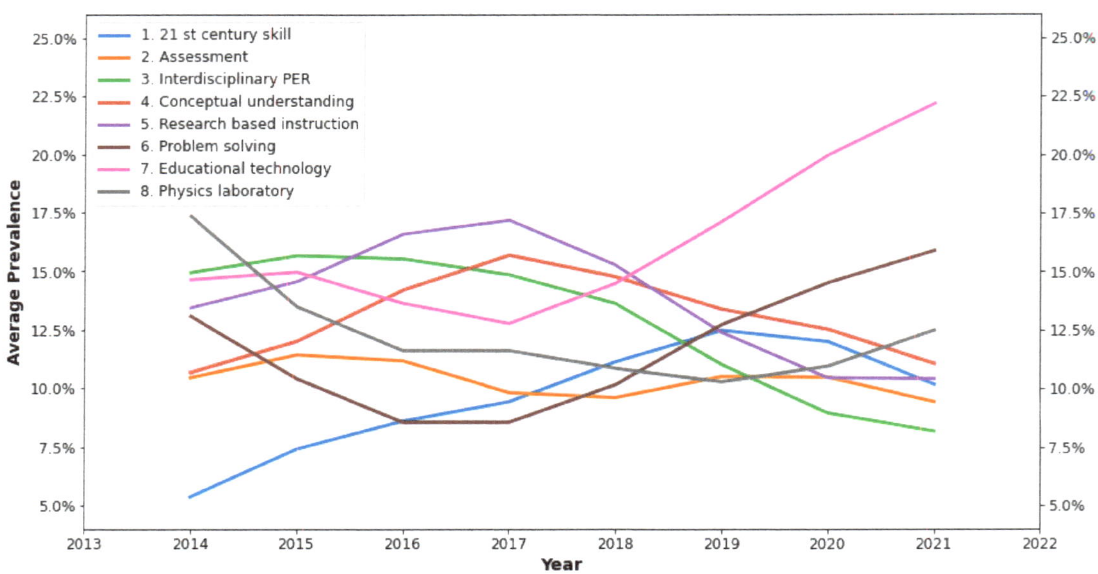

Figure 8. Average prevalence of the Indonesian PER topics development between 2014 and 2021.

The cumulative prevalence of eight Indonesian PER topics between 2014 and 2021 is illustrated in Figure 7. The cumulative prevalence of a topic in the y axis is provided as the number of "effective" papers disseminated in that year. For example, 25 cumulative prevalence of 21st century skill topics in 2018 (see Figure 7) means that there are equivalent to 25 "effective" articles discussing about 21st century skill topics in that year. This term "effective" is inspired by the previous research [11] because, keeping in mind, LDA results underlie the assumption of the mixed membership of topics. An individual article should be categorized into several topics (in varying weights) rather than a single topic.

We provide shaded areas in Figure 7 to describe the topical distribution within the annual topic development. The width of the shaded area in Figure 7 is the standard deviation (σ). We use as many as 3σ from the mean value represented by the solid line in the figure. We calculated this standard deviation using the jackknife resampling technique [340]. For certain topics and years, this procedure yielded a new sample of 100 cumulative prevalence values. Using this newly generated sample, the standard deviation is calculated to describe the distribution of a topic prevalence in each year. The jackknife resampling method described above produces the shaded areas that could be represented as the topical variation for a certain year. A shaded area of zero for one year would be produced if there is no difference among the cumulative prevalence of several topics during a single year. On the other hand, if there are several papers that are focused heavily on a certain topic, the shaded area (topical spread) would be larger.

Figure 7 illustrates that our whole topics have demonstrated relatively similar rise and fall between 2014 and 2021. There is a spike in 2018 and 2019 followed by a decrease in the subsequent year for all topics. We suspect that the apparent decrease can be driven by several publications in the year 2021 that are still progressing. Broadly speaking, the disruptive transition during the 2020 pandemic year has tremendously influenced the attendance of potential PER researchers from several parts of Indonesian institutions [341]. Moreover, our dataset for 2021 conference is merely sourced from the ISSE conference and the rest of the conferences are still progressing through publication processes. Figure 7 describes the lowest cumulative topic prevalence that occurred in the early year of 2014. The finding is not surprising because there were only two conferences that have been organized by UNY (through ICRIEMS) and UNNES (through ICMSE) in that year. A measure of cumulative topic prevalence is particularly dependent on the number of documents written for a particular year. There is stable cumulative prevalence particularly on 21st century skill and assessment topics even though the assessment topic has a lower prevalence. Educational technology has had the highest increased prevalence for the past few years. There are similar spikes described by the interdisciplinary aspect of physics education and conceptual understanding topics in 2018. However, for the following year after this, the interdisciplinary PER topic has a more substantial decrease than a conceptual understanding topic. Problem solving topics have the latest spike in 2019. Unfortunately, the physics laboratory seemed to be a minority within the Indonesian PER community due to the smallest topic prevalence among other Indonesian PER topics.

As described above, the cumulative measure of topic prevalence is merely dependent on the number of "effective" documents published in that year. Regarding the relative number of papers published in a certain year, an average measure should be defined. It could be fairly utilized to compare different topics from year to year. In the calculation of an average measure, we can employ the data-smoothing technique which dampens the effect of sample dependence in the year-to-year variation. In this study, we choose the three-year rolling windows that will average the prevalence values for each year with those of the former and the subsequent year. Figure 8 depicts our plot of average Indonesian PER topics prevalence over time.

Based on the average prevalence visualization in Figure 8, there is the relative stability of rising and falling for all the topics between 2014 and 2018. The most interesting topics within the literature are interchanged over years. In early of 2014, the physics laboratory topic emerged to dominate the movements, however, this topic follows a decreasing pattern through several subsequent years after that. In the next year, the interdisciplinary aspect of physics education has attracted our Indonesian PER scholars for their attention within the community. We suspect that the increasing pattern of the third topic must be motivated by the governmental policy of the 2013 curriculum. Moreover, there is a continuous pattern that research-based instruction topics lead the waves between 2016 and 2018. Nevertheless, this topic has substantially decreased in the subsequent years and the position is overtaken by educational technology topics after 2018 and problem-solving topics after 2019. We then notice that the assessment topics remain stable over time on average. The assessment of physics learning is inevitably a multidisciplinary field within educational science. Measurements of students' performance and validation studies using various methods, either from classical or modern theory, are still needed for the development of discipline-based educational research (DBER) including the Indonesian PER community. Furthermore, it then indicates that this PER topic has been studied through collective development to support the promotion of 21st century skill and other students' performance including interdisciplinary aspects of PER, conceptual understanding, and problem solving. In the early years, it is interesting that 21st century skills even had the lowest attention in 2014. Although we cannot conclude where this trend comes from. Looking at the representative papers on this topic (see Table 2), we argue that the lowest prevalence of 21st century skill in the early year of 2014 corresponded to the limited digital technology that has been approachable during this year. Eventually, this topic will continue to develop until 2019. It is likely to become greater in following the associated trends of increased educational technology until 2021.

5. Discussion

In this paper, we have demonstrated that the LDA algorithm from NLP, a subfield of ML studies, offers a potential tool to analyze the plethora of publications within the Indonesian PER community. For the answer to RQ1, we have extracted eight Indonesian PER topics using the LDA algorithm toward the selection of five publications on physics education research conferences organized, peer reviewed, and published by Indonesian PER members between 2014 and 2021 [1–5]. They are composed of (1) 21st century skills, (2) assessment, (3) interdisciplinary aspects of physics education, (4) conceptual understanding, (5) research-based instruction, (6) problem solving, (7) educational technology, and (8) physics laboratory. The description with the representative references to distinguish each of these emergent topics has been provided through Tables 1 and 2 above with a description of representative papers to emphasize our understanding of the topics.

Furthermore, Figures 7 and 8 above have been provided to enrich our insights about the development of Indonesian PER studies since the beginning of 2014 to date. For the answer to RQ2, the development of the Indonesian PER topics has dominated interchangeably over this timeframe. Nevertheless, we admit that several topics recommend that their development appear fair and stable between 2014 and 2021. In the early years of our analysis period, Indonesian PER members put their attention more towards studying how physics learning should be immersed through a physics laboratory. Thereafter, we discovered that it was overtaken by research-based instruction in transforming physics learning into several reforms to approach various forms of student performance that are constructed based on the interdisciplinary understanding of physics education. In more recent years, the Indonesian PER field has been encouraged by the demand for digital technology-enhanced learning that attracted Indonesian PER scholars to develop teaching aids for physics instruction using various technological approaches. This was also relevant to the movement of problem solving topics during the time to promote the increasing trends on 21st century learning since 2014.

We can discuss these current findings by comparing them to those previous works that have been published before our paper [9,11,12]. Table 3 summarizes PER themes that have been reported by Docktor and Mestre's review [9], Odden et al. study [11], and Yun's thematic analysis [12]. Some topics from our findings are found to be in common in these previous works, but some topics can be distinct. Using more traditional large-scale synthesis analysis, Docktor and Mestre have extracted PER topics into six primary topical areas of physics education research. Using the same method as the current study, Odden et al. have extracted PER topics into ten research themes based on 1302 individual papers published in the physics education research conference (PERC). Additionally, eight PER themes were also extracted by Yun [12] based on the data corpus from AJP and PRPER journals using a similar methodology to our paper (LDA algorithm). From these three references, we will discuss how our Indonesian PER findings show immediate points of overlap or several unique patterns different from the previous works.

Table 3. Previous works about characteristic and development of PER topics within the community.

Docktor and Mestre [9]	Odden et al. [11]	Yun [12]	
		AJP	PRPER
1. Conceptual understanding 2. Problem solving 3. Curriculum and instruction 4. Assessment 5. Cognitive psychology 6. Attitudes and beliefs about teaching and learning	1. Representation 2. Problem solving 3. Labs 4. Quantitative assessment of concept 5. K-12 6. Difficulties with quantum mechanics 7. Community, identity 8. Qualitative methodology and constructivist theory building 9. Research based instruction 10. Quantitative survey of demographic gap	1. Introductory physics 2. Teaching models 3. Force and motion 4. School program 5. Problem solving 6. Pedagogical content knowledge 7. Students' learning strategy 8. Experiment	1. Assessment 2. Gender 3. Student's concept 4. Teacher education 5. Students' reasoning process 6. School programs 7. Introductory physics 8. Problem solving

One can technically compare our topical findings in Table 1 to the previous works in Table 3. There are several topics or themes that are overlapped and are more distinctive. We have three similar findings precisely to Docktor and Mestre's [9] review on conceptual understanding, problem solving, and assessment topics. There are three topics overlapped with Odden, et al.'s [11] thematic analysis including problem solving, physics laboratory (labs), and research-based instruction. Yun's [12] results from AJP analysis exactly match our topical results on teaching models (research-based instruction), problem solving, and experiments (a physics laboratory). From PRPER findings of Yun's results, we demonstrate three relevant research themes including assessment, students' concept (conceptual understanding), and problem-solving topic.

These topical results are followed by three unique Indonesian PER topics that are missing from three previous studies. They are 21st century skills, interdisciplinary aspects of physics education, and educational technology. We argue that these immediate differences correspond to the different contexts according to the authors' point of view. If we review synthesis results of Docktor and Mestre [9], those three different topics might be categorized in the context of assessment or curriculum and instruction. Educational technology that has been developed by Indonesian PER members is assumed as a learning transformation within the PER community summarized in Docktor and Mestre's "curriculum and instruction" theme. Moreover, 21st century skills and interdisciplinary aspects of physics education are highly motivated by the Indonesian educational context, 2013 curriculum, and PISA results as explained above. They engage other forms of students' performance considered in the assessment topic of Docktor and Mestre's results. Moreover, this unique pattern derived from Indonesian PER literature can be understood as educational development within a certain country that should be determined through several social contexts and governmental policies [208,258,342,343].

Furthermore, based on Odden, et al. [11] topical findings, our unique findings can be illuminated by the topic of K-12 based education. In this scope of the theme, high school physics contributes to developing our third topic, the interdisciplinary aspect of physics learning. The scientific approach-based Indonesian 2013 curriculum inevitably directed physics educators to orient interdisciplinary high school (K-12) physics learning. The Indonesian PER community is tremendously conducted by the preparation of high school physics teachers on the national need for sustainable physics teaching and learning. Development of PER dissemination can be indirectly seen to respond to this national call. Several educational technologies have been developed by PER scholars to make the delivery of physics learning more engaging to all students from all backgrounds.

Moreover, we can discover other similar topics with different theoretical lenses from Yun's thematic analysis [12]. From her results, we highlight topics on force and motion, pedagogical content knowledge (PCK), students' reasoning process, and introductory physics. The latter is even reported by Yun both from AJP and PRPER journals. We found that force and motion is also the most interesting topic within the Indonesian PER community. In Table 1, we discover the keyword "motion" as the representative word to define the eighth topic, physics laboratory. Likewise, we discuss the relevant research on conceptual understanding and problem-solving topics addressing the concept of force and motion. We also believe that PCK and introductory physics can be related to each other to implement the transformation of physics learning. They are intended to deliver more effective physics learning for students. Therefore, we argue that these topics can have the same meaning as our fifth topic of the LDA results, research-based instruction.

For the open room of future projects, we argue that the Indonesian PER scholars should pay more attention to investigating physics education research more qualitatively. We argue that Indonesian PER topics should address research focused on qualitative aspects of physics teaching and learning as addressed by Docktor and Mestre's results as their fifth and sixth PER theme, Odden, et al.'s findings as their seventh and eighth PER theme, and Yun's inventions as their fourth topic from AJP results and their second and sixth theme from PRPER results. Compared to the Odden et al. thematic results, there are qualitative topics dealing with community and identity as well as qualitative methodology and constructivist theory building that are still missing within the Indonesian PER literature. Yun's topical results about gender and school program support these findings to grasp demographic factors within physics learning, including gender bias on physics assessment, students from underrepresented minorities or first generation, as well as supporting the vision of diversity in physics [344]. This methodological approach is also relevant to Docktor and Mestre's result to investigate cognitive psychology and attitudes and beliefs about physics education. Those trends still lack research within Indonesian PER literature and there is possible room for future study on this topic.

It is evident from our paper that the LDA algorithm has demonstrated several advantages in undertaking thematic analysis towards 852 Indonesian PER proceeding papers over time. We can describe its strength as two-fold explanations. First, the automation of the LDA algorithm inevitably has technically helped us to make classification of eight Indonesian PER topics without extra effort to manually scrutinize the data corpus. We also utilize almost the whole section of the body of the research paper. Thus, our current study can suggest that LDA considers the more comprehensive nature of thematic analysis rather than using the keywords from research titles as reported by Faisal [14] or selecting small parts of documents [13,15–17]. Second, the distributional hypothesis of topics and the mixed membership of topics have been satisfied through the LDA algorithm. These advantages have explained the existence of multidisciplinary aspects of physics education research. Categorization of a single topic in each document as reported by Faisal, et al. [14] and Bancong, et al. [13] fails to represent that each topic should be interchangeably in each document. Nevertheless, in nature, our paper is dedicated to the aim of exploration and attempts to deliver a promising tool to conduct a more efficient methodology of thematic analysis which successfully helps us to add dimensions of analysis and visualization. Traditional methods of thematic analysis must be worthwhile and cannot be replaced by the current methodology. Indeed, the LDA algorithm complements them to extract a more comprehensive understanding from thematic analysis.

On the other hand, we cannot forget the potential weaknesses after the implementation of the LDA model performed in this study. As discussed by previous work [11], there are admittedly several limitations of the LDA algorithm in the analysis of research literature.

First, LDA clearly neglects the sequence of words within sentences as clearly assumed in our theoretical review above. Our LDA results above are calculated based on the count of words occurring in the data corpus. Thereafter, the qualitative method of thematic analysis obviously can be more beneficial to address this issue. To address this first obstacle, evaluation methods through face validity with experts in specific domains (PER) should be attempted. Second, the instability of topical results is evident during the training of the most representative LDA model. This is driven by the random initialization of the computation of the LDA model. In order to address this second limitation, multiple LDA models should be trained across the mixture of several hyperparameters including a number of topics (K), alpha (α), and several filtering parameters to the most frequent and the rarest words. In this study, we trained a high number of LDA models within eleven numbers of topics (K), five different alphas (α), and we iterated ten selected different integers of our seed number. This produced 550 LDA models and then we chose the most optimum model based on the coherence measure using the elbow plot provided in Figure 5. Third, we discovered that LDA can be more sensitive to literature that has grown over a long period. Several specific topics that are not frequently mentioned within the data corpus cannot be detected in the results. Obviously, they are likely to be excluded based on our rule of filtering actions.

As a final mark, one can realize that our findings must be dedicated primarily to the Indonesian PER community. Since, to the best of our knowledge, similar research has never been attempted within the Indonesian PER community using the LDA to break down the growing size of Indonesian PER literature. Research institutions can adopt our topical findings to establish a solid definition for the research group of PER works. Subsequently, we hope it could encourage novice PER scholars to easily recognize the characteristic of the Indonesian PER and guide them to contribute to specific group within the community. Furthermore, our paper should recommend several topics that have been published and future directions that should be approached in the next research project within the community, particularly in the aspect of the qualitative methodology of physics education research. Through our LDA results, the Indonesian PER community can understand what valuable steps have been attempted and where the future Indonesian PER community must go. The LDA methodology demonstrated in this paper can inspire the wider Indonesian PER members to utilize this current method of thematic analysis. Admittedly, we cannot ignore that the results of this analysis may be interpreted as having a different meaning regarding other authors that accidentally did not publish their works at those conferences. The determination of five conferences that have been analyzed through our analysis might be an arguable position that has been selected by the authors. Ultimately, other PER researchers could look forward to using the LDA method for future explorations of the larger Indonesian PER literature in the next efforts.

6. Conclusions

In summary, Indonesian physics education research (PER) literature has been thematically analyzed using the LDA algorithm. Eight topics were attempted by our PER members including 21st century skills, assessment, interdisciplinary aspects of physics education, conceptual understanding, research-based instruction, problem solving, educational technology, and physics laboratory. In the early initiation of Indonesian PER conferences in 2014, our members placed more attention on approaching learning through physics laboratories. This brought us to the movement of the community in responding to the demands of 21st century learning experiences within physics lessons. Our educators then were encouraged to harness several educational technologies to promote several aspects of students' performance in physics and interdisciplinary aspects of physics education, including scientific literacy and higher order thinking skills (HOTS) based on the demand of 21st century learning. We can declare that the LDA algorithm has been demonstrated as a powerful computational tool to extract insights derived from Indonesian PER literature. The automation technology embedded in this algorithm made the literature review methodology through thematic analysis robust in terms of its findings for the merit of the research community. Furthermore, this paper could be the basis to understand the extent to which Indonesian PER scholars have made efforts to develop their community to date. Our results may recommend future work that should be conducted within the community, particularly about the qualitative aspect of physics learning and instruction, that is little known according to the results reported in this study.

Author Contributions: Conceptualization, P.H.S. and W.H.; methodology, P.H.S.; software, P.H.S. and W.H.; validation, E.I. and H.; formal analysis, P.H.S.; data curation, P.H.S.; writing—original draft preparation, P.H.S.; writing—review and editing, P.H.S.; visualization, W.H.; supervision, E.I. and H. All authors have read and agreed to the published version of the manuscript.

Funding: This research received no external funding.

Institutional Review Board Statement: Not applicable.

Informed Consent Statement: Not applicable.

Data Availability Statement: The cleaned data of this project (pickle format), the python code of this thematic analysis and metadata of the articles (excel format) can be downloaded at: https://github.com/santosoph/Indonesian-PER-thematic-analysis (accessed on 19 July 2022).

Acknowledgments: This article is an ongoing part of a doctoral study on the Graduate School of Educational Research and Evaluation (PEP) which the first author (P.H.S.) is currently pursuing at Universitas Negeri Yogyakarta (UNY), Indonesia. We would like to express the highest gratitude to The Ministry of Education, Culture, Research, and Technology (KEMENDIKBUDRISTEK), The Center for Education Financial Services (PUSLAPDIK), and The Indonesia Endowment Funds for Education (LPDP) of The Republic of Indonesia for providing the Indonesia Educational Scholarships (BPI) so that the first author (P.H.S.) is able to pursue this academic degree.

Conflicts of Interest: The authors declare no conflict of interest.

References

1. FMIPA UNY. International Conference on Research, Implementation, and Education of Mathematics and Science (ICRIEMS). Available online: http://icriems.uny.ac.id/ (accessed on 9 July 2022).
2. FMIPA UNY. International Seminar on Science Education (ISSE). Available online: http://isse.uny.ac.id/ (accessed on 9 July 2022).
3. FMIPA UNNES. International Conference on Mathematics, Science, and Education (ICMSE). Available online: https://icmseunnes.com/ (accessed on 9 July 2022).
4. FMIPA UPI. International Conference on Mathematics and Science Education (ICMScE). Available online: https://upiconf.org/2022/icmsce/kfz/ (accessed on 9 July 2022).
5. FMIPA UM. International Conference on Mathematics and Science Education (ICOMSE). Available online: http://icomse.fmipa.um.ac.id/ (accessed on 9 July 2022).
6. Choe, K.; Jung, S.; Park, S.; Hong, H.; Seo, J. Papers101: Supporting the Discovery Process in the Literature Review Workflow for Novice Researchers. In Proceedings of the 2021 IEEE 14th Pacific Visualization Symposium (PacificVis), Tianjin, China, 19–21 April 2021; pp. 176–180. [CrossRef]
7. Ameen, K.; Batool, S.H.; Naveed, M.A. Difficulties novice LIS researchers face while formulating a research topic. *Inf. Dev.* **2018**, *35*, 592–600. [CrossRef]
8. McDermott, L.C.; Redish, E.F. Resource Letter: PER-1: Physics Education Research. *Am. J. Phys.* **1999**, *67*, 755–767. [CrossRef]
9. Docktor, J.L.; Mestre, J.P. Synthesis of discipline-based education research in physics. *Phys. Rev. Spec. Top.-Phys. Educ. Res.* **2014**, *10*, 020119. [CrossRef]
10. Meltzer, D.E.; Otero, V.K. A brief history of physics education in the United States. *Am. J. Phys.* **2015**, *83*, 447–458. [CrossRef]
11. Odden, T.O.B.; Marin, A.; Caballero, M.D. Thematic analysis of 18 years of physics education research conference proceedings using natural language processing. *Phys. Rev. Phys. Educ. Res.* **2020**, *16*, 010142. [CrossRef]
12. Yun, E. Review of trends in physics education research using topic modeling. *J. Balt. Sci. Educ.* **2020**, *19*, 388–400. [CrossRef]
13. Bancong, H.; Nurazmi, N.; Fiskawarni, T.H.; Park, J. Trending Research Topics in the Field of Physics Education from 2017 to 2019 in Highly Reputable International Journals. *J. Ilm. Pendidik. Fis. Al-Biruni* **2021**, *10*, 29–36. [CrossRef]
14. Faisal, F.; Gi, G.M.; Martin, S.N. Analysis of Government-Funded Research in Indonesia from 2014-2018: Implications for Research Trends in Science Education. *J. Pendidik. IPA Indones.* **2020**, *9*, 146–158. [CrossRef]
15. Ni'Mah, F. Research trends of scientific literacy in Indonesia: Where are we? *J. Inov. Pendidik. IPA* **2019**, *5*, 23–30. [CrossRef]
16. Chusni, M.M.; Zakwandi, R. Trend Analysis of Physics Prospective Teachers' Research: An Effort to Improve The Academic Quality of Physics Study Program. *J. Ilm. Pendidik. Fis. Al-Biruni* **2018**, *7*, 11–19. [CrossRef]

17. Hari Kristiyanto, W.; Kardi, S. Trend of Research on Physics Learning Media and Its Findings. In Proceedings of the 2nd In-ternational Conference on Mathematics, Science, and Education (ICMSE), Semarang, Indonesia, 5–6 September 2015.
18. Harris, Z.S. Distributional Structure. *Word* **1954**, *10*, 146–162. [CrossRef]
19. Grimmer, J.; Stewart, B.M. Text as Data: The Promise and Pitfalls of Automatic Content Analysis Methods for Political Texts. *Political Anal.* **2013**, *21*, 267–297. [CrossRef]
20. Odden, T.O.B.; Marin, A.; Rudolph, J.L. How has Science Education changed over the last 100 years? An analysis using natural language processing. *Sci. Educ.* **2021**, *105*, 653–680. [CrossRef]
21. Blei, D.M.; Ng, A.Y.; Jordan, M.I. Latent Dirichlet Allocation. *J. Mach. Learn. Res.* **2003**, *3*, 993–1022. [CrossRef]
22. Hoffman, M.D.; Blei, D.M.; Bach, F. Online Learning for Latent Dirichlet Allocation. In *Advances in Neural Information Processing Systems*; Lafferty, J.D.A., Williams, C.K.I., Shawe-Taylor, J., Zemel, R.S., Eds.; Neural Information Processing Systems Foundation, Inc. (NeurIPS): La Jolla, CA, USA, 2010; Volume 23, pp. 856–864.
23. Snyder, H. Literature review as a research methodology: An overview and guidelines. *J. Bus. Res.* **2019**, *104*, 333–339. [CrossRef]
24. Vaismoradi, M.; Jones, J.; Turunen, H.; Snelgrove, S. Theme development in qualitative content analysis and thematic analysis. *J. Nurs. Educ. Pr.* **2015**, *6*, 100. [CrossRef]
25. Neuendorf, K.A. Content Analysis and Thematic Analysis. In *Advanced Research Methods for Applied Psychology*; Routledge: London, UK, 2019; pp. 211–223.
26. Creswell, J.W.; Poth, C.N. *Qualitative Inquiry & Research Design: Choosing Among Five Approaches*, 4th ed.; Sage: New York, NY, USA, 2017.
27. Braun, V.; Clarke, V. Using thematic analysis in psychology. *Qual. Res. Psychol.* **2006**, *3*, 77–101. [CrossRef]
28. Nowell, L.S.; Norris, J.M.; White, D.E.; Moules, N.J. Thematic Analysis: Striving to Meet the Trustworthiness Criteria. *Int. J. Qual. Methods* **2017**, *16*, 1–13. [CrossRef]
29. Tsafnat, G.; Glasziou, P.; Choong, M.K.; Dunn, A.; Galgani, F.; Coiera, E. Systematic review automation technologies. *Syst. Rev.* **2014**, *3*, 74. [CrossRef]
30. Gauthier, R.P.; Wallace, J.R. The Computational Thematic Analysis Toolkit. In Proceedings of the ACM on Human-Computer Interaction, New Orleans, LA, USA, 30 April–5 May 2022; Volume 6.
31. Syed, S.; Spruit, M. Full-Text or Abstract? Examining Topic Coherence Scores Using Latent Dirichlet Allocation. In Proceedings of the 2017 International Conference on Data Science and Advanced Analytics, DSAA 2017, Tokyo, Japan, 19–21 October 2017.
32. Agustina, A.; Pradina, R. *Analisis Dan Visualisasi Suara Pelanggan Pada Pusat Layanan Pelanggan Dengan Permodelan Topik Menggunakan Latent Dirichlet Allocation (LDA) Studi Kasus: PT*; Petrokimia Gresik, Institut Teknologi Sepuluh Nopember: Surabaya, Indonesia, 2017.
33. Nugroho, D.D.A.; Alamsyah, A. Analisis Konten Pelanggan Airbnb Pada Network Sosial Media Twitter. In *e-Proceeding of Management*; Telkom University: Bandung, Indonesia, 2018.
34. Hikmah, F.N.; Basuki, S.; Azhar, Y. Deteksi Topik Tentang Tokoh Publik Politik Menggunakan Latent Dirichlet Allocation (LDA). *J. Repos.* **2020**, *2*, 415–426. [CrossRef]
35. Prihatini, P.M.; Suryawan, I.K.; Mandia, I.N. Metode Latent Dirichlet Allocation Untuk Ekstraksi Topik Dokumen. *J. Log.* **2017**, *17*, 153–157.
36. Nurlayli, A.; Nasichuddin, M.A. Topik modeling penelitian dosen jptei uny pada google scholar menggunakan latent dirichlet allocation. *Elinvo (Electron. Inform. Vocat. Educ.)* **2019**, *4*, 154–161. [CrossRef]
37. Setijohatmo, U.T.; Rachmat, S.; Susilawati, T.; Rahman, Y.; Kunci, K. Analisis Metoda Latent Dirichlet Allocation Untuk Klasifikasi Dokumen Laporan Tugas Akhir Berdasarkan Pemodelan Topik. In *Prosiding Industrial Research Workshop and National Seminar*; Politeknik Negeri Bandung: Bandung, Indonesia, 2020; Volume 11.
38. Alfanzar, A.I.; Khalid, K.; Rozas, I.S. Topic modelling skripsi menggunakan metode latent diriclhet allocation. *J. Sist. Inf.* **2020**, *7*, 7–13. [CrossRef]
39. Riduwan, M.; Fatichah, C.; Yuniarti, A. Klasterisasi dokumen menggunakan weighted k-means berdasarkan relevansi topik. *JUTI J. Ilm. Teknol. Inf.* **2019**, *17*, 146. [CrossRef]
40. Arianto, B.W.; Anuraga, G. Topic Modeling for Twitter Users Regarding the "Ruangguru" Application. *J. Ilmu Dasar* **2020**, *21*, 149–154. [CrossRef]
41. Wilson, J.; Pollard, B.; Aiken, J.M.; Caballero, M.D.; Lewandowski, H.J. Classification of open-ended responses to a research-based assessment using natural language processing. *Phys. Rev. Phys. Educ. Res.* **2022**, *18*, 010141. [CrossRef]
42. Röder, M.; Both, A.; Hinneburg, A. Exploring the Space of Topic Coherence Measures. In Proceedings of the WSDM 2015—Proceedings of the 8th ACM International Conference on Web Search and Data Mining, Shanghai, China, 2–6 February 2015.
43. Hoffman, M.D.; Blei, D.M.; Bach, F. Online Learning for Latent Dirichlet Allocation. In Proceedings of the Advances in Neural Information Processing Systems 23: 24th Annual Conference on Neural Information Processing Systems 2010 (NIPS 2010), Vancouver, BC, Canada, 6–9 December 2010.
44. Erosheva, E.; Fienberg, S.; Lafferty, J. Mixed-membership models of scientific publications. *Proc. Natl. Acad. Sci. USA* **2004**, *101*, 5220–5227. [CrossRef] [PubMed]
45. FMIPA UNY. Proceedings of the 1st International Conference on Research, Implementation, & Education of Mathematics and Science 2014. Available online: https://eprints.uny.ac.id/view/subjects/ICRIEMS.html (accessed on 9 July 2022).

46. FMIPA UNY. Proceedings of the 6th International Conference on Research, Implementation, & Education of Mathematics and Science 2019. Available online: http://seminar.uny.ac.id/icriems/proceeding2019 (accessed on 9 July 2022).
47. FMIPA UNY. Proceedings of the 5th International Conference on Research, Implementation, & Education of Mathematics and Science 2018. Available online: http://seminar.uny.ac.id/icriems/proceeding2018 (accessed on 9 July 2022).
48. FMIPA UNY. Proceedings of the 4th International Conference on Research, Implementation, & Education of Mathematics and Science 2017. Available online: http://seminar.uny.ac.id/icriems/proceedings2017 (accessed on 9 July 2022).
49. FMIPA UNY. Proceedings of the 3rd International Conference on Research, Implementation, & Education of Mathematics and Science 2016. Available online: http://seminar.uny.ac.id/icriems/proceedings2016 (accessed on 9 July 2022).
50. FMIPA UNY. Proceedings of the 2nd International Conference on Research, Implementation, & Education of Mathematics and Science 2015. Available online: https://eprints.uny.ac.id/view/subjects/icriems2015.html (accessed on 9 July 2022).
51. FMIPA UNY. Proceedings of the 7th International Conference on Research, Implementation, and Education of Mathematics and Sciences 2020. Available online: https://www.atlantis-press.com/proceedings/icriems-20 (accessed on 9 July 2022).
52. FMIPA UNY. Proceedings of the 6th International Conference on Research, Implementation, and Education of Mathematics and Science 2019. *J. Phys. Conf. Ser.* **2019**, *1397*, 011001. [CrossRef]
53. FMIPA UNY. Proceedings of the 5th International Conference on Research, Implementation, & Education of Mathematics and Science 2018. *J. Phys. Conf. Ser.* **2018**, *1097*, 011001. [CrossRef]
54. FMIPA UNY. Proceedings of the 4th International Conference on Research, Implementation, and Education of Mathematics and Sciences (ICRIEMS). *AIP Conf. Proc.* **2017**, *1868*, 010001. [CrossRef]
55. FMIPA UNY. Proceedings of the 8th International Seminar on Science Education 2022. Available online: http://isse.uny.ac.id/presenter-2021 (accessed on 9 July 2022).
56. FMIPA UNY. Proceedings of the 6th International Seminar on Science Education 2020. Available online: https://www.atlantis-press.com/proceedings/isse-20 (accessed on 9 July 2022).
57. FMIPA UNY. Proceedings of the 5th International Seminar on Science Education 2019. *J. Phys. Conf. Ser.* **2020**, *1440*, 011001. [CrossRef]
58. FMIPA UNY. Proceedings of the 4th International Seminar on Science Education 2018. *J. Phys. Conf. Ser.* **2018**, *1233*, 011001. [CrossRef]
59. FMIPA UNY. Proceedings of the 3rd International Seminar on Science Education 2017. Available online: http://seminar.uny.ac.id/isse2017/ (accessed on 9 July 2022).
60. FMIPA UNY. Proceedings of the 2nd International Seminar on Science Education 2016. Available online: http://seminar.uny.ac.id/isse2016/?q=home (accessed on 9 July 2022).
61. FMIPA UNY. Proceedings of the 1st International Seminar on Science Education 2015. Available online: http://seminar.uny.ac.id/isse2015/ (accessed on 9 July 2022).
62. FMIPA UNNES. Proceedings of the 7th International Conference on Mathematics and Science Education 2020. Available online: https://iopscience.iop.org/volume/1742-6596/1918 (accessed on 9 July 2022).
63. FMIPA UNNES. Proceedings of the 6th International Conference on Mathematics and Science Education 2019. Available online: https://iopscience.iop.org/volume/1742-6596/1567 (accessed on 9 July 2022).
64. FMIPA UNNES. Proceedings of the 5th International Conference on Mathematics and Science Education 2018. Available online: https://iopscience.iop.org/volume/1742-6596/1321 (accessed on 9 July 2022).
65. FMIPA UNNES. Proceedings of the 4th International Conference on Mathematics and Science Education 2017. *J. Phys. Conf. Ser.* **2017**, *983*, 011001. [CrossRef]
66. FMIPA UNNES. Proceedings of the 3rd International Conference on Mathematics and Science Education 2016. *J. Phys. Conf. Ser.* **2017**, *824*, 011001. [CrossRef]
67. FMIPA UNNES. Proceedings of the 3rd International Conference on Mathematics and Science Education 2016. Available online: https://icmseunnes.com/2016/ (accessed on 9 July 2022).
68. FMIPA UNNES. Proceedings of the 2nd International Conference on Mathematics and Science Education 2015. Available online: https://icmseunnes.com/2015/ (accessed on 9 July 2022).
69. FMIPA UNNES. Proceedings of the 1st International Conference on Mathematics and Science Education 2014. Available online: https://icmseunnes.com/2015/?page_id=336 (accessed on 9 July 2022).
70. FMIPA UPI. Proceedings of the 5th International Conference on Mathematics and Science Education 2020. Available online: http://science.conference.upi.edu/proceeding/index.php/ICMScE/issue/view/5 (accessed on 9 July 2022).
71. FMIPA UPI. Proceedings of the 4th International Conference on Mathematics and Science Education 2019. Available online: http://science.conference.upi.edu/proceeding/index.php/ICMScE/issue/view/4 (accessed on 9 July 2022).
72. FMIPA UPI. Proceedings of the 3rd International Conference on Mathematics and Science Education 2018. Available online: http://science.conference.upi.edu/proceeding/index.php/ICMScE/issue/view/3 (accessed on 9 July 2022).
73. FMIPA UPI. Proceedings of the 5th International Conference on Mathematics and Science Education 2020. *J. Phys. Conf. Ser.* **2020**, *1806*, 011001. [CrossRef]
74. FMIPA UPI. Proceedings of the 4th International Conference on Mathematics and Science Education 2019. Available online: https://iopscience.iop.org/volume/1742-6596/1521 (accessed on 9 July 2022).

75. FMIPA UPI. Proceedings of the 3rd International Conference on Mathematics and Science Education 2018. Available online: https://iopscience.iop.org/volume/1742-6596/1157 (accessed on 9 July 2022).
76. FMIPA UPI. Proceedings of the 2nd International Conference on Mathematics and Science Education 2017. *J. Phys. Conf. Ser.* **2017**, *895*, 011001. [CrossRef]
77. FMIPA, UM. Proceedings of the 4th International Conference on Mathematics and Science Education 2020. *AIP Conf. Proc.* **2020**, *2330*, 010001. [CrossRef]
78. FMIPA, UM. Proceedings of the 3rd International Conference on Mathematics and Science Education 2019. *AIP Conf. Proc.* **2019**, *2215*, 010001. [CrossRef]
79. FMIPA, UM. Proceedings of the 2nd International Conference on Mathematics and Science Education 2018. *J. Phys. Conf. Ser.* **2018**, *1227*, 011001. [CrossRef]
80. FMIPA, UM. Proceedings of the 1st International Conference on Mathematics and Science Education 2017. *J. Phys. Conf. Ser.* **2018**, *1093*, 011001. [CrossRef]
81. Bird, S.; Loper, E.; Klein, E. *Natural Language ToolKit (NLTK) Book*; O'Reilly Media, Inc.: Sebastopol, CA, USA, 2009.
82. Denny, M.J.; Spirling, A. Text Preprocessing for Unsupervised Learning: Why It Matters, When It Misleads, and What to Do about It. *Political Anal.* **2018**, *26*, 168–189. [CrossRef]
83. Rehurek, R.; Sojka, P. Gensim–Python Framework for Vector Space Modeling. In *NLP Centre, Faculty of Informatics*; Masaryk University: Brno, Czech Republic, 2011; Volume 3.
84. Debortoli, S.; Müller, O.; Junglas, I.; Brocke, J.V. Text Mining for Information Systems Researchers: An Annotated Topic Modeling Tutorial. *Commun. Assoc. Inf. Syst.* **2016**, *39*, 110–135. [CrossRef]
85. Taylor, J.R. *An Introduction to Error Analysis: The Study of Uncertainties in Physical Measurements*; University Science Books: Sausalito, CA, USA, 1997.
86. Roberts, M.E.; Stewart, B.M.; Tingley, D. Navigating the Local Modes of Big Data: The Case of Topic Models. In *Computational Social Science: Discovery and Prediction*; Cambridge University Press: Cambridge, UK, 2016; ISBN 9781316257340.
87. Chuang, J.; Manning, C.D.; Heer, J. Termite: Visualization Techniques for Assessing Textual Topic Models. In Proceedings of the Workshop on Advanced Visual Interfaces AVI, Capri Island, Italy, 21-25 May 2012; pp. 74–77. [CrossRef]
88. Supahar, S. Applying Content Validity Ratios (CVR) to The Quantitative Content Validity of Physics Learning Achievement Tests. In Proceedings of the 2nd International Conference on Research, Implementation and Education of Mathematics and Sciences (ICRIEMS), Yogyakarta, Indonesia, 15–17 May 2015; pp. 61–65.
89. Ramalis, T.R.; Suwarma, I.R. Profile of Students' Critical Thinking Ability in Project Based Learning Integrated Science Technology Engineering and Mathematics. *J. Phys. Conf. Ser.* **2020**, *1521*, 022042.
90. Rizal, R.; Rusdiana, D.; Setiawan, W.; Siahaan, P.; Ridwan, I.M. Gender differences in digital literacy among prospective physics teachers. *J. Phys. Conf. Ser.* **2021**, *1806*, 012004. [CrossRef]
91. Masrifah, M.; Setiawan, A.; Sinaga, P. Profile of senior high school in-service physics teachers' technological pedagogical and content knowledge (TPACK). *J. Phys. Conf. Ser.* **2018**, *1097*, 012025. [CrossRef]
92. Anggraeni, D.M.; Sole, F.B. Developing creative thinking skills of STKIP weetebula students through physics crossword puzzle learning media using eclipse crossword app. *J. Phys. Conf. Ser.* **2020**, *1521*, 022045. [CrossRef]
93. Febriana, R.; Sinaga, P. Evaluation of critical thinking skills of class x high school students on the material of Newton's laws. *J. Phys. Conf. Ser.* **2021**, *1806*, 012012. [CrossRef]
94. Istiyono, E. *An Eight-Category Partial Credit Model As Very Appropriate For Four-Tier Diagnostic Test Scoring In Physics Learning. Proceedings of the 8th International Seminar on Science Education (ISSE)*; Universitas Negeri Yogyakarta: Yogyakarta, Indonesia, 2021.
95. Istiyono, E.; Dwandaru, W.S.B. Developing of Bloomian HOTS Physics Test: Content and Construct Validation of the PhysTeBio-HOTS. *J. Phys. Conf. Ser.* **2019**, *1397*, 012017. [CrossRef]
96. Ermansah; Muhammad, M.; Patria, Y.A.B.; Istiyono, E. Instrument Test Physics-Based Computer Adaptive Test to Meet the Slam Economic Community Literature Review. In Proceedings of the 2nd International Seminar on Science Education (ISSE). Yogyakarta, Indonesia,, 29 October 2016.
97. Asriadi, M.; Hadi, S. Implementation of Item Response Theory at Final Exam Test in Physics Learning: Rasch Model Study. In Proceedings of the 6th International Seminar on Science Education (ISSE). Yogyakarta, Indonesia,, 28–29 November 2020; Volume 541.
98. Jumadi; Wilujeng, I.; Prasetya, Z.K. Mapping of Professional, Pedagogical, Social, and Personal Competence of Senior High School Physics Teachers in Yogyakarta Special Region. In Proceedings of the 1st International Conference on Research, Implementation, and Education of Mathematics and Science (ICRIEMS). Yogyakarta, Indonesia,, 18–20 May 2014.
99. Arsyad, M.; Sopandi, W.; Chandra, D.T. Analysis of Scientific Literacy through PISA 2015 Framework. In Proceedings of the 1st Interna-tional Conference on Mathematics and Science Education (ICMScE), Bandung, Indonesia, 30 April 2016.
100. Napitupulu, N.D.; Munandar, A.; Redjeki, S.; Tjasyono, B. Shifting Attitude from Receiving to Characterisation as an Interdisciplinary Learning toward Ecological Phenomena. In Proceedings of the 3rd International Seminar on Science Education (ISSE), Yogyakarta, Indonesia, 28 October 2017; pp. 124–128.
101. Rahzianta, A.C.P. Promoting Metacognition and Students' Care Attitude Towards The Environment Through Learning Physics with STEM. In Proceedings of the 2nd International Seminar on Science Education (ISSE), Yogyakarta, Indonesia, 29 October 2016; pp. 85–88.

102. Maulita, S.R.; Sukarmin; Marzuki, A. Analysis of Senior High School Students' Higher Order Thinking Skills in Physics Learning. In Proceedings of the 5th International Conference on Research, Implementation, and Education of Mathematics and Science (ICRIEMS), Yogyakarta, Indonesia, 7–8 May 2018.
103. Wardiyah, K.; Suhandi, A.; Samsudin, A. Alternative Conception of High School Students Related to the Concepts in the Simple Electric Circuit Subject Matter. In Proceedings of the 2nd International Conference on Mathematics and Science Education (ICMScE), Bandung, Indonesia, 24 May 2017; pp. 183–187.
104. Saputra, O.; Setiawan, A.; Rusdiana, D. Identification of Student Misconception about Static Fluid. *J. Phys. Conf. Ser.* **2019**, *1157*, 032069. [CrossRef]
105. Handhika, J.; Cari, C.; Soeparmi, A.; Sunarno, W. External Representation to Overcome Misconception in Physics. In Proceedings of the 2nd International Conference on Mathematics, Science, and Education (ICMSE), Semarang, Indonesia, 5–6 September 2015; Volume 2015.
106. Anam, R.S.; Widodo, A.; Sopandi, W. Teachers, Pre-Service Teachers, and Students Understanding about the Heat Conduction. *J. Phys. Conf. Ser.* **2019**, *1157*, 022012. [CrossRef]
107. Wiyantara, A.; Widodo, A.; Prima, E.C. Identify students' conception and level of representations using five-tier test on wave concepts. *J. Phys. Conf. Ser.* **2021**, *1806*, 012137. [CrossRef]
108. Suastra, I.W. The Effectiveness of Local Culture-Based Physics Model of Teaching in Developing Physics Competence and National. In Proceedings of the 2nd International Conference on Research, Implementation and Education of Mathematics and Sciences (ICRIEMS), Yogyakarta, Indonesia, 15–17 May 2015.
109. Nurulsari, N.; Suyatna, A. Abdurrahman Cooperative Learning Model Design Based on Collaborative Game-Based Learning Approach as a Soft Scaffolding Strategy: Preliminary Research. In Proceedings of the 1st International Conference on Mathematics and Science Education (ICMScE), Bandung, Indonesia, 30 April 2016.
110. Kaleka, M. Effect of Free Inquiry Models to Learning Achievement and Character of Student Class XI. In Proceedings of the 5th International Conference on Research, Implementation, and Education of Mathematics and Science (ICRIEMS), Yogyakarta, Indonesia, 7–8 May 2018.
111. Ramayanti, S.; Utari, S.; Saepuzaman, D. Training Students' Science Process Skills through Didactic Design on Work and Energy. *J. Phys. Conf. Ser.* **2017**, *895*, 12110. [CrossRef]
112. Astra, I.M.; Susanti, D.; Sakinah, S. The Effects of Cooperative Learning Model Think Pair Share Assisted by Animation Media on Learning Outcomes of Physics in High School. *J. Phys. Conf. Ser.* **2020**, *1521*, 022005. [CrossRef]
113. Septiyono, W.E.; Prasetyo, Z.K.; Al Ihwan, M. The Effect of E-Learning Based Worksheet to Improve Problem Solving Ability of Senior High School Students. In Proceedings of the 6th International Seminar on Science Education (ISSE), Yogyakarta, Indonesia, 28–29 November 2020; Volume 541.
114. Yuliana, I.; Kusairi, S.; Taufiq, A.; Priyadi, R.; Rosyidah, N.D. The analysis of students' problem-solving ability in the 5E learning cycle with formative e-assessment. *AIP Conf. Proc.* **2020**, *2215*, 050015. [CrossRef]
115. Fitriadi, P.; Latumalukita, I.I.; Warsono, W. The Development of Physics E-Book Based on Contextual Teaching and Learning to Increase Student Problem-Solving Skill. In Proceedings of the 7th International Seminar on Science Education (ISSE), Yogyakarta, Indonesia, 19–20 November 2021.
116. Himawan, N.A.; Wilujeng, I. Improving students' problem-solving skills through quick on the draw model assisted by the optical learning book integrated the Pancasila. *J. Phys. Conf. Ser.* **2020**, *1440*, 012031. [CrossRef]
117. Sakti, A.O.P. Profile of Problem Solving Ability of Islamic Senior High School Students on Momentum and Impuls. In Proceedings of the 8th International Seminar on Science Education (ISSE), Yogyakarta, Indonesia, 11–12 November 2022.
118. Anggraini, A.I.; Warsono; Hamidiyah, H.; Jatmika, S. Developing Whiteboard Animation Video Through Local Wisdom on Work and Energy Materials as Physics Learning Solutions During the COVID-19 Pandemic. In Proceedings of the 6th International Seminar on Science Education (ISSE 2020), Yogyakarta, Indonesia, 28–29 November 2020; pp. 394–400. [CrossRef]
119. Rahayu, M.S.I.; Kuswanto, H.; Pranowo, C.Y. Android-Based Carrom Game Comics Integrated with Discovery Learning for Physics Teaching. In Proceedings of the 7th International Conference on Research, Implementation, and Education of Mathematics and Sciences (ICRIEMS), Yogyakarta, Indonesia, 25–26 September 2020; Volume 528.
120. Aji, S.H.; Jumadi; Saputra, A.T.; Tuada, R.N. Development of physics mobile learning media in optical instruments for senior high school student using android studio. *J. Phys. Conf. Ser.* **2020**, *1440*, 012032. [CrossRef]
121. Nadhiroh, N.; Wilujeng, I.; Sa'diyah, A.; Erlangga, S.Y. Smartphone-Based Learning Media on Microscope Topic for High School Students. In Proceedings of the 6th International Seminar on Science Education (ISSE), Yogyakarta, Indonesia, 28–29 November 2020; Volume 541.
122. Adi, N.P.; Yulianto, R.A.; Irwan, M.; Endris, W.M. Android For The 21st Century Learning Media and Its Impact on Students. In Proceedings of the 2nd International Seminar on Science Education (ISSE), Yogyakarta, Indonesia, 29 October 2016; pp. 173–178.
123. Janah; Ishafit. Dwandaru Simple Vertical Upward Motion Experiment Using Smartphone Based Phyphox App for Physics Learning. In Proceedings of the 7th International Seminar on Science Education (ISSE), Yogyakarta, Indonesia,, 19–20 November 2021.
124. Listiaji, P.; Darmawan, M.S.; Dahnuss, D. The Atwood machine experiment assisted by smartphone acceleration sensor for enhancing classical mechanics experiments. *J. Phys. Conf. Ser.* **2021**, *1918*, 022009. [CrossRef]
125. Maisyaroh, S.; Mariyo, H.; Supahar; Kuswanto, H. Development of sound wave experimentation tools influenced by wind velocity. *J. Phys. Conf. Ser.* **2020**, *1440*, 012021. [CrossRef]

126. Mu'iz, M.S.; Lestari, K.M.; Yulianawati, D.; Rusdiana, D.; Hasanah, L. Analysis of simple harmonic spring motion using tracker software. In Proceedings of the 2nd International Conference on Mathematics and Science Education (ICMScE), Bandung, Indonesia, 24 May 2017.
127. Ristanto, S.; Novita, E.S. Real Laboratory Based Learning Using Video Tracker on Terminal Velocity. In Proceedings of the 2nd International Seminar on Science Education (ISSE), Yogyakarta, Indonesia, 29 October 2016.
128. Keane, T. Leading with Technology: 21st Century Skills = 3Rs + 4Cs. *Aust. Educ. Lead.* **2012**, *34*, 44.
129. Rahmawati, L.; Wilujeng, I. Feasibility of STEM Teaching Kit for Heat Material Through Simple Technology Design. In Proceedings of the 7th International Conference on Research, Implementation, and Education of Mathematics and Sciences (ICRIEMS), Yogyakarta, Indonesia, 25–26 September 2020; Volume 528.
130. Putri, I.E.; Sinaga, P. Collaborative problem-solving: How to implement and measure it in science teaching and learning. *J. Phys. Conf. Ser.* **2021**, *1806*, 012018. [CrossRef]
131. Tania, R.; Jumadi. The Application of Physics Learning Media Based on Android with Learning Problem Based Learning (PBL) to Improve Critical Thinking Skills. In Proceedings of the 7th International Conference on Research, Implementation, and Education of Mathematics and Sciences (ICRIEMS), Yogyakarta, Indonesia, 25–26 September 2020; Volume 528.
132. Purwita, T.D.; Rosana, D. Bringing Indigenous Knowledge into Physics Learning Instruments for Enhancing Students' Data Literacy: Its feasibility and practicality. In Proceedings of the 7th International Conference on Research, Implementation, and Education of Mathematics and Sciences (ICRIEMS 2020), Yogyakarta, Indonesia, 25–26 September 2020; pp. 600–607. [CrossRef]
133. Rahmawati, L.; Wilujeng, I.; Satriana, A. Application of STEM learning approach through simple technology to increase data literacy. *J. Phys. Conf. Ser.* **2020**, *1440*, 012047. [CrossRef]
134. Herliandry, L.D.; Kuswanto, H.; Hidayatulloh, W. Improve Critical Thinking Ability Through Augmented Reality Assisted Worksheets. In Proceedings of the 6th International Seminar on Science Education (ISSE), Yogyakarta, Indonesia, 28–29 November 2020; Volume 541.
135. Rahmawati, R.G.; Wilujeng, I.; Kamila, A.U. The Effectiveness of STEM-Based Student Worksheets to Improve Students' Data Literacy. In Proceedings of the 6th International Seminar on Science Education (ISSE), Yogyakarta, Indonesia, 28–29 November 2020; Volume 541.
136. Kartika, E.; Ariswan; Suban, M.E.; Arafah, Z.U. Students' Data Literacy Ability in Physics Using the Physics E-Module Integrated with the Values of Pancasila During the COVID-19. In Proceedings of the 6th International Seminar on Science Education (ISSE), Yogyakarta, Indonesia, 28–29 November 2020; Volume 541.
137. Sugita, M.I. Implementation of Creative Physics Experiment on the Creativity of Students' Ability. *J. Phys. Conf. Ser.* **2021**, *1918*, 022007. [CrossRef]
138. Tiyaswati, I. Students' Creative and Innovation Skill on Chapter of Newton's Law Using SSCS Learning Model. *J. Phys. Conf. Ser.* **2021**, *1806*, 012120. [CrossRef]
139. Rufaida, S.; Nurfadilah, N. The effectiveness of hypercontent module to improve creative thinking skills of prospective physics teachers. *J. Phys. Conf. Ser.* **2021**, *1918*, 022022. [CrossRef]
140. Azmy, W.N.; Damayanti, A.E.; Kuswanto, H.; Susetyo, B. Learning optics with android-assisted comics: The impacts on students critical thinking. *J. Phys. Conf. Ser.* **2020**, *1440*, 012055. [CrossRef]
141. Yusuf, I.; Widyaningsih, S.W. HOTS profile of physics education students in STEM-based classes using PhET media. *J. Phys. Conf. Ser.* **2019**, *1157*, 032021. [CrossRef]
142. Rahayu, E.C. The Critical Thinking Ability Profile of Grade X SMA N 2 Kudus. *J. Phys. Conf. Ser.* **2020**, *1567*, 032086. [CrossRef]
143. Rusnayati, H.; Saepuzaman, D.; Karim, S.; Feranie, S. Correlation of Cognitive Ability Relevance to the Ability of Scientific Creative Thinking and Scientific Critical Thinking Skills of Students of Work and Energy Concept. In Proceedings of the 3rd International Conference on Mathemarics and Science Education (ICMScE), Bandung, Indonesia, 5 May 2018; Volume 3.
144. Edie, S.S.; Krismonika, Z. Analysis of the combination aspects of creativity level in product design for physics students in basic physics learning. *J. Phys. Conf. Ser.* **2021**, *1918*, 022008. [CrossRef]
145. Andriani, R.; Hidayat, A.; Supriana, E.; Anantanukulwong, R. Examining the relationship between students' motivation and critical thinking skills in learning torque and static equilibrium. *J. Phys. Conf. Ser.* **2020**, *1567*, 032087. [CrossRef]
146. Utami, S.N.; Siahaan, P.; Setiawan, A. Development of Instrument Critical and Creative Thinking Skills on Fluids Motion. In Proceedings of the 3rd International Conference on Mathemarics and Science Education (ICMScE), Bandung, Indonesia, 5 May 2018; Volume 3.
147. Istiyono, E.; Dwandaru, W.S.B.; Asysyifa, D.S.; Viana, R.V. Development of computer-based test in critical thinking skill assessment of physics. *J. Phys. Conf. Ser.* **2020**, *1440*, 012062. [CrossRef]
148. Saputri, D.I.; Sunarno, W.; Supriyanto, A. Measurement of Critical Thinking in Physics: The Identification of Students' Criti-cal Thinking Skill through the Work Report on Momentum Conservation? *J. Phys. Conf. Ser.* **2020**, *1567*, 032073. [CrossRef]
149. Eveline, E.; Suparno, S.; Ardiyati, T.K.; DaSilva, B.E. Development of Interactive Physics Mobile Learning Media for Enhancing Students' HOTS in Impulse and Momentum with Scaffolding Learning Approach. *J. Penelit. Pengemb. Pendidik. Fis.* **2019**, *5*, 123–132. [CrossRef]
150. Muhajir, S.N.; Utari, S.; Suwarma, I.R. How to Develop Test for Measure Critical and Creative Thinking Skills of the 21st Century Skills in POPBL? *J. Phys. Conf. Ser.* **2019**, *1157*, 032051. [CrossRef]

151. Silvianty, A.; Suhandi, A.; Setiawan, W. Video supported critical thinking test in the kinetic theory of gases: Validity and reliability. *J. Phys. Conf. Ser.* **2019**, *1157*, 032052. [CrossRef]
152. Agustina, E.; Nabila, S.A.; Rahayu, P. Developing Physics Digital Literacy Skill Diagnostic Test Assisted by Google Form for Senior High School Students. In Proceedings of the 6th International Seminar on Science Education (ISSE), Yogyakarta, Indonesia, 28–29 November 2020; Volume 541.
153. Mulhayatiah, D.; Sinaga, P.; Rusdiana, D.; Kaniawati, I.; Suhendi, H.Y. Pedagogical and Professional Physics Teacher Training: Why Hybrid Learning Is Important? *J. Phys. Conf. Ser.* **2021**, *1806*, 012036. [CrossRef]
154. Ma'Ruf, M.; Setiawan, A.; Suhandi, A.; Siahaan, P. Profile of early ICT capabilities of prospective physics teachers through basic physics learning in Makassar. *J. Phys. Conf. Ser.* **2021**, *1806*, 012044. [CrossRef]
155. Efwinda, S.; Mannan, M.N. Technological pedagogical and content knowledge (TPACK) of prospective physics teachers in distance learning: Self-perception and video observation. *J. Phys. Conf. Ser.* **2021**, *1806*, 012040. [CrossRef]
156. Rizal, R.; Rusdiana, D.; Setiawan, W.; Siahaan, P. Creative Thinking Skills of Prospective Physics Teacher. *J. Phys. Conf. Ser.* **2020**, *1521*, 022012. [CrossRef]
157. Erwin, E.; Rustaman, N.Y.; Firman, H.; Ramalis, T.R. Profile of the prospective teachers response to the development of scientific communication skills through physics learning. *J. Phys. Conf. Ser.* **2019**, *1157*, 032040. [CrossRef]
158. Istiyono, E. Constructing Reasoning Multiple Choice Test to Measure Bloomian Higher Order Thinking Skills in Physics of XI Grade Students. *J. Phys. Conf. Ser.* **2019**, *1233*, 012037. [CrossRef]
159. Istiyono, E. The Analysis of Senior High School Students' Physics HOTS in Bantul District Measured Using PhysReMChoTHOTS. *AIP Conf. Proc.* **2017**, *1868*, 070008.
160. Maulidiansyah, D.; Meutia, I.; Istiyono, E. Computer-Based Two-Tier Diagnostic Test to Identify Critical Thinking Skills in Optical Instrument. In Proceedings of the 6th International Seminar on Science Education (ISSE), Yogyakarta, Indonesia, 28–29 November 2020; Volume 541.
161. Adawiyah, R.; Istiyono, E. Assessment Instrument on Measuring Physics Verbal Representation Ability of Senior High School Students. In Proceedings of the 7th International Conference on Research, Implementation, and Education of Mathematics and Sciences (ICRIEMS 2020), Yogyakarta, Indonesia, 25–26 September 2020; pp. 591–599. [CrossRef]
162. Syarif, A.N.; Kuswanto, H. Developing an Essay Test Instrument for Measuring Diagram Representation and the Capability of Argumentation on Newton's Law. *J. Phys. Conf. Ser.* **2019**, *1227*, 012030. [CrossRef]
163. Nirmala, M.F.T.; Sundari, S. Dissemination of symbolic representation ability in high school physics subjects. *J. Phys. Conf. Ser.* **2020**, *1440*, 012056. [CrossRef]
164. Sari, L.P.; Istiyono, E. Developing Assessment Instrument to Measure Senior High School Student's Mathematical Representation Ability in Physics Learning. In Proceedings of the 7th International Conference on Research, Implementation, and Education of Mathematics and Sciences (ICRIEMS), Yogyakarta, Indonesia, 25–26 September 2020; Volume 528.
165. Istiyono, E.; Fenditasari, K. Physics Graphical Representation Test of Straight Motion Kinematics Based on Boti Boat Local Wisdom: Development and Validity. In Proceedings of the 6th International Seminar on Science Education (ISSE), Yogyakarta, Indonesia, 28–29 November 2020; Volume 541.
166. Adawiyah, R.; Istiyono, E.; Wilujeng, I.; Hardiyanti, S. Development of an instrument measuring the multi representation ability of senior high school students. *J. Phys. Conf. Ser.* **2020**, *1440*, 012028. [CrossRef]
167. Larasati, P.E.; Supahar; Yunanta, D.R.A. Validity and Reliability Estimation of Assessment Ability Instrument for Data Literacy on High School Physics Material. *J. Phys. Conf. Ser.* **2020**, *1440*, 012020. [CrossRef]
168. Perdana, R.; Yani, R.; Jumadi, J.; Rosana, D. The Multiple Choice and Open Ended Test to Measure Students' Digital Literacy Skill in Physics Simulation Learning. In Proceedings of the 6th International Conference on Research, Implementation, and Education of Mathematics and Sciences (ICRIEMS), Yogyakarta, Indonesia, 12–13 July 2019.
169. Efendi, R. Systematic review of physic laboratory skills assessment instruments based on PhysPort with Nvivo. *J. Phys. Conf. Ser.* **2021**, *1806*, 012037. [CrossRef]
170. Hall, J.; Setiawan, A. Assessment Inside Assessment: Developing Course Embedded Assessment to Measure Science Process Skills and Scientific Reasoning in Simple Harmonic Motion Labwork. In Proceedings of the 4th International Conference on Research, Implementation, and Education of Mathematics and Sciences (ICRIEMS), Yogyakarta, Indonesia, 15–16 May 2017; pp. 43–48.
171. Astuti, A.T.; Istiyono, E. Development of assessment instruments to measure problem solving skills in senior high school. *J. Phys. Conf. Ser.* **2020**, *1440*, 012063. [CrossRef]
172. Yusal, Y.; Suhandi, A.; Setiawan, W.; Kaniawati, I. Construction and Testing of Decision-Problem Solving Skills Test Instruments Related Basic Physics Content. *J. Phys. Conf. Ser.* **2020**, *1521*, 022007. [CrossRef]
173. Halim, A.; Ayunda, D.S.; Syukri, M. Development and validation of students' achievement, ability to ask and inductive thinking instruments in the static fluid course. *J. Phys. Conf. Ser.* **2021**, *1806*, 012023. [CrossRef]
174. Agustina, E.; Supahar. Development of Visual Literacy Test Instrument on High School Physics Material. In Proceedings of the 7th International Conference on Research, Implementation, and Education of Mathematics and Sciences (ICRIEMS), Yogyakarta, Indonesia, 25–26 September 2020; Volume 528.
175. Oktasari, D.; Siahaan, S.M. Validation Construct: Confirmatory Factor Analysis (CFA) Instruments Scientific Communication Skills Students in Learning Physics. *J. Phys. Conf. Ser.* **2020**, *1567*, 032095. [CrossRef]

176. Sekarini, Y.P.; Adiningsih, E.T.; Anisa, Z.L.; Setiaji, B. A New Alternative to Measure Students' Analytical Thinking Skill: A Validity Test for Mechanics Problem Based Learning Module. In Proceedings of the 7th International Conference on Research, Implementation, and Education of Mathematics and Sciences (ICRIEMS 2020), Yogyakarta, Indonesia, 25–26 September 2020; pp. 618–626. [CrossRef]
177. Wati, M. The Development of Scientific Literacy Test Instruments on Newton's Law Materials for High School Students. In Proceedings of the 8th International Seminar on Science Education (ISSE), Yogyakarta, Indonesia, 11–12 November 2021.
178. Halim, A.; Elim; Elisa; Wahyuni, A.; Balqis, N.N. Development of concept maps diagnostic test for identification of students' misconceptions. *AIP Conf. Proc.* **2020**, *2215*, 050003. [CrossRef]
179. Handhika, J.; Cari, C.; Suparmi, A.; Sunarno, W.; Purwandari, P. Development of diagnostic test instruments to reveal level student conception in kinematic and dynamics. *J. Phys. Conf. Ser.* **2018**, *983*, 012025. [CrossRef]
180. Lengkong, M.; Istiyono, E.; Rampean, B.A.O.; Tumanggor, A.M.R.; Nirmala, M.F.T Development of Two-Tier Test Instruments to Detect Student's Physics Misconception. In Proceedings of the 7th International Conference on Research, Implementation, and Education of Mathematics and Sciences (ICRIEMS), Yogyakarta, Indonesia, 25–26 September 2020; Volume 528.
181. Pramesti, Y.S.; Mahmudi, H.; Setyowidodo, I. Using Three-Tier Test to Diagnose Students' Level of Understanding. *J. Phys. Conf. Ser.* **2021**, *1806*, 012013. [CrossRef]
182. Janah, A.F.; Mindyarto, B.N. Developing Four-Tier Diagnostic Test to Measure Students' Misconceptions on Simple Harmonic Motion Material. *J. Phys. Conf. Ser.* **2021**, *1918*, 052050. [CrossRef]
183. Nirmala, M.F.T.; Tumanggor, A.M.R. Analysis of Validity and Reliability of Diagnostic Test of Picture Representation Ability in High School Physics Learning. In Proceedings of the 7th International Conference on Research, Implementation, and Education of Mathematics and Sciences (ICRIEMS), Yogyakarta, Indonesia, 25–26 September 2020; Volume 528.
184. Meutia, I.; Maulidiansyah, D.; Istiyono, E. Identifying the Drawbacks of the Problem-Solving Skills by Using a Three-Tier Diagnostic Test with Google Form Assistant. In Proceedings of the 6th International Seminar on Science Education (ISSE), Yogyakarta, Indonesia, 28–29 November 2020; Volume 541.
185. Tumanggor, A.M.R.; Nirmala, M.F.T. The Development of Diagnostic Test Instrument for Verbal Representation Ability in High School Physics Learning. In Proceedings of the 7th International Conference on Research, Implementation, and Education of Mathematics and Sciences (ICRIEMS 2020), Yogyakarta, Indonesia, 25–26 September 2020; pp. 471–476. [CrossRef]
186. Istiyono, E. The Developing and Calibration of PhysEDiTHOTS Based on IRT and IQF for Students' HOTS Diagnostic. *J. Phys. Conf. Ser.* **2019**, *1233*, 012038. [CrossRef]
187. Suban, M.E.; Kartika, E.; Arafah, Z.U. A Diagnostic Test to Measure Students Physics Data Literacy Skills During the COVID-19 Pandemic. In Proceedings of the 6th International Seminar on Science Education (ISSE), Yogyakarta, Indonesia, 28–29 November 2020; Volume 541.
188. Utari, G.P.; Liliawati, W.; A Utama, J. Design and validation of six-tier astronomy diagnostic test instruments with Rasch Model analysis. *J. Phys. Conf. Ser.* **2021**, *1806*, 012028. [CrossRef]
189. Supahar, S. The Estimation of Inquiry Performance Test Items of High School Physics Subject With Quest Program. In Proceedings of the 1st International Conference on Research, Implementation and Education of Mathematics and Sciences (ICRIEMS), Yogyakarta, Indonesia, 18–20 May 2014.
190. Yusuf, I.; Widyaningsih, S.W.; Prasetyo, Z.K.; Istiyono, E. The Analysis of Self Directed Learning (SDL) through Rasch Modeling: Case Study on Prospective Teachers during the Use of e-Learning with HOTS-Oriented in the Period of COVID-19 Pandemic. *AIP Conf. Proc.* **2021**, *2330*, 050006.
191. Rosa, I.S.; Liliawati, W.; Efendi, R. Design and validation smart teaching materials oriented multiple intelligences on global warming (STM2I-GLOW): Rasch model analysis. *J. Phys. Conf. Ser.* **2021**, *1806*, 012010. [CrossRef]
192. Istiyono, E. The Application of GPCM on MMC Test as a Fair Alternative Assessment Model in Physics Learning. In Proceedings of the 3rd International Conference on Research, Implementation and Education of Mathematics and Science (ICRIEMS), Yogyakarta, Indonesia, 16–17 May 2016.
193. Istiyono, E.; Dwandaru, W.S.B.; Erfianti, L.; Astuti, W. Applying CBT in physics learning to measure students' higher order thinking skills. *J. Phys. Conf. Ser.* **2020**, *1440*, 012061. [CrossRef]
194. Patria, Y.A.B.; Istiyono, E. The development of CAT-MARZANO as an assessment media in the industrial revolution 4.0. *J. Phys. Conf. Ser.* **2020**, *1440*, 012024. [CrossRef]
195. Mindyarto, B.N.; Mardapi, D.; Bastari. Development of a Testlet Generator in Re-Engineering the Indonesian Physics National-Exams. *AIP Conf. Proc.* **2017**, *1868*, 070002.
196. Sulsilah, H.; Utari, S.; Saepuzaman, D. The application of scientific approach to improve scientific literacy on domain competency at secondary school on dynamic electricity topic. *J. Phys. Conf. Ser.* **2019**, *1157*, 032056. [CrossRef]
197. Damayanti, N.; Subali, B.; Nugroho, S.E.; Sureeporn, K. Items Analysis of Physics Assessment Based on Cognitive Level of High Order Thinking Skills in Bloom Taxonomy. *J. Phys. Conf. Ser.* **2020**, *1521*, 022022. [CrossRef]
198. Malik, A.; Setiawan, A.; Suhandi, A.; Permanasari, A. Enhancing pre-service physics teachers' creative thinking skills through HOT lab design. *AIP Conf. Proc.* **2017**, *1868*, 070001. [CrossRef]
199. Trisnawaty, W. Analyze of Student's Higher Order Thinking Skills to Solve Physics Problem on Hooke's Law. In Proceedings of the 4th International Conference on Research, Implementation, & Education of Mathematics and Science (ICRIEMS), Yogyakarta, Indonesia, 14–17 May 2017.

200. Yulianti, E.; Pratiwi, N.; Mustikasari, V.R.; Putri, A.P.; Hamimi, E.; Rahman, N.F.A. Evaluating the Effectiveness of Problem-Based Learning in Enhancing Students' Higher Order Thinking Skills. *AIP Conf. Proc.* **2020**, *2215*, 050017.
201. Nurjannah, N.; Setiawan, A.; Rusdiana, D.; Muslim, M. Students' critical thinking skills toward analyzing argumentation on heat conductivity concept. *J. Phys. Conf. Ser.* **2019**, *1157*, 032053. [CrossRef]
202. Puspita, I.; Kaniawati, I.; Suwarma, I.R. Analysis of Critical Thinking Skills on the Topic of Static Fluid. *J. Phys. Conf. Ser.* **2017**, *895*, 012100. [CrossRef]
203. Yulianti, D.; Rusilowati, A.; Nugroho, S.E.; Pangesti, K.I. Science, Technology, Engineering, and Mathematics (STEM) Based Learning of Physics to Develop Senior High School Student's Critical Thinking. *J. Phys. Conf. Ser.* **2019**, *1321*, 022029. [CrossRef]
204. Susilowati, E.; Mayasari, T.; Winarno, N.; Rusdiana, D.; Kaniawati, I.; Santoso, P.H. Correlation between Increasing Mastery Concepts of Wave and Optics and Habits of Mind Prospective Physics Teacher Students. *J. Phys. Conf. Ser.* **2019**, *1397*, 012011. [CrossRef]
205. Sutarno, S.; Setiawan, A.; Suhandi, A.; Kaniawati, I.; Malik, A. The development and validation of critical thinking skills test on photoelectric effect for pre-service physics teachers. *J. Phys. Conf. Ser.* **2019**, *1157*, 032032. [CrossRef]
206. Juliyanto, E.; Siswanto, S. An Analysis of Thinking Patterns of Natural Sciences Teacher Candidate Students in Understanding Physics Phenomena Using P-Prims Perspective. *J. Phys. Conf. Ser.* **2021**, *1918*, 022043. [CrossRef]
207. OECD. *Programme for International Student Assessment (PISA) Results from PISA 20187*; OECD: Paris, France, 2019.
208. Amini, S.; Sinaga, P. Inventory of scientific literacy ability of junior high school students based on the evaluation of PISA framework competency criteria. *J. Phys. Conf. Ser.* **2021**, *1806*, 012017. [CrossRef]
209. Riskawati, A.A.; Aqil, M.; Sitti, R.; Yunus, R. Analysis Student's Level of Science Literacy in Class x Sman Khusus Jeneponto. In Proceedings of the 2nd International Conference on Mathematics and Science Education (ICMSE), Semarang, Indonesia, 5–6 September 2015; Volume 2015.
210. Hidayat, Y.A.; Siahaan, P.; Liliawati, W. Profile of Scientific Literacy Temperature and Heat Matter Competence Student on. In Proceedings of the International Conference on Mathematics and Science Education, Malang, Indonesia, 28–29 August 2018; Volume 3.
211. Wasis, W. Analyzing Physics Item of UN, TIMSS, and PISA (Based on Higher-Order Thinking and Scientific Literacy). In Proceedings of the 1st International Conference on Research, Implementation and Education of Mathematics and Sciences (ICRIEMS). Yogyakarta, Indonesia,, 18–20 May 2014.
212. Rusilowati, A.; Yulianto, A.; Astuti, B.; Huda, N. Developing an instrument of scientific literacy assessment to measure natural science teacher candidates in force subject. *J. Phys. Conf. Ser.* **2019**, *1321*, 022027. [CrossRef]
213. Astuti, B.; Suryaningsih, I.; Rusilowati, A.; Kusuma, H.H. Science Literacy Profile of Student on Landslide Disaster Mitigation in Semarang City. *J. Phys. Conf. Ser.* **2021**, *1918*, 022017. [CrossRef]
214. Widodo, E. The Effect of Virtual Laboratory Application of Problem-Based Learning Model to Improve Science Literacy and Problem-Solving Skills. In Proceedings of the 7th International Conference on Research, Implementation, and Education of Mathematics and Sciences (ICRIEMS), Yogyakarta, Indonesia, 25–26 September 2020; Volume 528.
215. Nasution, I.B.; Liliawati, W.; Hasanah, L. Development of scientific literacy instruments based on pisa framework for high school students on global warming topic. *J. Phys. Conf. Ser.* **2019**, *1157*, 032063. [CrossRef]
216. Hartini, S.; Mahtari, S. Developing of Physics Learning Material Based on Scientific Literacy to Train Scientific Process Skills. *J. Phys. Conf. Ser.* **2018**, *1097*, 012032. [CrossRef]
217. Nugroho, S.E. Preparing prospective physics teachers to teach integrated science in junior high school. *J. Phys. Conf. Ser.* **2018**, *983*, 012053. [CrossRef]
218. Pellegrino, J.W. Teaching, Learning and Assessing 21st Century Skills. In *Pedagogical Knowledge and the Changing Nature of the Teaching Profession*; Guerriero, S., Ed.; OECD: Paris, France, 2017; pp. 223–251.
219. Pellegrino, J.W.; Hilton, M.L. (Eds.) *National Research Council Education for Life and Work: Developing Transferable Knowledge and Skills in the 21st Century*; National Academy Press: Washington, DC, USA, 2013.
220. Dewi, T.S.; Suresman, E.; Ramalis, T.R. The exploration of character education contents in the physics textbooks about newton's law. *J. Phys. Conf. Ser.* **2020**, *1521*, 022021. [CrossRef]
221. Yulianci, S.; Adiansha, A.A.; Kaniawati, I.; Liliawati, W. The Development of Character and Scientific Knowledge of Students through Inquiry-Based Learning Neuroscience Approach. *J. Phys. Conf. Ser.* **2021**, *1806*, 012019.
222. Hindarto, N.; E Nugroho, S. Using history of physics as a media to introduce and internalize characters values in physics instruction. *J. Phys. Conf. Ser.* **2018**, *983*, 012002. [CrossRef]
223. Faizah, R.; Taqwa, M.R.A.; Istiyono, E. Senior High School Student's Higher Order Thinking Skills Based on Gender and Grade. *J. Phys. Conf. Ser.* **2021**, *1918*, 022031. [CrossRef]
224. Prastyaningrum, I.; Pratama, H. Student Conception of Ohm's Law. *J. Phys. Conf. Ser.* **2019**, *1321*, 022028. [CrossRef]
225. Taqwa, M.R.A.; Zainuddin, A.; Riantoni, C. Multi Representation Approach to Increase the Students' Conceptual Understanding of Work and Energy. *J. Phys. Conf. Ser.* **2020**, *1567*, 032090. [CrossRef]
226. Cahyaningrum, R.; Hidayat, A. Contrasting-Cases Problems: Learning Material to Improve Students' Conceptual Understanding on Magnetism. *J. Phys. Conf. Ser.* **2018**, *1097*, 012028. [CrossRef]
227. Shodiqin, M.I.; Taqwa, M.R.A. Identification of Student Difficulties in Understanding Kinematics: Focus of Study on the Topic of Acceleration. *J. Phys. Conf. Ser.* **2021**, *1918*, 022016. [CrossRef]

228. E Saputro, D.; Sarwanto, S.; Sukarmin, S.; Ratnasari, D. Pre-services science teachers' conceptual understanding level on several electricity concepts. *J. Phys. Conf. Ser.* **2019**, *1157*, 032018. [CrossRef]
229. Saputra, O.; Setiawan, A.; Dan Muslim, D.R. Teacher's Conception about Static Fluid. *J. Phys. Conf. Ser.* **2020**, *1521*, 022010. [CrossRef]
230. Hermita, N.; Suhandi, A.; Syaodih, E.; Samsudin, A.; Johan, H.; Rosa, F.; Setyaningsih, R.; Safitri, D. Constructing and Implementing a Four Tier Test about Static Electricity to Diagnose Pre-Service Elementary School Teacher' Misconceptions. *J. Phys. Conf. Ser.* **2017**, *895*, 012167. [CrossRef]
231. Rahmawati, D.U.; Kuswanto, H.; A Oktaba, I. Identification of students' misconception with isomorphic multiple choices test on the force and newton's law material. *J. Phys. Conf. Ser.* **2020**, *1440*, 012052. [CrossRef]
232. Iewaru, A.A.L.; Kusairi, S.; Pramono, N.A. Senior high school students understanding of vector concepts in mathematical and physical representations. *AIP Conf. Proc.* **2021**, *2330*, 050024. [CrossRef]
233. Latif, F.H.; Buhungo, T.J.; Odja, A.H. Analysis of Students' Misconceptions Using the Certainty of Response Index (CRI) on the Concept of Work and Energy in SMA Negeri 1 Gorontalo Utara After Online Learning. In Proceedings of the 7th International Conference on Research, Implementation, and Education of Mathematics and Sciences (ICRIEMS 2020), Yogyakarta, Indonesia, 25–26 September 2020; pp. 511–515. [CrossRef]
234. Rivaldo, L.; Taqwa, M.R.A.; Zainuddin, A.; Faizah, R. Analysis of Students' Difficulties about Work and Energy. *J. Phys. Conf. Ser.* **2020**, *1567*, 032088. [CrossRef]
235. Makiyah, Y.S.; Utari, S.; Samsudin, A. The Effectiveness of Conceptual Change Texts in Reducing Pre-Service Physics Teachers' Misconceptions in Photoelectric Effect. *J. Phys. Conf. Ser.* **2019**, *1157*, 022055. [CrossRef]
236. Sianturi, I.N. Exploring Multiple Representation Preference to Develop Students Misconception Inventory in Measuring of Students Science Conception Awareness. *J. Phys. Conf. Ser.* **2019**, *1233*, 012039. [CrossRef]
237. Siswanto, J.; Susantini, E.; Jatmiko, B. Multi-Representation Based on Scientific Investigation for Enhancing Students' Representation Skills. *J. Phys. Conf. Ser.* **2018**, *983*, 012034. [CrossRef]
238. Ulum, A.S.; Basori, H.; Suhandi, A.; Samsudin, A. Improving the Mental Model of High School Students Related to the Concept of Global Warming through the Implementation of the Context Based Learning (CBL) Model Combined with the CM2RA Strategy. *J. Phys. Conf. Ser.* **2020**, *1521*, 022008. [CrossRef]
239. Iffa, U.; Supriana, E. Drawing Ability of Force Diagram with Modeling Instruction Based Free-Body Diagram Learning. *AIP Conf. Proc.* **2020**, *2215*, 050005.
240. Pratiwi, I.K.; Kusairi, S. Analyzing Students' Skill in Drawing a Free-Body Diagram. *AIP Conf. Proc.* **2021**, *2330*, 050009.
241. Trisniarti, M.D.; Aminah, N.S.; Sarwanto, S. How Interpersonal and Generic Science Skills Influence Students' Alternative Conceptions in Learning Physics? *J. Phys. Conf. Ser.* **2020**, *1521*, 022051. [CrossRef]
242. Sari, D.R.; Ramdhani, D.; Surtikanti, H.K. Analysis of elementary school students' misconception on force and movement concept. *J. Phys. Conf. Ser.* **2019**, *1157*, 022053. [CrossRef]
243. Haryono, H.E.; Aini, K.N.; Samsudin, A.; Siahaan, P. Reducing the students' misconceptions on the theory of heat through cognitive conflict instruction (CCI). *AIP Conf. Proc.* **2021**, *2330*, 050001. [CrossRef]
244. Rusilowati, A.; Susanti, R.; Sulistyaningsing, T.; Asih, T.S.N.; Fiona, E.; Aryani, A. Identify misconception with multiple choice three tier diagnostik test on newton law material. *J. Phys. Conf. Ser.* **2021**, *1918*, 052058. [CrossRef]
245. Listianingrum, S.A. A Review of Various Misconceptions in Physics Learning. In Proceedings of the 8th International Seminar on Science Education (ISSE), Yogyakarta, Indonesia, 11–12 November 2022.
246. Suhendi, H.Y.; Ardiansyah, R. Development of HATRADI: A four tier test for diagnostic student misconception in heat transfer concept. *J. Phys. Conf. Ser.* **2021**, *1918*, 022015. [CrossRef]
247. Cahyaningsih, S.; Suhandi, A.; Maknun, J. Application of Predict-Discuss-Explain-Observed-Discuss-Explore-Explain (PDEODE*E) Strategy to Remediate Students' Misconceptions on Hydrostatic Pressure. In Proceedings of the 4th International Conference on Research, Implementation, & Education of Mathematics and Science (ICRIEMS 2017), Yogyakarta, Indonesia, 14–17 May 2017; pp. 71–76.
248. Suhandi, A.; Samsudin, A.; Suhendi, E.; Basori, H. Using CCOText assisted by dynamic model and analogy to fostering students' misconception about the concept of heat conduction. *J. Phys. Conf. Ser.* **2020**, *1521*, 022044. [CrossRef]
249. Haryono, H.E.; Aini, K.N. Diagnosis misconceptions of junior high school in Lamongan on the heat concept using the three-tier test. *J. Phys. Conf. Ser.* **2021**, *1806*, 012002. [CrossRef]
250. Odja, A.H. Misconception Analysis to Know the Understanding of Static Electrical Concept at SMK Bina Taruna Gorontalo by Using Certainty of Response Index (CRI). In Proceedings of the 7th International Conference on Research, Implementation, and Education of Mathematics and Sciences (ICRIEMS 2020), Yogyakarta, Indonesia, 25–26 September 2020; pp. 608–611. [CrossRef]
251. Hindarto, N. Study on Latent Misunderstanding on Electrical Current Concept and Its Impact. *J. Phys. Conf. Ser.* **2017**, *824*, 012012. [CrossRef]
252. Basori, H.; Suhandi, A.; Kaniawati, I.; Rusdiana, D. Concept progression of high school students related to the concept of parallel electric circuits as the effect of applying CCROI integrated with T-ZPD strategy. *J. Phys. Conf. Ser.* **2020**, *1521*, 022009. [CrossRef]
253. Mahmudiah, E.; Suhandi, A.; Samsudin, A. Learning progression of madrasah aliyah-students in remedial teaching about interaction of an electrically charged object with a neutral object concept using CSCCText. *J. Phys. Conf. Ser.* **2019**, *1157*, 032067. [CrossRef]

254. Putri, A.R.; Yuliati, L. Analysis of Conceptual Changes of Static Fluid Topic through Authentic Learning Based on Phenomena. *AIP Conf. Proc.* **2020**, *2215*, 050018.
255. Setiono, I.A.; Suhandi, A.; Liliawati, W. Conceptual progression of K-10 student on the free-falling objects acceleration concept as an effect of E-CDCCText. *J. Phys. Conf. Ser.* **2021**, *1806*, 012026. [CrossRef]
256. Surtiana, Y.; Suhandi, A.; Samsudin, A.; Siahaan, P.; Setiawan, W. The Preliminary Study of the Application of the Conceptual Change Laboratory (CC-Lab) for Overcoming High School Students Misconception Related to the Concept of Floating, Drifting and Sinking. *J. Phys. Conf. Ser.* **2020**, *1521*, 022018. [CrossRef]
257. Suyanto, S. The Implementation of the Scientific Approach through "5M" of the Revised Curriculum 2013 in Indonesia. *Cakrawala Pendidik.* **2018**, *37*, 22–29.
258. Mukminin, A.; Habibi, A.; Prasojo, L.D.; Idi, A.; Hamidah, A. Curriculum Reform in Indonesia: Moving from an Exclusive to Inclusive Curriculum. *Cent. Educ. Policy Stud. J.* **2019**, *9*, 53–72. [CrossRef]
259. Dewanti, N.K.; Wilujeng, I.; Kuswanto, H. Application of Outdoor Inquiry Learning Model on Cognitive Learning Outcomes of Class XI Senior High School Students. *J. Phys. Conf. Ser.* **2019**, *1233*, 012070. [CrossRef]
260. Hasanah, U.; Hamidah, I.; Utari, S. Trained Inquiry Skills on Heat and Temperature Concepts. *J. Phys. Conf. Ser.* **2017**, *895* 012103. [CrossRef]
261. Hariadi, M.H.; Wilujeng, I.; Kuswanto, H. Improving Mathematical Representation Ability of Student's Senior High School by Inquiry Training Model with Google Classroom. *J. Phys. Conf. Ser.* **2019**, *1233*, 012043. [CrossRef]
262. Nisyah, M.; Gunawan, G.; Harjono, A.; Kusdiastuti, M. Inquiry Learning Model with Advance Organizers to Improve Students' Understanding on Physics Concepts. *J. Phys. Conf. Ser.* **2020**, *1521*, 022057. [CrossRef]
263. Viana, R.V.; Wilujeng, I.; Kuswanto, H. The Influence of Project Based Learning based on Process Skills Approach to Student's Creative Thinking Skill. *J. Phys. Conf. Ser.* **2019**, *1233*, 012033. [CrossRef]
264. Hidayah, A.; Yulianto, A.; Marwoto, P. Effect of Project Based Learning Approach Contextual to Creativity of Student of Madrasah. In Proceedings of the 2nd International Conference on Mathematics, Science and Education (ICMSE), Semarang, Indonesia, 5–6 September 2015; Volume 2015.
265. Umamah, C.; Andi, H.J. The Effect Of Project Based Learning As Learning Innovation in Applied Physics. In Proceedings of the 5th International Conference on Research, Implementation, and Education of Mathematics and Science (ICRIEMS), Yogyakarta, Indonesia, 7–8 May 2018.
266. Setiaji, B. Developing Physics Subject-Spesific Pedagogy on Problem Based Learning Model Assisted by E-Learning to Enhance Student's Scientific Literacy Skill. *Int. J. Sci. Basic Appl. Res.* **2018**, *37*, 255–268.
267. Putry, A.A.; Pratama, A.C.; Delima, E. Enhancing Physics Student's Achievement Through Problem Based Learning Assisted PhET on High School. In Proceedings of the 3rd International Seminar on Science Education (ISSE), Yogyakarta, Indonesia, 28 October 2017; pp. 189–192.
268. Karmila, N.; Wilujeng, I.; Sulaiman, H. The Effectiveness of Problem Based Learning (PBL) Assisted Google Classroom to Scientific Literacy in Physics Learning. In Proceedings of the 6th International Seminar on Science Education (ISSE), Yogyakarta, Indonesia, 28–29 November 2020; Volume 541.
269. Halsyar, I.R. The Effectiveness of Cooperative Learning Model with Time Token Arends Type with Respect to Increasing of Students' Physics Concept Understanding and Communication Skill. In Proceedings of the 2nd International Conference on Mathematics, Science, and Education (ICMSE), Semarang, Indonesia, 5–6 September 2015; pp. 9–11.
270. Sukariasih, L.; Ato, A.S.; Fayanto, S.; Nursalam, L.O.; Sahara, L. Application of SSCS model (Search, Solve, Create and Share) for improving learning outcomes: The subject of optic geometric. *J. Phys. Conf. Ser.* **2019**, *1321*, 032075. [CrossRef]
271. Buhungo, T.J.; Prastowo, T. Description of Problem Solving Ability Students in Physics Lesson. In Proceedings of the 2nd International Seminar on Science Education (ISSE), Yogyakarta, Indonesia, 29 October 2016; pp. 480–483.
272. Ekasari, A.; Diantoro, M. Problem Solving and Metacognition Abilities on Heat and Temperature of Three Different Groups of High School Students in Tulungagung. In Proceedings of the International Conference on Mathematics and Science Education, Malang, Indonesia, 29–30 August 2017; pp. 55–57.
273. Ma'Ruf, M.; Setiawan, A.; Suhandi, A.; Siahaan, P. Identification of the ability to solve the problem of contextual physics possessed by prospective physics teachers related to basic physics content. *J. Phys. Conf. Ser.* **2020**, *1521*, 022011. [CrossRef]
274. Putra, A.A.I.A.; Aminah, N.S.; Marjuki, A.; Pamungkas, Z.S. The profile of student's problem solving skill using analytical problem solving test (apst) on the topic of thermodynamic. *J. Phys. Conf. Ser.* **2020**, *1567*, 032082. [CrossRef]
275. Yuliati, L.; Putri, E.G.; Taufiq, A.; Purwanigsih, E.; Affriyenni, Y.; Halim, L. Exploration of Problem-Solving Skill with Inquiry-Based Authentic Learning for the Stem Program. *AIP Conf. Proc.* **2020**, *2215*, 050019.
276. Lestari, I.F. Experiential learning using STEM approach in improving students' problem solving ability. *J. Phys. Conf. Ser.* **2021**, *1806*, 012005. [CrossRef]
277. Rusilowati, A.; Hidayah, I.; Abidin, Z. Development of Simulation Integrated Learning Model with Mikir Approach to School for Disaster Mitigation. *J. Phys. Conf. Ser.* **2021**, *1918*, 052056. [CrossRef]
278. Ringo, E.S.; Kusairi, S.; Latifah, E.; Tumanggor, A.M.R. Student's Problem Solving Skills in Collaborative Inquiry Learning Supplemented by Formative E-Assessment: Case of Static Fluids. *J. Phys. Conf. Ser.* **2019**, *1397*, 012012. [CrossRef]
279. Setyowidodo, I.; Jatmiko, B.; Susantini, E.; Handayani, A.D.; Pramesti, Y.S. The role of science project based peer interaction on improving collaborative skills and physical problem solving: A mini review. *J. Phys. Conf. Ser.* **2020**, *1521*, 022032. [CrossRef]

280. Malik, A.; Yuningtias, U.A.; Mulhayatiah, D.; Chusni, M.M.; Sutarno, S.; Ismail, A.; Hermita, N. Enhancing problem-solving skills of students through problem solving laboratory model related to dynamic fluid. *J. Phys. Conf. Ser.* **2019**, *1157*, 032010. [CrossRef]
281. Sari, L.P.; Purwita, T.D.; Wilujeng, I. Application of TTW (Think-Talk-Write) Learning Model Using Pictorial Riddle Worksheet to Improve Students's Conceptual Undertstanding Abilities. *J. Phys. Conf. Ser.* **2020**, *1440*, 012057. [CrossRef]
282. Al Ihwan, M.; Prasetyo, Z.K.; Septiyono, W.E. Student Worksheet Based on E-Learning Development to Improve Problem-Solving Skills of Class X MAN 3 Yogyakarta Students in 2019/2020. In Proceedings of the 6th International Seminar on Science Education (ISSE), Yogyakarta, Indonesia, 28–29 November 2020; Volume 541.
283. Yuliana, A.S.; Parno; Taufiq, A. Application of teaching materials based on 7E-STEM learning cycle to improve student's problem solving skills. *AIP Conf. Proc.* **2020**, *2215*, 050014. [CrossRef]
284. Bontinge, S.; Sutopo; Taufiq, A. Epistemic Games of Students Grade X IPA SMAN 5 Malang in Solving Newton Law Problems. *AIP Conf. Proc.* **2021**, *2330*, 050023.
285. Rahmawati, L.; Labibah, U.N.; Kuswanto, H. The implementation of android-based physics learning media integrated with landslide disaster education to improve critical thinking ability and disaster preparedness. *J. Phys. Conf. Ser.* **2020**, *1440*, 012042. [CrossRef]
286. Ujulu, I.; Umar, M.K.; Odja, A.H. Students' Problem-Solving Profile in Overcoming Sound Wave Concepts Based Students' Academic Abilities on Online Class. In Proceedings of the 7th International Conference on Research, Implementation, and Education of Mathematics and Sciences (ICRIEMS 2020), Yogyakarta, Indonesia, 25–26 September 2020; pp. 516–520. [CrossRef]
287. Zulaikha, D.F. STEM-PBL with Integration of Local Wisdom in Physics Learning: Teachers' Perspective. In Proceedings of the 7th International Seminar on Science Education (ISSE), Yogyakarta, Indonesia, 19–20 November 2021.
288. Abdillah, A.J.; Rany, T.D.; Kuswanto, H.; Riyadi, I. Implementation of physics learning media based on android integrated earthquake disaster education to enhance problem solving abilities and natural disaster preparedness. *J. Phys. Conf. Ser.* **2020**, *1440*, 012027. [CrossRef]
289. Labibah, U.N.; Kuswanto, H. Integrated Landslide Disaster Education in Physics Subject Viewed from High School Students Preparedness in Kulon Progo, Yogyakarta. *J. Phys. Conf. Ser.* **2020**, *1440*, 012026. [CrossRef]
290. Jannah, M.M. Integration of Volcanic Eruption Disaster Education with Physics Learning Process to Improve Students' Disaster Preparedness in Magelang Regency. In Proceedings of the 6th International Seminar on Science Education (ISSE), Yogyakarta, Indonesia, 28–29 November 2020; Volume 541.
291. Prakasiwi, L.R.; Gusemanto, T.G. Development of Adobe Animate Assisted Physics Learning Media as Online Learning Aid. In Proceedings of the 6th International Seminar on Science Education (ISSE), Yogyakarta, Indonesia, 28–29 November 2020; Volume 541.
292. Delima, E. The Importance of Multimedia Learning Modules (Mlms) Based on Local Wisdom as an Instructional Media of 21st Century Physics Learning. *J. Phys. Conf. Ser.* **2018**, *1097*, 012018. [CrossRef]
293. Kurniawan, R.B.; Mujasam, M.; Yusuf, I.; Widyaningsih, S.W. Development of physics learning media based on Lectora Inspire Software on the elasticity and Hooke's law material in senior high school. *J. Phys. Conf. Ser.* **2019**, *1157*, 032022. [CrossRef]
294. Luliyarti, D.S.; Prasetyo, Z.K. Development of Inquiry-Based Multimedia Learning Module with PhET Simulation in New-ton's Law of Motion. In Proceedings of the 7th International Conference on Research, Implementation, and Education of Mathematics and Sciences (ICRIEMS), Yogyakarta, Indonesia, 25–26 September 2020; Volume 528.
295. Liliana, R.A.; Raharjo, W.; Jauhari, I. The development of interactive learning media with lectora inspire in gas kinetic theory subject to improve the result and students' interest of the eleventh grade students of senior high school. *J. Phys. Conf. Ser.* **2020**, *1567*, 032092. [CrossRef]
296. Mahardika, I.K.; Delftana, R.E.; Rasagama, I.G.; Rasyid, A.N.; Sugiartana, I.W. Practicality of physics module based on contextual learning accompanied by multiple representations in physics learning on senior high school. *J. Phys. Conf. Ser.* **2020**, *1521*, 022064. [CrossRef]
297. Ratnaningtyas, L.; Wilujeng, I.; Kuswanto, H. Android-Based Physics Comic Media Development on Thermodynamic Experiment for Mapping Cooperate Attitude for Senior High School. *J. Phys. Conf. Ser.* **2019**, *1233*, 012054. [CrossRef]
298. Sari, A.M. Development of Integrated Physics Learning E-Module with Pancasila Character Values in Work and Energy Subjects. In Proceedings of the 7th International Conference on Research, Implementation, and Education of Mathematics and Sciences (ICRIEMS), Yogyakarta, Indonesia, 25–26 September 2020; Volume 528.
299. Saehana, S.; Wahyono, U.; Darmadi, I.W.; Kendek, Y.; Widyawati, W. Development of Website for Studying Modern Physics. *J. Phys. Conf. Ser.* **2018**, *983*, 012052.
300. Astuti, I.A.D.; Sulisworo, D.; Firdaus, T. What Is the Student Response to Using the Weblogs for Learning Resources? *J. Phys. Conf. Ser.* **2019**, *1157*, 032012. [CrossRef]
301. Nabila, S.A.; Agustina, E. Development of Interactive Physics Learning Media Using Smartphone Integrated with Pancasila Values on Optical Instrument. In Proceedings of the 6th International Seminar on Science Education (ISSE), Yogyakarta, Indonesia, 28–29 November 2020; Volume 541.
302. Dasilva, B.E. Development of the Android-Based Interactive Physics Mobile Learning Media (IPMLM) to Improve Higher Order Thinking Skills (HOTS) of Senior High School Students. *J. Phys. Conf. Ser.* **2019** *1397*, 012010. [CrossRef]

303. Damayanti, A.E.; Kuswanto, H. Developing Android-Based Marbles Game Comics Using Group Investigation Model in Physics Learning. In Proceedings of the 7th International Conference on Research, Implementation, and Education of Mathematics and Sciences (ICRIEMS 2020), Yogyakarta, Indonesia, 25–26 September 2020; pp. 440–445. [CrossRef]
304. Rahayu, M.S.I.; Kuswanto, H. Development of android-based comics integrated with scientific approach in physics learning. *J. Phys. Conf. Ser.* **2020**, *1440*, 012040. [CrossRef]
305. Kurniawan, H.D.; Kuswanto, H. CAKA as Physics Learning Media Based on Android Apps on Smartphones. *J. Phys. Conf. Ser.* **2019**, *1227*, 012032. [CrossRef]
306. Saputra, M.R.D.; Kuswanto, H. Development of Physics Mobile (Android) Learning Themed Indonesian Culture Hombo Batu on the Topic of Newton's Law and Parabolic Motion for Class X SMA/MA. *J. Phys. Conf. Ser.* **2018**, *1097*, 012023. [CrossRef]
307. Maghfiroh, A.; Kuswanto, H.; Susetyo, B. The development of android-based physics comic on optical devices for high school students. *J. Phys. Conf. Ser.* **2020**, *1440*, 012023. [CrossRef]
308. Mahfudz, A.Z.; Billah, A. The development of android-based learning media on vibrations and waves topic for junior high school students. *J. Phys. Conf. Ser.* **2020**, *1567*, 042009. [CrossRef]
309. Permana, A.H.; Muliyati, D.; Bakri, F.; Dewi, B.P.; Ambarwulan, D. The Development of an Electricity Book Based on Augmented Reality Technologies. *J. Phys. Conf. Ser.* **2019**, *1157*, 032027. [CrossRef]
310. Agustina, E. The Teacher's Necessity for a Diagnostic Test to Detect Student Weaknesses in Learning Physics in Offline and Online Classes During the COVID-19 Pandemic. In Proceedings of the 6th International Seminar on Science Education (ISSE), Yogyakarta, Indonesia, 28–29 November 2020; Volume 541.
311. Putri, R.Z. Moodle as E-Learning Media in Physics Class. *J. Phys. Conf. Ser.* **2020**, *1567*, 032075. [CrossRef]
312. Darmawan, A.S.; Setyani, W.A. Development of Audio Visual Media for Distance Learning. In Proceedings of the 6th International Seminar on Science Education (ISSE), Yogyakarta, Indonesia, 28–29 November 2020; Volume 541.
313. Sari, F.P.; Nikmah, S.; Kuswanto, H.; Wardani, R. Developing Physics Comic Media a Local Wisdom: Sulamanda (Engklek) Traditional Game Chapter of Impulse and Momentum. *J. Phys. Conf. Ser.* **2019**, *1397*, 012013. [CrossRef]
314. Haroky, F.; Amirta, P.D.; Handayani, D.P.; Kuswanto, H.; Wardani, R. Creating physics comic media dol (a Bengkulu local wisdom musical instrument) in sound wave topic. *AIP Conf. Proc.* **2020**, *2215*, 050004. [CrossRef]
315. Azmy, W.N.; Kuswanto, H. Comic Indigenous (Bola Kasti) Based Android: The Development Integrate Problem Based Learning. In Proceedings of the 7th International Conference on Research, Implementation, and Education of Mathematics and Sciences (ICRIEMS), Yogyakarta, Indonesia, 25–26 September 2020; Volume 528.
316. Wardani, Y.R.; Mundilarto. Development of Android-Based Physics e-Book to Local Wisdom of Traditional Games Nekeran. *AIP Conf. Proc.* **2021**, *2330*, 050011.
317. Anggraini, R.; Kuswanto, H. Karapan Sapi as Android-Based Learning Module Material of Physics. *J. Phys. Conf. Ser.* **2019**, *1233*, 012063. [CrossRef]
318. Nikmah, S.; Sari, F.P.; Kuswanto, H.; Wardani, R. Development of Android Physics Comics Based on Local Wisdom Pak-Pak Dor for Sound Wave Material. In Proceedings of the 6th International Conference on Research, Implementation, and Education of Mathematics and Sciences (ICRIEMS), Yogyakarta, Indonesia, 12–13 July 2019.
319. Hartini, S.; Dewantara, D. The effectiveness of physics learning material based on South Kalimantan local wisdom. *AIP Conf. Proc.* **2017**, *1868*, 070006. [CrossRef]
320. Handayani, D.P.; Wilujeng, I.; Kuswanto, H. Development of Comic Integrated Student Worksheet to Improve Critical Thinking Ability in Microscope Material. *J. Phys. Conf. Ser.* **2019**, *1233*, 012069. [CrossRef]
321. Sa'Diyah, A.; Wilujeng, I.; Nadhiroh, N. The Effect of Using Smartphone Based Learning Media to Improve Students' Critical Thinking Skills During COVID-19 Pandemic. In Proceedings of the 6th International Seminar on Science Education (ISSE 2020), Yogyakarta, Indonesia, 28–29 November 2020; pp. 374–379. [CrossRef]
322. Malik, A.; Mardianti, D.; Nurlutfiah, D.; Izzah, D.W.; Mulhayatiah, D.; Nasrudin, D.; Suhendi, H.Y. Determination of Refractive Index on Three Mediums Based on the Principle of Refraction of Light. *J. Phys. Conf. Ser.* **2021**, *1806*, 012003. [CrossRef]
323. Festiana, I.; Herlina, K.; Kurniasari, L.S.; Haryanti, S.S. Damping Harmonic Oscillator (DHO) for learning media in the topic damping harmonic motion. *J. Phys. Conf. Ser.* **2019**, *1157*, 032062. [CrossRef]
324. Imtinan, N.; Rahmawati, I.; Hidayah, H.; Linuwih, S.; Aji, M.P. Demonstration of a New Collision Phenomenon Using Air Track. *J. Phys. Conf. Ser.* **2021**, *1918*, 022036. [CrossRef]
325. Nursulistiyo, E. Design and development of multipurpose Kundt's tube as physics learning media. *J. Phys. Conf. Ser.* **2018**, *983*, 012011. [CrossRef]
326. Uskenat, K.; Iswanto, B.H.; Indrasari, W. Spring Oscillator as Case Based Learning (CBL) Device. *J. Phys. Conf. Ser.* **2021**, *1806*, 012008. [CrossRef]
327. Wijaya, P.A.; Widodo, A. Virtual Experiment of Simple Pendulum to Improve Student's Conceptual Understanding. *J. Phys. Conf. Ser.* **2021**, *1806*, 012133. [CrossRef]
328. Rani, S.A.; Dwandaru, W.S.B. Physics virtual laboratory: An innovative media in 21st century learning. *J. Phys. Conf. Ser.* **2019**, *1321*, 022026. [CrossRef]
329. Saprudin, S.; Liliasari, S.; Prihatmanto, A.S.; Setiawan, A.; Viridi, S.; Safitri, H.; Yulina, I.K.; Rochman, C. Gamified Experimental Data on Physics Experiment to Measuring the Acceleration Due to Gravity. *J. Phys. Conf. Ser.* **2020**, *1567*, 032079. [CrossRef]

330. Firmansyah, J.; Suhandi, A.; Setiawan, A.; Permanasari, A. Development of Augmented Reality in the Basic Physics Practicum Module. *J. Phys. Conf. Ser.* **2020**, *1521*, 022003. [CrossRef]
331. Swandi, A.; Amin, B.D.; Viridi, S.; Eljabbar, F.D. Harnessing technology-enabled active learning simulations (TEALSim) on modern physics concept. *J. Phys. Conf. Ser.* **2020**, *1521*, 022004. [CrossRef]
332. Saputra, H.; Suhandi, A.; Setiawan, A.; Permanasari, A.; Putra, R.A. Real-Time Data Acquisition of Dynamic Moving Objects. *J. Phys. Conf. Ser.* **2021**, *1806*, 012046. [CrossRef]
333. Akhlis, I.; Syaifurrozaq, M.; Marwoto, P.; Iswari, R. The determination of fluid viscosity using tracker-assisted falling ball viscosimeter. *J. Phys. Conf. Ser.* **2020**, *1567*, 042102. [CrossRef]
334. Susilawati, S.; Satriawan, M.; Rizal, R.; Sutarno, S. Fluid Experiment Design Using Video Tracker and Ultrasonic Sensor Devices to Improve Understanding of Viscosity Concept. *J. Phys. Conf. Ser.* **2020**, *1521*, 022039. [CrossRef]
335. Fahrunnisa, S.A.; Rismawati, Y.; Sinaga, P.; Rusdiana, D. Experiments of the law of conservation of mechanical energy using video tracker in high school learning. *J. Phys. Conf. Ser.* **2021**, *1806*, 012035. [CrossRef]
336. Yakob, M.; Wahyuni, A.; Saputra, H.; Putra, R.A.; Mustika, D. Development of measuring instrument based on microcontroller for physics laboratory. *J. Phys. Conf. Ser.* **2020**, *1521*, 022028. [CrossRef]
337. Anissofira, A.; Latief, F.D.E.; Kholida, L.; Sinaga, P. Newton's Cradle Experiment Using Video Tracking Analysis with Multiple Representation Approach. *J. Phys. Conf. Ser.* **2017**, *895*, 012107. [CrossRef]
338. Iradat, R.D.; Alatas, F. The Implementation of Problem-Solving Based Laboratory Activities to Teach the Concept of Simple Harmonic Motion in Senior High School. *J. Phys. Conf. Ser.* **2017**, *895*, 12014. [CrossRef]
339. Susanti, D.; Nilawati, W.; Fitri, U.R.; Kurniawati, H. The contribution of physics media laboratory management towards physics education courses. *J. Phys. Conf. Ser.* **2020**, *1521*, 022031. [CrossRef]
340. Efron, B.; Stein, C. The Jackknife Estimate of Variance. *Ann. Stat.* **1981**, *9*, 586–596. [CrossRef]
341. Houston, S. Lessons of COVID-19: Virtual conferences. *J. Exp. Med.* **2020**, *217*, e20201467. [CrossRef]
342. Hardiono, H.; Umar, M.A.; Hidyantari, E. Zonation System Policy Implementation in the Admission of New Students in the City of Surabaya, East Java, Indonesia. *Int. J. Bus. Manag.* **2020**, *8*, 12. [CrossRef]
343. Purba, M.; Saad, M.Y.; Falah, M. (Eds.) *Center of Curriculum and Instruction Council of Educational Standard Curriculum and Assessment Academic Manuscript: Principles of Developing Differentiated Instruction for Flexible Curriculum as Independent Learner Manifestation*; Ministry of Education Culture Research and Technology: Jakarta, Indonesia, 2021.
344. Santoso, P.H.; Istiyono, E.; Haryanto, H. Physics Teachers' Perceptions about Their Judgments within Differentiated Learning Environments: A Case for the Implementation of Technology. *Educ. Sci.* **2022**, *12*, 582. [CrossRef]

data

MDPI

Article

Advances in Contextual Action Recognition: Automatic Cheating Detection Using Machine Learning Techniques

Fairouz Hussein [1,*], Ayat Al-Ahmad [2], Subhieh El-Salhi [1], Esra'a Alshdaifat [1] and Mo'taz Al-Hami [1]

1 Department of Computer Information System, Faculty of Prince Al-Hussein Bin Abdallah II for Information Technology, The Hashemite University, P.O. Box 330127, Zarqa 13133, Jordan
2 Department of Computer Science and Applications, Faculty of Prince Al-Hussein Bin Abdallah II for Information Technology, The Hashemite University, P.O. Box 330127, Zarqa 13133, Jordan
* Correspondence: fairouzf@hu.edu.jo; Tel.: +962-791329214

Abstract: Teaching and exam proctoring represent key pillars of the education system. Human proctoring, which involves visually monitoring examinees throughout exams, is an important part of assessing the academic process. The capacity to proctor examinations is a critical component of educational scalability. However, such approaches are time-consuming and expensive. In this paper, we present a new framework for the learning and classification of cheating video sequences. This kind of study aids in the early detection of students' cheating. Furthermore, we introduce a new dataset, "actions of student cheating in paper-based exams". The dataset consists of suspicious actions in an exam environment. Five classes of cheating were performed by eight different actors. Each pair of subjects conducted five distinct cheating activities. To evaluate the performance of the proposed framework, we conducted experiments on action recognition tasks at the frame level using five types of well-known features. The findings from the experiments on the framework were impressive and substantial.

Keywords: action recognition; machine learning; cheating; computer vision; feature extraction; video surveillance

Citation: Hussein, F.; Al-Ahmad, A.; El-Salhi, S.; Alshdaifat, E.; Al-Hami, M. Advances in Contextual Action Recognition: Automatic Cheating Detection Using Machine Learning Techniques. *Data* **2022**, *7*, 122. https://doi.org/10.3390/data7090122

Academic Editors: Antonio Sarasa Cabezuelo and Ramón González del Campo Rodríguez Barbero

Received: 1 August 2022
Accepted: 29 August 2022
Published: 31 August 2022

1. Introduction

Interest in monitoring examinations and their mechanisms is increasing. Universities and academic institutions around the world are racing to obtain the latest technologies to monitor cheating in exam halls and secure a cheat-free environment. Typically, to ensure the management of examinations and detect cheating in exams, professional proctors are employed to supervise the entire examination process. In conjunction with the change in the examination control system worldwide due to COVID-19, all universities and institutes are now seeking to work with an electronic mechanism to monitor paper and electronic exams in order to provide safe and secure exams. They are also keen to use the latest mechanisms to detect cheating methods in exams. This is what universities and academic institutes around the world have been planning in recent years, but COVID-19 has definitely sped up their schedule. There is no doubt that cheating is a dangerous phenomenon and disgraceful behavior. Exam cheating is a concern in the educational industry. For this purpose, we focus on automatic cheating detection in exams, as many teachers and educators complain about the spread of cheating and failure of detection methods. Cheating, in fact, has begun to spread not only at the university level, but also at the secondary and primary levels. Action recognition in videos has been a fruitful topic in computer vision in recent years. Its significance is demonstrated in many diverse applications, including remote sensing applications, video surveillance, video recovery, human–computer interactions, sports video analysis, home intelligence, and feature extraction. Action recognition is a challenging field due to the inherent noisy nature of interpretations captured by sensors, which are frequently subject to viewpoint occlusion, scaling, illumination, cluttered background, camera motion, variation, and brightness. The importance of action recognition is substantiated in machine learning and data mining applications through the use of eligible metrics for choosing features and structure in these applications. The action recognition task is usually classified into two main categories: long-range recognition and short-range recognition. The former, long-range recognition, focuses on videos that span more than a minute. From this, it infers the future action based on the current action. The latter, short-range recognition, focuses on short-duration video sequences that consist of just a few seconds, such as video sequences in MSR DailyActivity and MSR-II [1]. The objective of this work is to infer the current action labels founded upon temporally unfinished video sequences. In this work, we present a comprehensive framework to detect and classify the strange actions and behaviors that occur in exam halls and lead to cheating. This is achieved by examining the exam by video and observing the students through the camera. The acquired model is optimized through renowned feature extraction. Another main contribution of this study is presenting a novel dataset on exam cheating. We generated and compiled the dataset ourselves because there is no open source dataset for identifying cheating in paper tests. The dataset was created to depict actions that students could take during a paper-based exam to allow them to cheat. It includes the most common cheating methods, such as exchanging exam papers, looking at another student's exam paper, using a cheat sheet, using a cellular device, and not cheating. The following is the order in which the manuscript was written. The sections "Introduction" and "Related Works" contain the introduction and literature review, respectively. The detailed description of the dataset and how the dataset was acquired is explained in Section 3. The key terms and the feature extractions of the proposed method are discussed in Section 4. Section 5 introduces the results and discusses the experiments in detail. Finally, the conclusion and an outlook are presented in Section 6.

2. Related Works

The significance of recognizing a human action from a video containing a complete action execution is dramatically increasing. The basic steps of action recognition are the preprocessing of raw data, feature extraction and training, and classification [2]. The work in [3] presented a survey of popular algorithms, existing models, popular action databases, technical difficulties, and evolution protocols for action recognition and prediction from videos, which represent the mainstay for real-world applications such as autonomous driving vehicles, video retrievals, etc. Deep learning algorithms and sensors embedded within smartphones and smartwatches were exploited in [4] to recognize eight human activities such as walking, jogging, sitting in a car, etc. The results of the study showed that a combination of data from wrist and pocket sensors can be used to accurately recognize many human activities. In [5], the authors developed techniques to control home appliances using multimodal interaction such as speech, gestures, and smartphone applications. The accuracy of control home appliances using gesture action was 79.25%. For few-shot action recognition, the researchers in [6] suggested a temporal-relation cross-transformation novel approach (TRX). The contribution was the construction of class prototypes using the CrossTransformer attention mechanism. The method proposed by [7] utilizes convolutional neural networks paired with temporal layers for video sequence classification tasks. The researchers in [8] introduced the Action for Cooking Eggs dataset (ACE). The ACE dataset contains activities that occurred in a kitchen, and action label and action recognition methods for analyzing scene contexts were provided for each frame. The use of Kinect devices improves the effectiveness of the application with an in-depth video for intelligent monitoring.

Image processing is still in its infancy, and requires many manual inputs to provide computers with the instructions they need to access the result. These computers were programmed to recognize images [9]. Many studies concentrate on tackling cheating action recognition and all aspects related to it [10]. Ref. [11] organized eight online exam control procedures to detect cheating without employing human proctors or robotic proctors. The essential reasons for cheating actions were investigated by [12]. They found that the most influential factors are the papers exchanged and the environment in which the exam was held. However, Ref. [13] realized the danger of online exams with the tremendous development of technology, allowing for students to master cheating. Weka is used as a tool to identify student behavior that can be classified as cyber-cheating. Ref. [14] introduced computational methods involving a support vector machine (SVM) and text-mining to detect plagiarism. The used computational methods succeed with an accuracy and precision above 90% in determining the original author of the submitted document. Data-mining algorithms, hierarchical clustering, and dendrogram trees have been used to detect patterns in multiple-choice online exam responses that indicate cheating during an exam [15]. Human proctoring is the most prevalent methodology to control cheating in exams. The authors in [16] presented a multimedia analytical system for online exam proctoring. The system is composed of two inexpensive cameras and one microphone. The system's results hinted at future robust behavior-recognition educational applications. The work developed by [17] offered a system that functioned by capturing the data regarding head pose estimates and eye gaze using an internet connection and webcam. The visual focus of attention system (VFOA) was implemented using a hybrid classifier approach and machine learning to classify the students' actions as either malpractice or a momentary lapse in concentration. The COVID-19 pandemic imposed a rapidly invented system to prevent fraud during remote online exams [18]. This took advantage of CNN-based technologies and a new method to provide software that guaranteed more protection during e-exams. This technology was used during the COVID-19 pandemic and was recommended by the majority of governments around the world. Ref. [19] collected sensor data from the iPhone 7's accelerometer and gyroscope during movements, and machine learning was suggested as a candidate for detecting cheat behaviors in physical activities. The work offered by [20] proposed a framework based on deep learning to distinguish suspicious activities during exams held at halls. The proposed model was tested using the CIFAR-100 dataset. The developed system in [21] utilized 3D convolutional neural networks (3D CNN) for image recognition and processing. The system aims to monitor movements and gestures during exams. A recent study [22] reviewed 58 publications about online exams published from 2010 to 2021. The comprehensive review is a very useful resource to obtain an understanding of cheating mitigation, detection, and prevention for educators and academic workers. In the literature, the objectives for preventing and detecting cheating varied, including: (1) strengthening the morality and ethics of students; (2) limiting the possibilities of cheating, e.g., by assessment environment design optimization; and (3) detecting the students caught cheating. However, such approaches are time-consuming and expensive. To fill the gaps in the literature, this study proposes a new framework for the early detection of students' cheating practiced on exams.

3. Data Preparation and Acquisition

One of the main contribution of this study is providing a dataset that will soon be available for public use. Since there is no open source dataset related to detecting cheating in paper exams, we designed and prepared the dataset ourselves. We designed the dataset to contain actions that students may perform during the paper-based exam that will enable them to cheat. It covers most cheating techniques, including: exchanging exam papers, looking at another student's exam paper, using cheat sheets, using cellular devices, and not cheating. Figure 1 depicts several activity classes. A Canon 70D sensor camera was used to capture scenes. The scenes were captured in a classroom in the information technology faculty at the Hashemite University. The sensor recorded 24 frames per second, and the image size was 1920 × 1080 pixels. This period is very appropriate to determine the actions and not to ignore any movement, even if it is simple. The Canon sensor also captured the hand area of a subject. The distance between the sensor and the recorded scene was approximately 3 m. Video clips were grouped into five action types, as shown in Figure 1.

The presented dataset is a challenging one, as many activities appear very to be similar and offer actions that do not depend only on the movement of the body. For example, additional information such as "using cheat sheet" or "use of cellular device" should be taken into account to make a final decision on action recognition. Therefore, it is important to focus not only on the movement of the body but also the adjacent objects. Our dataset consisted of five classes. The total number of video sequences was 37, and the average number of images in each class was 1650. Table 1 shows the number of sequences and frames per class. For action recognition, not all frames are equally crucial; only a few are critical. Therefore, we asked annotators to select a subset of 300 images from each class such that they best depict the class. Overall, we recorded eight unique subjects: four female students and four male students. Each pair of subjects conducted five distinct cheating activities, that is, 1000 images for training were available for each class. In addition, 500 images were also captured as testing images for each class.

Table 1. Details of the actions of the student cheating dataset.

Action	No. of Sequences	No. of Frames
1 Use of cellular device	13	3192
2 Exchange exam paper	4	744
3 Looking at another student's exam paper	8	1734
4 Using cheats sheet	8	1626
5 Not cheating	14	954

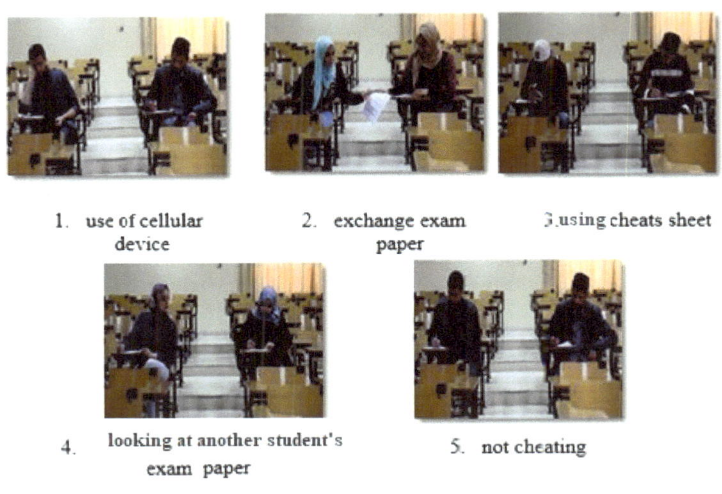

Figure 1. Example shots of each class.

Our task is to classify five kinds of exam cheating actions at frame-level, including: exchanging exam papers, looking at another student's exam paper, using cheat sheets, using cellular devices, and not cheating. It is an attractive dataset since most of the classes involve human–object interaction and share the same body movements.

4. Proposed Method

The proposed work is being developed for a computer vision-based system. The goal of this work is to create a multimedia analysis system that can detect and classify various actions indicative of cheating during an exam. The model includes scaling all of the frames in the dataset and the extraction of five renowned features. For each type of feature, a visual vocabulary codebook is created with different-sized words to encode the visual occurrences in each frame. Finally, a support vector machine is used to classify the specified features. The proposed approach proves its effectiveness using the proposed dataset.

4.1. Definition of Key Terms

In our research, we want to infer the class label y for each frame in the video. More formally, a video V is represented by a set of frames $V = x_1, x_2, \ldots, x_T$, where x_t is an element from some input domain X (e.g., a video frame) and T is the length of a video sequence. Suppose we are given set of N samples (x_i, y_i), $i = 1, \ldots, N$, such that x_i is the feature vector of the i-th sample and y_i is its class label that falls from some discrete set of classes Y. The task is to produce a function F (classifier) that will work well on unseen samples. Mathematically, the frame label y is selected to maximize the scoring function F:

$$Y = argmax_y F(V) \tag{1}$$

Here, in Equation (1), let V represent the space of all possible inputs and y represent the set of identifiable actions such as "using cheats sheet", "use of cellular device", "no cheating", etc. $F(V)$ is a function that measures how well a sequence is presented. The task is to assign a class label Y at frame level. At test time, the maximizer function $F : X\text{->}Y$ assigns a predictive label to the real vector space x. To find F, we used a multiclass Support Vector Machine (SVM) classifier [23]. This kind of classification is used in many action recognition applications. The formulation to solve multi-class SVM can be carried out by building (assuming Y classes) $Y(Y-1)/2$ multiple binary SVM classification problems. The objective of SVM is to learn the optimal separating hyperplane w, which can be found by:

$$argmin \frac{1}{2}||w||^2 + C\sum_{n=1}^{N} \xi_i \tag{2}$$

For each sample, one slack variable ξ_i is introduced to measure the loss of misclassification. The upper bound on the empirical risk is measured by the summation of the slack variables on the training set. For general purposes, a non-differentiable regularization parameter C is introduced to equilibrium the trade-off between complexity and loss. For example, we are given video sequences of "Exchange exam paper" and "no cheating"; each sequence is represented by frames that are considered our measurements and we want to correctly classify an unseen frame as either of these two classes. Each frame is digitized as 1920 × 1080 pixels, so we have measurement vectors $\xi_i \in R_d$, where d = 2,073,600. The positive label could indicate the "Exchange exam paper" class, and the negative label may indicate the "no cheating" class. Then, a new frame is given, which we want to classify: is it an "Exchange exam paper" or a "no cheating"?

4.2. Feature Extraction

Feature extraction is a kind of dimensionality reduction that professionally identifies informative parts of an image as a compressed feature vector. It is recommended to adapt this technique to large images to reduce processing time during tasks such as image retrieval and matching. In our experiments, to evaluate the effectiveness of the proposed method, we extracted five well-known features that are described as follows:

- BRISK: For each frame, we extracted the Binary Robust Invariant Scalable Key-points (BRISK) multi-scale corner features [24]. BRISK is a scale-invariant and rotation-invariant feature point detection and description technique. The BRISK features contain information about points and objects detected in a 2D gray-scale input image. An example of the detected key-points in the "use a cellular device" class is shown in Figure 2. Brisk accomplishes rotation in-variance by attempting to rotate the sample pattern by the measured orientation of the key-points. For clarity, the radials of the circles represent the orientation of the detected key-points while their size represents their scale. In our experiments, to extract BRISK features, we set the scale to 12 and specified the minimum accepted quality of corners as 10% within the designated region of interest (rectangular region for the detected corner). The minimum accepted quality of corners denotes a fraction of the maximum corner measured value in the frame. Note that increasing this value will remove inaccurate corners.

- MSER: We extracted MSER features from the proposed dataset. The maximally stable extremal regions (MSER) technique was used to extract co-variant regions from images [25]. The word "extremal" means that all pixels within a certain region have a higher or lower intensity (brightness) than those outside their boundaries. This process is achieved by arranging the pixels in ascending order according to their intensity and then assigning pixels to regions. The region boundaries were specified by applying a series of thresholds, one for each gray-scale level. Almost all the producing regions resembled an ellipse shape. The resulting region descriptors are considered MSER features. For parameters, we set the step size between intensity threshold levels at 2. Increasing this value will return fewer regions. We also considered the vector [30, 14,000] for the size of the region in pixels. The vector $[minimum_area maximum_area]$ allows for the selection of regions whose total pixels are within the vector. An example of the detected keypoints in the "exchange exam paper" class is shown in Figure 2. It depicts MSER regions, which are designated by pixel lists and are kept in the regions object. Figure 2 displays centroids and ellipses that fit into the MSER regions.
- HOG: The Histogram of Oriented Gradient is one of the most famous feature-extraction algorithms for object detection, proposed by [26]. It extracts features from a region of interest in the frame or from all locations in the frame. The shape of objects in the region is captured by collecting information about gradients. The image is divided into cells, and each group (grid) of adjacent cells forms spatial regions called blocks. The block is the foundation for the normalization and grouping of histograms. The cell is represented by angular bins according to the gradient orientation. Each pixel in the cell participates in a weighted gradient to its corresponding bin; this means that each cell's pixel polls for a gradient bin with a vote proportional to the gradient amount at that pixel (e.g., if a pixel has a gradient orientation of 85 degrees, it will poll with a weighted gradient of 0.9 for the 85-to-95 degree bin and a weighted gradient of 0.9 for the 75-to-85 degree bin). In the experiments, we extract HOG features from blocks specified by [16, 16] cells and 9 orientation histogram bins to encode finer orientation details. However, an increasing number of bins increases the length of the feature vector, which then requires more time to access. A close-up of a HOG detection example is shown in Figure 2.
- SURF: Speeded-Up Robust Features (SURF) is a detector–descriptor scheme used in the fields of computer vision and image analysis [27]. The SURF detector finds distinctive interest points in the image (blobs, T-junctions, corners) based on the Hessian detector. The idea behind the Hessian detector is that it searches for strong derivatives in two orthogonal directions, thereby reducing the computational time. The Hessian detector also uses a multiple-scale iterative algorithm to localize the interest points. The SURF descriptor recaps the pixel information within a local neighborhood called "block". The block calculates directional derivatives of the frame's intensity. The SURF descriptor describes features unrelated to the positioning of the camera or the objects [28]. This rotational in-variance property allows for the objects to be accurately identified regardless of their perspectives or their different locations within the frame. The region of interest (ROI) is presented as a vector with the form [x y width height]. As parameters, we set the region size to [1 1$size$(I, 2) $size(I, 1)$], where the [1 1] elements specify the left upper corner of the rectangular region of size $[size(I, 2) size(I, 1)]$. An example of the ROIs in the "using cheat sheet" class is shown in Figure 2.
- SURF&HOG: We used two of the aforementioned features, SURF and HOG, in the extraction process [29]. First, we used the SURF detector to obtain objects that contain information about the interest points in the images We created a regular-spaced grid

of interest point locations over each image. This permitted dense feature extraction. Then, we computed the HOG descriptors centered on the point locations produced by the SURF detector. For clear visualization, we selected 100 points with the strongest metrics. Figure 2 shows the SURF interest points and the HOG descriptors in the "using cheat sheet" class. Bulleted lists look like this:

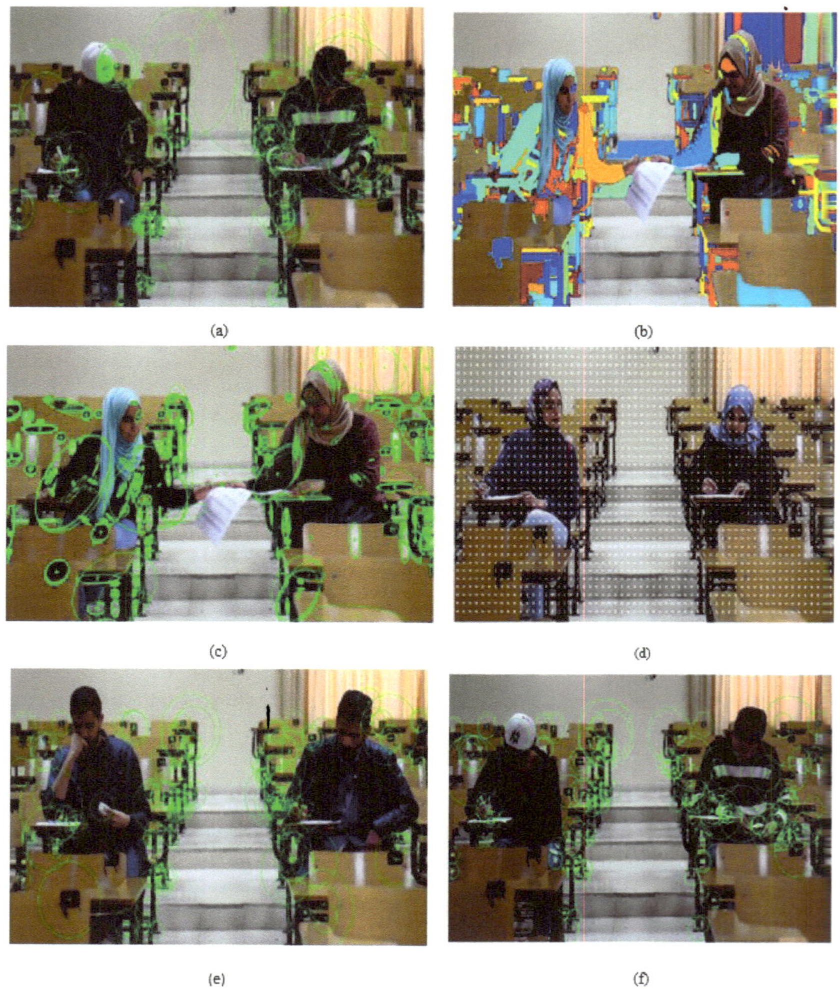

Figure 2. Here some sampling patterns of the (**a**) BRISK features, (**b**) MSER regions, (**c**) MSER ellipses and centroids, (**d**) HOG blocks around the strongest corners, (**e**) SURF locations of interest, and (**f**) SURF detectors and HOG descriptors.

5. Experiments

To evaluate the performance of the proposed method, we conducted experiments on action recognition tasks in the proposed dataset at the frame level using five kinds of well-known features. The dataset was made up of a set of short video sequences representing exam cheating actions. The dataset included a total of 37 sequences at a resolution of 1920 $\times$ 1080 pixels. Due to some issues with feature extraction in terms of the high dimensionality of the feature vector, we cropped the frames from the sides to 960 × 540 pixels without affecting their contents or affecting the main objective of the classification task. We prepared training and validation frame sets. Since the frame sets contained an unequal number of frames per action, we adjusted this so that the number of frames in the training set was balanced. Note that each action set has exactly the same number of images. We separated the frames of classes into training and validation data. We chose 30% of the frames from each class for the training data and the remainder and 70%, for the validation data, and randomized the fragments to avoid biasing the results. Note that this ratio is not easy and is a challenge in the field of classification.

For each of the features listed in Section 4.2, we created a visual vocabulary code-book by using the bag of words technique. Bag of words (BOW) is a natural language processing technique adapted to computer vision. Additionally, the bag of words technique offers an encoded method to count the visual vocabulary occurrences in an image. BOW produces a histogram that becomes a reduced representation of an image. The vocabularies are constructed by reducing the number of features through a quantization of feature space using K-means clustering. In our experiments, to establish the code-book, an unsupervised learning clustering K-mean is used with $k = 400, 500, 600, 700$, where the clusters' centers are characterized as the video's vocabulary.

Tables 2 and 3 show the classification performance of the validation data for each class, with different types of features and different values for vocabulary. On the one hand, in these experiments, we used multiple vocabulary (k) values for each type of feature. It is good to note that the change in the number of vocabulary significantly affected the classification performance. Perhaps there are other vocabulary values that may increase accuracy, but this is beyond the scope of this research. In short, the experiments perform best in this classification task by leveraging SURF descriptors when the vocabulary size was 500. Additionally, Figure 3 displays a comparison of visual word occurrences using $k = 400$ and $k = 500$. On the other hand, based on the classification performance, we can categorize the accuracy of the cheat classes into four categories, and illustrate the results in Tables 2 and 3.

Table 2. Accuracy of classifying the validation dataset using BRISK and HOG features with multiple values for vocabularies.

Features	BRISK				HOG			
Vocabulary	400	500	600	700	400	500	600	700
1 Use of cellular device	65%	86%	69%	69%	80%	69%	51%	67%
2 Exchange exam paper	63%	84%	86%	92%	69%	94%	86%	88%
3 looking at another student's exam paper	57%	73%	78%	94%	75%	80%	92%	94%
4 Using cheats sheet	84%	55%	84%	80%	80%	80%	82%	88%
5 Not cheating	100%	98%	100%	96%	98%	86%	98%	100%
Average Accuracy	74%	79%	84%	86%	80%	82%	82%	87%

Table 3. Accuracy of classifying the validation dataset using MSER, SURF, and SURF&HOG features with multiple values for vocabularies.

Features	MSER				SURF				SURF & HOG			
Vocabulary	400	500	600	700	400	500	600	700	400	500	600	700
1 Use of cellular device	75%	73%	73%	92%	65%	90%	82%	69%	61%	75%	69%	67%
2 Exchange exam paper	82%	98%	94%	96%	75%	96%	73%	75%	92%	84%	94%	86%
3 looking at another student's exam paper	98%	78%	78%	84%	75%	86%	86%	82%	67%	98%	94%	92%
4 Using cheats sheet	94%	67%	78%	80%	73%	82%	94%	78%	82%	82%	90%	100%
5 Not cheating	98%	96%	100%	76%	94%	98%	100%	90%	100%	100%	96%	76%
Average Accuracy	89%	82%	85%	86%	76%	91%	87%	79%	80%	88%	89%	84%

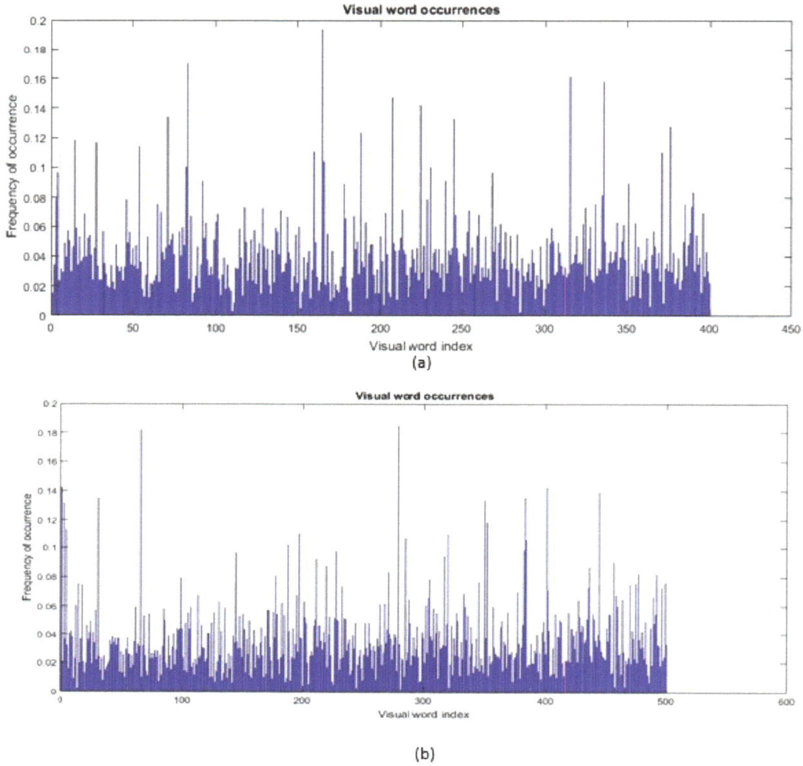

Figure 3. The comparison of visual word occurrences using SURF features at k = 400 (**a**) and k = 500 (**b**).

First, for the classification of the (looking at another student's exam paper, using cheat sheet) classes, the accuracy ranged from 86% to 98% for the "look at the student paper", and 84% to 100% for the "use cheat cheat" class. This is because the classes contain extremely varied kinds of cheating, which lead to huge variations in the feature space.

Second, for the classification of the "exchange exam paper," the accuracy ranged from 92% to 98%. The classifier maintained high accuracy despite choosing varying vocabulary values from different features. These accuracy values are considered reasonably high and are welcome in the classification world. Typically, the "exchange exam paper" class is triggered by specific object interactions in specific scene settings. As a result, it must include not only actions but also the interpretation of objects, situations, and their temporal arrangements with actions, as this knowledge might provide a valuable indication as to "what's going on now".

Third, when classifying the "using a cellular device" class, the accuracy varied between 75% and 92%. The results were not encouraging, and this could be for several reasons, including using a phone of a dark color, the same color as men's clothing; phones are also different shapes and sizes, which requires the system to be trained enough to be able to distinguish and classify them.

Fourth, in the classification of the "not cheating" class, we note that all the selected features were able to classify frames with a very encouraging accuracy of 100%. This is expected: classifying a class that contains very simple movements without interacting with objects is considered a difficult task in the classification process. From this, we conclude that the results are better for the classification of non-cheating than for the classification of cheating.

Figure 4 highlights the best results. The results were achieved with different features. Note that choosing various features does not significantly reduce the recognition performance. Given the results shown in Figure 4, we were looking at the types of features from which the classifier was able to infer the best results. The results obtained from BRISK and HOG features were reasonable. For the BRISK features, the best was 94%, for the "looking at another student's exam paper" class, and the lowest was 69%, for the "use cellular device" class. For the HOG features, the best was 100%, for the "not cheating" class, while the lowest was, again, 67% for the "use cellular device" class. The average accuracy when classifying the validation dataset was 86% and 87% for BRISK and HOG, respectively. There may be a good opportunity to improve these results by increasing the number of detected keypoints in the descriptors, combining the BRISK and HOG descriptors with other detectors, or just tuning some of the parameters. There were encouraging results when using the MSER and SURF and HOG features. An identical average accuracy of 89% was obtained from both features. The MSER features distinguished "looking at another student's exam paper" and "not cheating" with an accuracy of 98%, and "use cheat sheet" with an accuracy of 94%. This is not strange, because the detected regions are well-defined by the intensity function. This leads to the regions having many key properties that make them valuable. Additionally, the significant results obtained by SURF and HOG features for the classification of "no cheating", "looking at another student's exam paper" and "exchange exam paper" cannot be avoided. HOG demonstrated its positive effects in detecting texture information and the edge of the image. However, SURF is the fastest, and comparable to SIFT in terms of performance.

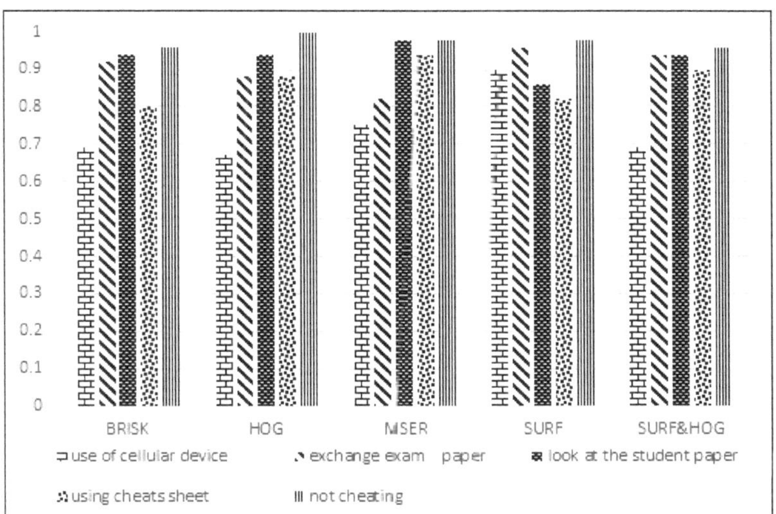

Figure 4. The accuracy obtained when classifying the validation dataset using different features.

Typically, the results obtained from the SURF features are remarkable. The average accuracy was 91%; see Table 4. It had the distinct ability to distinguish between the features of "using a cellular device" with an accuracy of up to 90%. This notable accuracy could not be reached by the other features most of the time. The SURF technique is well-known for its quick computation of operators utilizing box filters. Figure 5 shows a comparison of the different correlations between the five cheating classes using SURF features.

Table 4. The average accuracy of classifying the validation dataset.

Features	Accuracy
BRISK	86%
HOG	87%
MSER	89%
SURF	91%
SURF&HOG	89%

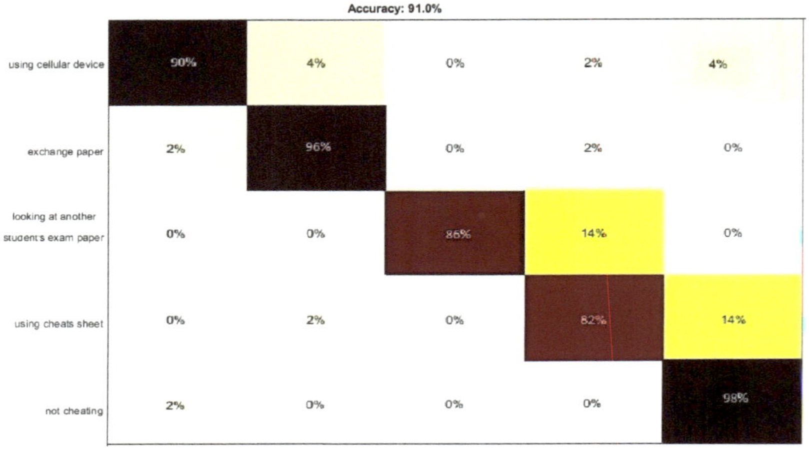

Figure 5. Comparing the accuracy of different correlations between the five cheating classes using SURF features.

6. Conclusions

In this research, we created a cheating video sequence dataset that detects cheating actions in paper-based exams. The dataset contains very challenging video sequences, since many activities appear to be quite similar and include actions that are not solely dependent on body movement. The results from the experiments on the framework were impressive and substantial. The cheating recognition model correctly recognized the cheating actions with an accuracy of 91%. As the results of the work were encouraging and distinct, there are several ways in which our work might be enhanced. For example, more complex algorithms could be used, such as deep learning for learning and more appropriate features and classifiers for classification. The system can also be expanded in the future to detect cheating in online exams with more than one subject. Moreover, the proposed dataset was captured in one country, and the examination environment is different in every country. Therefore, the dataset can be expanded by recording more videos and taking more dynamic factors such as: different environments; lighting (dim, normal, bright); camera angle (low angle, face-level, on-looking, top-down); presence of various motions; blurriness; resolution (SD, HD, 4K); etc.

Author Contributions: F.H. proposed the research framework, conceptualization, methodology, formal analysis and data curation. A.A.-A. worked on formal analysis and writing—original draft preparation. S.E.-S., E.A. and M.A.-H. worked on the writing—review, supervision and editing. All authors have read and agreed to the published version of the manuscript.

Funding: This research received no external funding.

Institutional Review Board Statement: Not applicable.

Informed Consent Statement: Informed consent was obtained from all individual participants included in this study.

Data Availability Statement: Data are available from the authors upon request.

Acknowledgments: The authors would like to thank the Hashemite University for its encouragement.

Conflicts of Interest: The authors declare no conflict of interest.

References

1. Oreifej, O.; Liu, Z. Hon4d: Histogram of oriented 4d normals for activity recognition from depth sequences. In Proceedings of the IEEE Conference on Computer Vision and Pattern Recognition, Portland, OR, USA, 23–28 June 2013; pp. 716–723.
2. Alshdaifat, E.; Alshdaifat, D.; Alsarhan, A.; Hussein, F.; El-Salhi, S.M.F.S. The effect of preprocessing techniques, applied to numeric features, on classification algorithms' performance. *Data* **2021**, *6*, 11. [CrossRef]
3. Kong, Y.; Fu, Y. Human action recognition and prediction: A survey. *Int. J. Comput. Vis.* **2022**, *130*, 1366–1401. [CrossRef]
4. Alam, A.; Das, A.; Tasjid, M.; Al Marouf, A. Leveraging Sensor Fusion and Sensor-Body Position for Activity Recognition for Wearable Mobile Technologies. *Int. J. Interact. Mob. Technol.* **2021**, *15*, 141–155. [CrossRef]
5. Fakhrurroja, H.; Machbub, C.; Prihatmanto, A.S. Multimodal Interaction System for Home Appliances Control. *Int. J. Interact. Mob. Technol.* **2020**, *14*, 44. [CrossRef]
6. Perrett, T.; Masullo, A.; Burghardt, T.; Mirmehdi, M.; Damen, D. Temporal-relational crosstransformers for few-shot action recognition. In Proceedings of the IEEE/CVF Conference on Computer Vision and Pattern Recognition, Nashville, TN, USA, 20–25 June 2021; pp. 475–484.
7. Fernando, B.; Gould, S. Learning end-to-end video classification with rank-pooling. In Proceedings of the International Conference on Machine Learning, PMLR, New York, NY, USA, 19–24 June 2016; pp. 1187–1196.
8. Shimada, A.; Kondo, K.; Deguchi, D.; Morin, G.; Stern, H. Kitchen scene context based gesture recognition: A contest in ICPR2012. In *International Workshop on Depth Image Analysis and Applications*; Springer: Berlin/Heidelberg, Germany, 2012; pp. 168–185.
9. Hussein, F.; Piccardi, M. V-JAUNE: A framework for joint action recognition and video summarization. *ACM Trans. Multimed. Comput. Commun. Appl. (TOMM)* **2017**, *13*, 1–19. [CrossRef]
10. Liu, X.; Li, Y.; Li, Y.; Yu, S.; Tian, C. The study on human action recognition with depth video for intelligent monitoring. In Proceedings of the 2019 Chinese Control And Decision Conference (CCDC), Nanchang, China, 3–5 June 2019; pp. 5702–5706.
11. Cluskey, G., Jr.; Ehlen, C.R.; Raiborn, M.H. Thwarting online exam cheating without proctor supervision. *J. Acad. Bus. Ethics* **2011**, *4*, 1–7.
12. Wang, J.; Tong, Y.; Ling, M.; Zhang, A.; Hao, L.; Li, X. Analysis on test cheating and its solutions based on extenics and information technology. *Procedia Comput. Sci.* **2015**, *55*, 1009–1014. [CrossRef]
13. Hernándeza, J.A.; Ochoab, A.; Muñozd, J.; Burlaka, G. Detecting cheats in online student assessments using Data Mining. In Proceedings of the Conference on Data Mining | DMIN, Las Vegas, NV, USA, 26–29 June 2006; Volume 6, p. 205.
14. Diederich, J. Computational methods to detect plagiarism in assessment. In Proceedings of the 2006 7th International Conference on Information Technology Based Higher Education and Training, Ultimo, Australia, 10–13 July 2006; pp. 147–154.
15. Chen, M. Detect multiple choice exam cheating pattern by applying multivariate statistics. In Proceedings of the International Conference on Industrial Engineering and Operations Management, Bogota, Colombia, 25–26 October 2017; Volume 2017, pp. 173–181.
16. Atoum, Y.; Chen, L.; Liu, A.X.; Hsu, S.D.; Liu, X. Automated online exam proctoring. *IEEE Trans. Multimed.* **2017**, *19*, 1609–1624. [CrossRef]
17. Indi, C.S.; Pritham, K.; Acharya, V.; Prakasha, K. Detection of Malpractice in E-exams by Head Pose and Gaze Estimation. *Int. J. Emerg. Technol. Learn.* **2021**, *16*, 47. [CrossRef]
18. Sharma, N.K.; Gautam, D.K.; Rathore, S.; Khan, M. CNN implementation for detect cheating in online exams during COVID-19 pandemic: A CVRU perspective. *Mater. Today Proc.* **2021**. [CrossRef]
19. Kock, E.; Sarwari, Y.; Russo, N.; Johnsson, M. Identifying cheating behaviour with machine learning. In Proceedings of the 2021 Swedish Artificial Intelligence Society Workshop (SAIS), Stockholm, Sweden, 14–15 June 2021; pp. 1–4.
20. Genemo, M.D. Suspicious activity recognition for monitoring cheating in exams. *Proc. Indian Natl. Sci. Acad.* **2022**, *88*, 1–10. [CrossRef]
21. El Kohli, S.; Jannaj, Y.; Maanan, M.; Rhinane, H. Deep Learning: New Approach for Detecting Scholar Exams Fraud. *Int. Arch. Photogramm. Remote Sens. Spat. Inf. Sci.* **2022**, *46*, 103–107. [CrossRef]
22. Noorbehbahani, F.; Mohammadi, A.; Aminazadeh, M. A systematic review of research on cheating in online exams from 2010 to 2021. *Educ. Inf. Technol.* **2022**, *27*, 8413–8460. [CrossRef] [PubMed]
23. Hsu, C.W.; Lin, C.J. A comparison of methods for multiclass support vector machines. *IEEE Trans. Neural Netw.* **2002**, *13*, 415–425. [PubMed]
24. Leutenegger, S.; Chli, M.; Siegwart, R.Y. BRISK: Binary robust invariant scalable keypoints. In Proceedings of the 2011 International Conference on Computer Vision, Barcelona, Spain, 6–13 November 2011; pp. 2548–2555.

25. Matas, J.; Chum, O.; Urban, M.; Pajdla, T. Robust wide-baseline stereo from maximally stable extremal regions. *Image Vis. Comput.* **2004**, *22*, 761–767. [CrossRef]
26. Dalal, N.; Triggs, B. Histograms of oriented gradients for human detection. In Proceedings of the 2005 IEEE Computer Society Conference on Computer Vision and Pattern Recognition (CVPR'05), San Diego, CA, USA, 20–25 June 2005; Volume 1, pp. 886–893.
27. Bay, H.; Tuytelaars, T.; Gool, L.V. Surf: Speeded up robust features. In *European Conference on Computer Vision*; Springer: Berlin/Heidelberg, Germany, 2006; pp. 404–417.
28. Jegham, I.; Khalifa, A.B.; Alouani, I.; Mahjoub, M.A. Safe driving: Driver action recognition using surf keypoints. In Proceedings of the 2018 30th International Conference on Microelectronics (ICM), Sousse, Tunisia, 16–19 December 2018; pp. 60–63.
29. Madan, R.; Agrawal, D.; Kowshik, S.; Maheshwari, H.; Agarwal, S.; Chakravarty, D. Traffic Sign Classification using Hybrid HOG-SURF Features and Convolutional Neural Networks. In Proceedings of the ICPRAM, Prague, Czech Republic, 19–21 February 2019; pp. 613–620.

data

MDPI

Data Descriptor

Student Dataset from Tecnologico de Monterrey in Mexico to Predict Dropout in Higher Education

Joanna Alvarado-Uribe [1,2,*], Paola Mejía-Almada [1], Ana Luisa Masetto Herrera [3], Roland Molontay [4,5], Isabel Hilliger [6], Vinayak Hegde [7], José Enrique Montemayor Gallegos [3], Renato Armando Ramírez Díaz [3] and Hector G. Ceballos [1,2]

1 Institute for the Future of Education, Tecnologico de Monterrey, Monterrey 64849, Mexico
2 School of Engineering and Sciences, Tecnologico de Monterrey, Monterrey 64849, Mexico
3 Analytics and Business Intelligence Department, Tecnologico de Monterrey, Monterrey 64849, Mexico
4 Department of Stochastics, Institute of Mathematics, Budapest University of Technology and Economics, 1111 Budapest, Hungary
5 ELKH-BME Stochastics Research Group, 1111 Budapest, Hungary
6 School of Engineering, Pontificia Universidad Católica de Chile, Santiago 7820436, Chile
7 Department of Computer Science, Mysuru Campus, Amrita Vishwa Vidyapeetham, Mysore 570026, India
* Correspondence: joanna.alvarado@tec.mx

Citation: Alvarado-Uribe, J.; Mejía-Almada, P.; Masetto Herrera, A.L. Molontay, R.; Hilliger, I.; Hegde, V.; Montemayor Gallegos, J.E.; Ramírez Díaz, R.A.; Ceballos, H.G. Student Dataset from Tecnologico de Monterrey in Mexico to Predict Dropout in Higher Education. *Data* 2022, 7, 119. https://doi.org/10.3390/data7090119

Academic Editors: Antonio Sarasa Cabezuelo and Ramón González del Campo Rodríguez Barbero

Received: 7 July 2022
Accepted: 20 August 2022
Published: 25 August 2022

Abstract: High dropout rates and delayed completion in higher education are associated with considerable personal and social costs. In Latin America, 50% of students drop out, and only 50% of the remaining ones graduate on time. Therefore, there is an urgent need to identify students at risk and understand the main factors of dropping out. Together with the emergence of efficient computational methods, the rich data accumulated in educational administrative systems have opened novel approaches to promote student persistence. In order to support research related to preventing student dropout, a dataset has been gathered and curated from Tecnologico de Monterrey students, consisting of 50 variables and 143,326 records. The dataset contains non-identifiable information of 121,584 High School and Undergraduate students belonging to the seven admission cohorts from August–December 2014 to 2020, covering two educational models. The variables included in this dataset consider factors mentioned in the literature, such as sociodemographic and academic information related to the student, as well as institution-specific variables, such as student life. This dataset provides researchers with the opportunity to test different types of models for dropout prediction, so as to inform timely interventions to support at-risk students.

Dataset: https://doi.org/10.57687/FK2/PWJRSJ.

Dataset License: CC0

Keywords: dropout prediction; student attrition; machine learning; educational data mining; learning analytics; educational innovation; higher education

1. Introduction

High dropout rates and delayed completion in higher education are associated with considerable personal and social costs. Dropping out from higher education represents a cost for the government and society, an unnecessary expense for the family, and an experience of failure for the university student [1,2]. Therefore, the early identification of at-risk students and understanding of the main factors of dropping out have recently attracted a great deal of research interest [3–5]. Early detection of at-risk students allows higher education institutions to offer individualized assistance in varied forms, including remedial courses and tutoring sessions to mitigate academic failure.

The rich data accumulated in educational administrative systems together with the emergence of efficient Statistical and Machine Learning methods have opened novel approaches to address the problem of student attrition, generating a new line of research. In the last few years, a high number of predictive analytical educational research papers have been published and Artificial Intelligence-based decision support systems have been developed to assist stakeholders in higher education [6–8]. For example, the application of Learning Analytics in higher education institutions can provide quality and actionable information to implement educational interventions, such as timely support for at-risk students of dropping out [9,10].

Most dropout prediction studies rely on pre-enrollment achievement measures (high school grades, assessment tests) and personal details [11–14]; some also consider first-semester university performance indicators [12,15], such as course grades [16]. On the other hand, other factors were also found to have incremental predictive power on academic performance and retention such as on/off-campus housing [17], socioeconomic status [11,18,19], psychological factors such as coping and emotional intelligence [20], and schooling background of parents [21]. Notwithstanding, ethical principles on the collection and use of educational data should be proposed and applied with the aim of protecting the privacy of students, such as the ethical principle of considering student performance as a dynamic variable [22].

In Latin America, college access grew dramatically in the early 2000s, and particularly for those students from middle and low-income segments [23]. Most of these 'new students' enrolled in new private programs, relying on the recent growth of middle-class family incomes, student loans, and scholarships [24]. Although the coverage expansion of higher education systems was crucial for knowledge production and social mobility, it generated major challenges regarding quality and equity. According to Lemaitre [25], 50% of students drop out and only 50% of the remaining ones graduate on time. Considering that low-income students are the ones at higher risk of dropping out and being disfavoured by disparities in lifetime earnings [26], there is an urgent need to improve higher education quality in the region and reduce dropout rates [23,27]. In this context, data-based strategies are seen as an opportunity to tackle issues related to these problems, such as providing personalized feedback and support to an increasing number of learners [27].

Therefore, in order to support the prediction of student dropout and increase student retention rates, a student dataset has been gathered and curated based on the related work and the retention prediction model developed for Tecnologico de Monterrey within the early alerts program. This program is a project whose purpose is to provide timely and reliable information in the follow-up process to high school and undergraduate students according to their information and their retention indicator. Although retention rates at the institution have increased from 91.2% in High School and 89.9% in Undergraduate in 2014 to 94.5% in High School and 92.1% in Undergraduate in 2020, new or disruptive models are needed to identify all at-risk students in an effective and timely manner. A call for proposals was launched to research and develop solutions based on this dataset using Machine Learning algorithms [28]. According to the proposals received, the dataset was enriched with more variables related to student life and dropout time. Resulting in a dataset of 50 variables and 143,326 records.

The rest of the descriptor is organized as follows. Section 2 provides the context and detail description of the student dataset. Then, Section 3 provides the methodology carried out to collect, preprocess, preserve, and explore the proposed dataset, mentioning the materials and methods used as well as presenting a brief exploratory analysis of the dataset. Finally, Section 4 gives the conclusions.

2. Data Description

The Tecnologico de Monterrey is a university in Mexico made up of 29 campuses and 18 offices around the world. The institution has a total current population of 94,424 students, of which 26,794 are in High School, 60,169 in Undergraduate, and 7461 in Postgraduate programs [29]. In the dataset given through this descriptor, non-identifiable information is provided for 121,584 High School and/or Undergraduate students who have enrolled at Tecnologico de Monterrey. The information corresponds to seven admission cohorts to the institution from 2014 to 2020; that is, August–December 2014 (AD14), August–December 2015 (AD15), August–December 2016 (AD16), August–December 2017 (AD17), August–December 2018 (AD18), August–December 2019 (AD19), and August–December 2020 (AD20).

The dropout rates in the institution have decreased from 8.8% in High School and 10.1% in Undergraduate in 2014 to 5.5% in High School and 7.9% in Undergraduate in 2020. However, in the 2015–2016 period, the dropout rates increased from 7.3% to 7.6% for High School, as well as in the 2018–2019 period from 7.5% to 9.4% for Undergraduate. Therefore, it is necessary to continue researching and developing models and strategies for student retention.

Among the categories of information available in this dataset are:

- Sociodemographic information, such as age, gender, and type of zone to which the student's address belongs.
- Enrollment information, such as program, school, and educational model.
- Academic information related to the student, such as the average of the previous level, the average in the first term or midterm of the first semester, and the number of failed subjects.
- Information associated with scores on admission tests, such as the admission test, standardized English proficiency test, and Mathematics grade.
- Academic history, such as type of school from provenance, national/international student, and relationship with the Tecnologico de Monterrey system.
- Student life, such as participation in sports, cultural, and leadership activities.
- Scholarship and financial aid information, such as type of scholarship, percentage of scholarship, and percentage of scholarship loan.
- Academic information related to the student's parents, such as educational level and whether the parents were students of the Tecnologico de Monterrey.
- Information on the student's retention or dropout in the first year.

Tables 1–3 provide a detailed description of the variables constituting the student dataset.

It is relevant to mention that this student dataset provides information on two educational models implemented at Tecnologico de Monterrey. The previous model, corresponding to the AD14–AD18 generations, is based on the teaching-learning process while the current model called "TEC21 Model", corresponding to the AD19–AD20 generations, is based on challenges and competencies [29]. In this dataset, information on the average obtained in the first term or midterm, the number of subjects failed, and the number of subjects dropped out by the student is only provided for the AD19–AD20 generations. Hence, this data is interesting to analyze from this perspective as well.

In the same way, co-curricular activities related to the integrated learning of students have also evolved in accordance with the new educational model ("TEC21 Model"). The AD14–AD17 generations of students contemplated enrolling in one type of activity or the three categories of activities offered: (1) physical education, (2) cultural diffusion, and (3) student society. For the AD18–AD20 generations, the offer of activities increased since they are now part of the well-rounded education of the student to contribute to the development of transversal skills for all students [30,31]. This evolution is called the LiFE (Leadership and Student Education) program, which goes hand in hand with the TEC21 educational model [31] and is made up of the following categories: athletic or sports activities, art or culture activities, student society activities, life or work mentoring, and wellness activities.

Table 1. Description of the attributes of the student dataset (Part I).

No.	Attribute	Data Type	Description	Values
1	student.id	Integer	Masked enrollment number of the student. There are duplicate student identifiers (IDs) as one identifier may be related to a different educational level: High School or Undergraduate. In addition, there are some student IDs that are repeated three times due to those students have additional information related to different generations.	1-121584
2	generation	String	Unique indicator that denotes the generation to which the student belongs.	AD14, AD15, AD16, AD17, AD18, AD19, AD20
3	educational.model	Binary	Educational model to which the student belongs.	1: TEC21 Model, 0: Previous educational model
4	level	String	Educational level to which the student belongs.	High School, Undergraduate
5	gender	String	Student gender.	Male, Female
6	age	Integer	Student's age.	Range from 13 to 55 years
7	zone.type	String	Description of the type of zone to which the student's address belongs.	Rural, Semiurban, Urban, No information
8	socioeconomic.level	String	Socioeconomic level of the student.	Level 1, Level 2, Level 3, Level 4, Level 5, Level 6, Level 7, No information
9	social.lag	String	It indicates the level of social backwardness at the level of urban areas of the student's address according to the zip code.	Low, Medium, High, No information
10	id.school.origin	String	Masked identifier of the school where the student comes from.	Range from "School 0" to "School 10242".
11	school.cost	String	Classification of the tuition cost of the student's school of origin.	Public, Low cost, Medium cost, Medium-high cost, High cost, Not defined
12	tec.no.tec	String	Indicator that denotes if the student comes from a school that belongs to Tecnologico de Monterrey.	TEC, NO TEC
13	max.degree.parents	String	Highest educational level obtained by the student's parents.	No information, No degree, Undergraduate degree, Master degree, PhD
14	father.education.complete	String	Description of the last educational level completed by the father.	Attended university, but did not graduate; Graduated from elementary or middle school; Graduated from high school; None educational degree; Received master degree; Received PhD; Received technical or commercial degree; Received undergraduate degree; No information
15	father.education.summary	String	Classification of the last educational level completed by the father.	No information, No degree, Undergraduate degree, Master degree, PhD
16	mother.education.complete	String	Description of the last educational level completed by the mother.	Attended university, but did not graduate; Graduated from elementary or middle school; Graduated from high school; None educational degree; Received master degree; Received PhD; Received technical or commercial degree; Received undergraduate degree; No information
17	mother.education.summary	String	Classification of the last educational level completed by the mother.	No information, No degree, Undergraduate degree, Master degree, PhD
18	parents.exatec	String	Indicator that denotes if either of the parents is an exatec (was a student at Tecnologico de Monterrey).	Yes, No, No information
19	father.exatec	String	Indicator that denotes if the student's father is an exatec (was a student at Tecnologico de Monterrey).	Yes, No, No information
20	mother.exatec	String	Indicator that denotes if the student's mother is an exatec (was a student at Tecnologico de Monterrey).	Yes, No, No information
21	first.generation	String	It indicates if the student is the first person in the family to study for a professional career.	Yes, No, No information, Does not apply

Table 2. Description of the attributes of the student dataset (Part II).

No.	Attribute	Data Type	Description	Values
22	school	String	Acronyms of the school to which the student's academic program belongs.	High school, EN = Business School, EMCS = School of Medicine and Health Sciences, EIC = School of Engineering and Sciences, ECSG = School of Social Sciences and Government, EHE = School of Humanities and Education, EAAD = School of Architecture, Art and Design
23	program	String	Acronyms of the academic program to which the student belongs.	The meaning of the acronyms is found in Appendix A
24	region	String	Code of the region to which the campus where the student is enrolled belongs.	RM = Monterrey Region, RO = West Region, RCM = Mexico City Region, RCS = South/Central Region, DR = Regional Development Region
25	foreign	String	Indicator to identify if the student is a foreigner (Yes: Foreigner), if the Mexican student's birthplace is different from the location of the school campus (Yes: National), or if the student belongs to the same location (Local).	Local, Yes: National, Yes: Foreigner
26	PNA	Float	Previous level score (average)	Range from 0 to 100
27	english.evaluation	Integer	Level of English obtained from a standardized test of English language proficiency.	Level 0: No information, Level 1: Beginner, Level 2: Basic, Level 3: Basic, Level 4: Intermediate, Level 5: Intermediate, Level 6: Upper Intermediate, Level 7: Advanced
28	admission.test	Integer and String	Admission test score. There are two scoring scales depending on how the test is applied: (1) Academic Aptitude Test (Prueba de Aptitud Académica-PAA): admission test applied face-to-face for all generations of students before the closure due to the COVID-19 pandemic. The range of scores is from 400 to 1600. (2) Online Aptitude Test (Prueba de Aptitud en Línea-PAL): admission test that, as a consequence of the closure due to COVID-19, is applied online. The range of scores is from 0 to 100.	Ranges from 1 to 100 and from 400 to 1600, Does not apply
29	online.test	Binary	It indicates if the student took the online admission test.	1: Yes, 0: No
30	general.math.eval	Float and String	Mathematics score from the admission test or from the school of origin.	Range from 0 to 100, Does not apply, No information
31	admission.rubric	Integer	Score generated from the student's profile where 50 is outstanding and 0 is average.	Range from 0 to 50
32	scholarship.type	String	Type of scholarship.	Academic talent, Army/Navy scholarship, Child of Professor/Employee/Director, Contingency scholarship, Cultural talent, Entrepreneurial talent, Leaders of Tomorrow Scholarship, Leadership talent, No scholarship, Sports Talent, Traditional
33	scholarship.perc	Integer	Scholarship percentage.	Range from 0 to 100
34	loan.perc	Integer	Percentage of the educational loan.	Range from 0 to 50
35	total.scholarship.loan	Integer	Total percentage of financial support provided to the student for education (scholarship + educational loan).	Range from 0 to 100
36	FTE	Float	It indicates if the student is a full-time student at Tecnologico de Monterrey according to the number of subjects enrolled.	Range from 0.04 to 1.44
37	average.first.period	Float	Average obtained in the first term (five weeks–Undergraduate) or the first midterm (six weeks–High School) of the student's first semester. This data corresponds only to the AD19 and AD20 generations (TEC21 Model).	Range from 0 to 100

Table 3. Description of the attributes of the student dataset (Part III).

No.	Attribute	Data Type	Description	Values
38	failed.subject.first.period	Integer	Number of subjects failed in the first term (five weeks–Undergraduate) or the first midterm (six weeks–High School) of the student's first semester. This data corresponds only to the AD19 and AD20 generations (TEC21 Model).	Range from 0 to 8
39	dropped.subject.first.period	Integer	Number of subjects dropped out in the first term (five weeks–Undergraduate) or the first midterm (six weeks–High School) of the student's first semester. This data corresponds only to the AD19 and AD20 generations (TEC21 Model).	Range from 0 to 9
40	retention	Binary	Value that indicates if the student continues studying at Tecnologico de Monterrey.	1: Retention, 0: Dropout
41	dropout.semester	Integer	Value indicating the semester when the student dropped out. Where 0 = the student continues studying, 1 = the student dropped out during the first semester, 2 = the student did not enroll in the second semester, 3 = the student dropped out during the second semester, and 4 = the student did not enroll in the third semester.	0, 1, 2, 3, 4
42	physical.education	Binary and String	Value that indicates if the student was enrolled in any physical education activities during the first semester. This data corresponds only to the AD14, AD15, AD16, and AD17 generations.	0: No, 1: Yes, Does not apply, No information
43	cultural.diffusion	Binary and String	Value that indicates if the student was enrolled in any cultural diffusion activities during the first semester. This data corresponds only to the AD14, AD15, AD16, and AD17 generations.	0: No, 1: Yes, Does not apply, No information
44	student.society	Binary and String	Value that indicates if the student was enrolled in any student society activities during the first semester. This data corresponds only to the AD14, AD15, AD16, and AD17 generations.	0: No, 1: Yes, Does not apply, No information
45	total.life.activities	Integer and String	Number of LiFE (Leadership and Student Education) activities in which the student was enrolled during the first semester. This data corresponds only to the AD18, AD19, and AD20 generations.	0, 1, 2, 3, 4, 5, Does not apply, No information
46	athletic.sports	Binary and String	Value that indicates if the student was enrolled in any athletic or sports activities during the first semester. This data corresponds only to the AD18, AD19, and AD20 generations.	0: No, 1: Yes, Does not apply, No information
47	art.culture	Binary and String	Value that indicates if the student was enrolled in any artistic or cultural activities during the first semester. This data corresponds only to the AD18, AD19, and AD20 generations.	0: No, 1: Yes, Does not apply, No information
48	student.society.leadership	Binary and String	Value that indicates if the student was enrolled in any student society activities and a leadership program during the first semester. This data corresponds only to the AD18, AD19, and AD20 generations.	0: No, 1: Yes, Does not apply, No information
49	life.work.mentoring	Binary and String	Value that indicates if the student received advice on life and work plans during the first semester. This data corresponds only to the AD18, AD19, and AD20 generations.	0: No, 1: Yes, Does not apply, No information
50	wellness.activities	Binary and String	Value that indicates if the student was enrolled in any integral wellness activities during the first semester. This data corresponds only to the AD18, AD19, and AD20 generations.	0: No, 1: Yes, Does not apply, No information

3. Materials and Methods

The methodology used in this research is based on the Data Life Cycle used in the field of Research Data Management shown in Figure 1. The Data Life Cycle illustrates the research process and its different phases, as well as the stages associated with the data generation, use, and dissemination [32].

Figure 1. Data Life Cycle Diagram, based on [33].

3.1. Data Planning

The first 40 variables shown in Tables 1–3 were defined according to the related work cited in this descriptor, as well as the Analytics and Business Intelligence Department of Tecnologico de Monterrey due to its experience in the early alerts program (student retention). The following nine variables (listed from 41 to 50 in Table 3) related to the student's dropout semester and the student's co-curricular activities were gathered after receiving the proposals of the researchers participating in the call for proposals. The dataset along with its data dictionary were built in Excel files to allow downloading them through the Tecnologico de Monterrey's Data Hub (https://datahub.tec.mx/dataverse/tec (accessed on 24 August 2022)). Taking into account the sensitivity of the data, the dataset will be made available to researchers who request it through the Data Hub.

3.2. Data Collection

The data was extracted in two phases. Firstly, data was collected from the Tecnologico de Monterrey's Data Warehouse by the Analytics and Business Intelligence Department through the SAP BusinessObjects Web Intelligence (WebI) tool. This first dataset includes personal and academic information on Undergraduate and High School students, such as gender, age, tests, schooling background of parents, among others. The variables related to retention and the socioeconomic level of the students were calculated by the same department with the purpose of designing a model to identify students at risk, used in the early alerts program. Secondly, the co-curricular activities of the students from 2014 to 2020 were obtained from the Tecnologico de Monterrey's LiFE Department.

3.3. Data Assurance

For the dataset that was extracted from the WebI tool, the following preprocessing steps were performed:

1. Considering the privacy of students and faculty, it is important to emphasize that the data must be de-identified before it is made available for institutional use and research purposes [22]. Therefore, the student's enrollment identifier (student.id)

and the name of the previous level school (id.school.origin) became non-identifiable values as they represent sensitive information.
2. All records were translated into the English language.
3. An exhaustive exploration was carried out to find inconsistencies in the values of variables 1 to 40 (described in Tables 1–3) and in the relationships among them.
4. Spelling and typographical errors were checked for the categorical values of each variable.
5. Missing values for the variables socioeconomic.level and social.lag were filled in with "No information".
6. The empty values corresponding to admission.test for the Undergraduate level were replaced by "Does not apply" when the variable tec.no.tec has the value "TEC". That is, the student is a graduate of the Tecnologico de Monterrey's High School.
7. The variable dropout.semester was categorized according to the period in which the student dropped out: before or during the semester.
8. The values of the variables scholarship.perc, loan.perc, and total.scholarship.loan were multiplied by 100 to represent a percentage.

3.4. Data Description

The dataset was described in detail in Section 2.

3.5. Data Preservation

This dataset will be available upon request through the Tecnologico de Monterrey's Data Hub repository for its long-term preservation. The metadata was properly described and a specific Digital Object Identifier (DOI) was assigned in order that the data can be easily traceable and correctly cited. This dataset is protected by the Creative Commons Zero (CC0) waiver and is governed by Tecnologico de Monterrey's Terms of Use and a Data Policy.

3.6. Data Discovery

Based on the proposals received by the researchers, information on co-curricular activities and dropout semester were identified as potential data that could be valuable for the student dropout prediction model and were added to the original dataset.

3.7. Data Integration

The first dataset consisting of 40 variables was merged with the co-curricular activities database and semester dropout information based on the variables *student.id* and *generation* to create a single data file. As a result, the final dataset is made up of 50 attributes to test and predict student dropout at the High School and Undergraduate levels.

3.8. Data Analysis

Firstly, a descriptive analysis of dataset variables was performed using the Pandas library version 1.4.3 and the Scikit-learn library version 1.1.2 in Python 3 shown in Tables 4 and 5. Secondly, a data visualization was carried out using Tableau Desktop Professional Edition 2021.4.4.

On the one hand, Table 4 describes the numerical variables of the dataset through their unique, mean, minimum, and maximum values. The identifier of each variable corresponds to the identifier assigned in Tables 1–3. Similarly, the gain information is integrated to demonstrate the dependency between each feature in the dataset and the target variable retention. The information gain was calculated using a mutual information classifier, the values "Does not apply" and "No information" were excluded from the calculation of the statistical variables *admission.test*, *general.math.eval*, and *total.life.activities* since they do not represent numerical values, and the records containing null values were also not considered in the information gain calculation. It is important to remember that for the variables *average.first.period*, *failed.subject.first.period*, and *dropped.subject.first.period* the data is only available for AD19 and AD20.

Table 4. Description of the numerical attributes of the student dataset.

No.	Attribute	Unique	Mean	Min	Max	Information Gain
6	age	32	17	13	55	0.0086
26	PNA	2881	88.15	0	100	0.0068
28	admission.test	907	1259	1	1600	0.0026
30	general.math.eval	423	68.50	0	100	0.0062
31	admission.rubric	51	33	0	50	0.0025
33	scholarship.perc	26	17	0	100	0.0066
34	loan.perc	14	4	0	50	0.0010
35	total.scholarship.loan	3066	21	0	100	0.0064
36	FTE	64	1.02	0.04	1.44	0.0154
37	average.first.period	545	87.26	0	100	0.0321
38	failed.subject.first.period	9	0	0	8	0.0039
39	dropped.subject.first.period	10	0	0	9	0.0006
45	total.life.activities	8	1.74	0	8	0.0061

Table 5. Description of the categorical attributes of the student dataset .

No.	Attribute	Unique	Mode	Frequency	Information Gain
2	generation	7	AD20	21,962	0.0047
3	educational model	2	0	99,534	0.0029
4	level	2	Undergraduate	77,517	0.0089
5	gender	2	Male	75,285	0.0081
7	zone.type	4	No information	101,920	0.0058
8	socioeconomic.level	8	No information	124,041	0.0174
9	social.lag	4	No information	119,327	0.0208
10	id.school.origin	10,243	School 5,328	3106	0.0080
11	school.cost	6	High cost	67,135	0.0057
12	tec.no.tec	2	NO TEC	102,481	0.0026
13	max.degree.parents	5	Undergraduate degree	52,494	0.0128
14	father.education.complete	9	Received undergraduate degree	49,888	0.0110
15	father.education.summary	5	Undergraduate degree	49,888	0.0124
16	mother.education.complete	9	Received undergraduate degree	53,453	0.0119
17	mother.education.summary	5	Undergraduate degree	53,453	0.0130
18	parents.exatec	3	No	94,020	0.0056
19	father.exatec	3	No	97,845	0.0047
20	mother.exatec	3	No	104,787	0.0039
21	first.generation	4	Does not apply	65,809	0.0064
22	school	7	High School	65,809	0.0100
23	program	76	PBB	38,506	0.0074
24	region	5	RCM	36,678	0.0078
25	foreign	3	Local	116,933	0.0020
27	english.evaluation	8	6	49,296	0.0070
29	online.test	2	0	142,204	0.0004
32	scholarship.type	11	No scholarship	71,866	0.0165
40	retention	2	1	131,687	Target
41	dropout.semester	5	0	131,687	0.2819
42	physical.education	4	1	58,701	0.0243
43	cultural.diffusion	4	1	40,768	0.0233
44	student.society	4	0	52,710	0.0235
46	athletic.sports	4	1	36,908	0.0176
47	art.culture	4	0	43,566	0.0174
48	student.society.leadership	4	0	42,987	0.0175
49	life.work.mentoring	4	0	51,553	0.0176
50	wellness.activities	4	0	44,364	0.0175

In addition, a correlation matrix is provided in Figure 2 to show the correlation coefficients between each numerical attribute in the dataset. Due to the considerations mentioned above, the dataset used for these analyzes resulted in 25,061 records. From this matrix, it can be seen that the degree of linear relationship between the variable *total.scholarship.loan* and the variable *scholarship.perc* is 0.94, which means that these variables are strongly correlated. While between the variables *average.first.period* and *failed.subject.first.period* the coefficient is −0.43, which indicates that they are associated in the opposite direction.

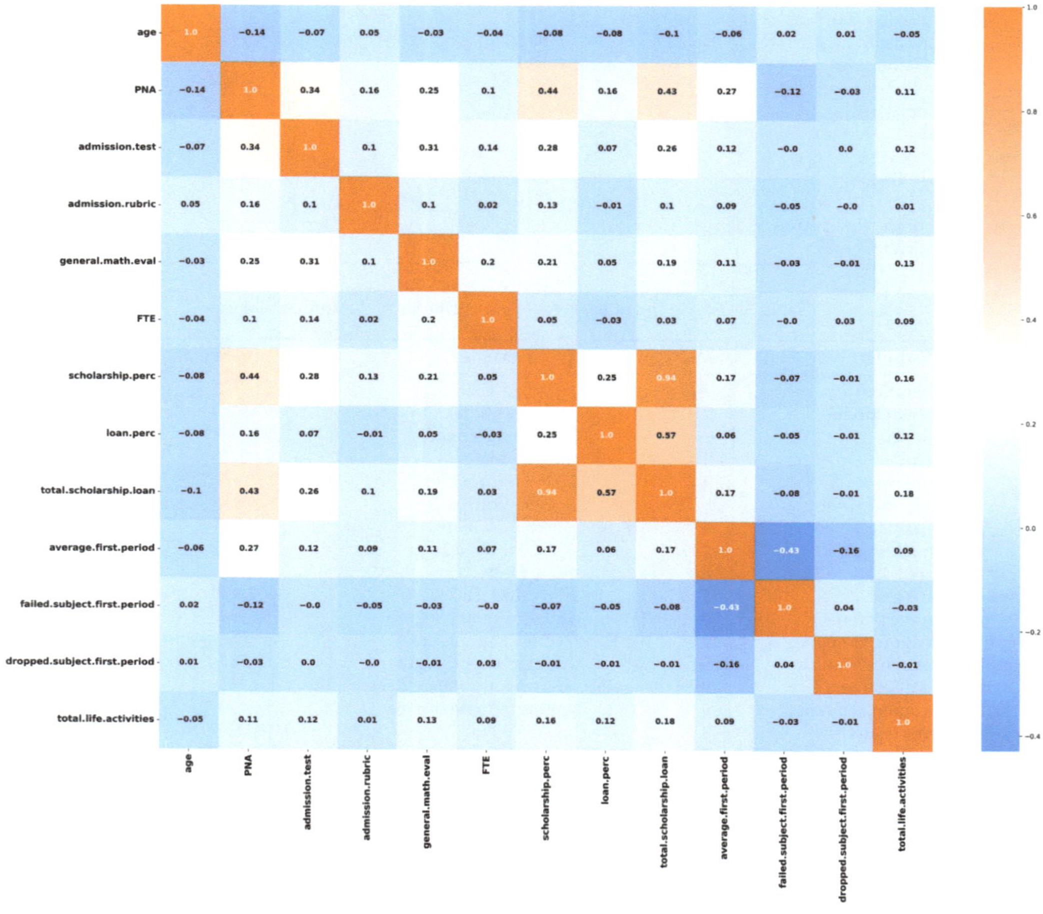

Figure 2. Correlation matrix of the numerical attributes shown in Table 4.

On the other hand, Table 5 describes the categorical variables of the dataset through their unique and mode values, and the frequency of the mode. The identifier of each variable corresponds to the identifier assigned in Tables 1–3. Regarding the co-curricular activities, the mode and frequency were calculated according to the generation to which they correspond. For example, for the variables *physical.education*, *cultural.diffusion*, and *student.society*, only the values corresponding to the generations AD14 to AD17 were considered. Similarly, for the LiFE activities, only the values of the generations AD18 to AD20 were contemplated. Furthermore, the "Does not apply" value was ignored for all generations. In the same way, the gain information is integrated to demonstrate the dependency between each feature in the dataset and the target variable: retention. The information gain was calculated using a mutual information classifier, it was necessary to encode the features using an *OrdinalEncoder* while the target variable, in this case, "retention" was encoded with a *LabelEncoder*. From this calculation, it can be deduced that the retention variable is more dependent on the students' co-curricular activities, such as *cultural.diffusion*, *student.society*, and *physical.education*, while the variables *online.test* and *dropped.subject.first.period* have less dependency on retention.

It is worth mentioning that it is recommended to carry out a greater analysis of the factors since the gain values may vary depending on the data preprocessing and the approach that each researcher considers in their experiments.

Subsequently, graphical representations were performed with the variables related to the dropout rates and the specific variables of the institution (student life). Figure 3 illustrates the number of High School and Undergraduate students who dropped out during their first year of study from AD14 to AD20. In general, the number of students enrolled increased over time for both levels. Figure 3 shows that in AD14 the number of High School students who dropped out is higher compared to other generations. It is also found that in AD15 there is a slight decrease in student dropout of 7.28% but during the following three generations, from AD16 to AD18, the dropout rates increased and ranged between 7.61% and 7.98%. In AD19, when the Tec21 model started, this rate started to decrease from 6.48% to 5.51% in AD20, which is the lowest dropout rate of the seven generations.

Although at the Undergraduate level the number of students enrolled seems to increase year after year, the number of dropouts does not behave the same. It is observed in the orange line of Figure 3 that the year with the highest student dropout is also found in the AD14 generation with a dropout rate of 10.09%. According to the graph, there was a downward trend starting from the AD15 generation with a dropout rate of 9.20%, then between the AD16 and AD17 generations, the dropout rates decreased and had a minimum variation with percentages of 8.82% and 8.71%, respectively. In AD18, the dropout rate continued to decrease with a percentage of 7.53%. Although there was a decreasing trend in dropout rates during the past generations, in AD19, despite the number of students enrolled increased, the dropout rate rose to 9.43% but in AD20 this rate decreased to 7.95%.

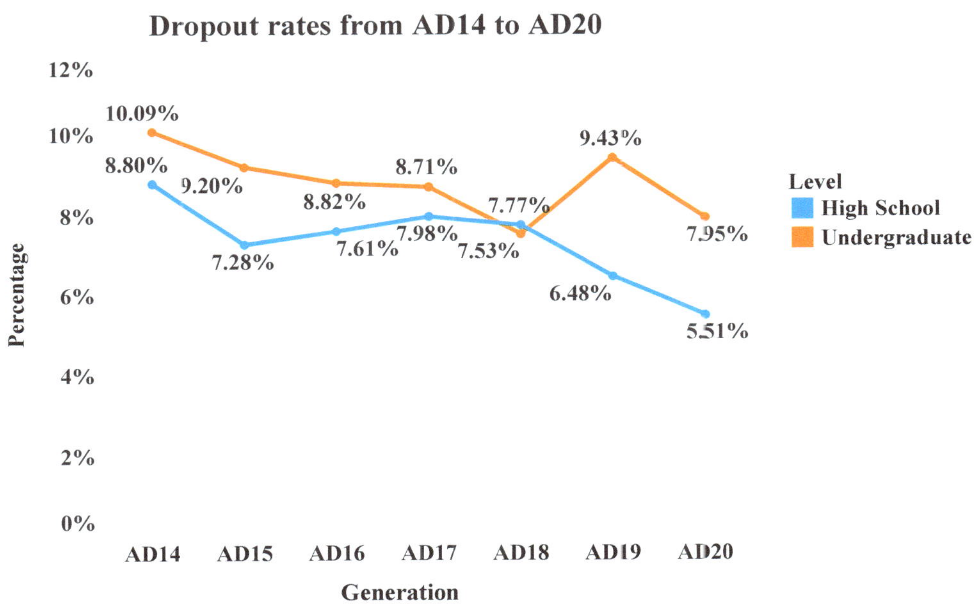

Figure 3. High School and Undergraduate dropout rates.

Moreover, Figure 4 presents information on the number of High School and Undergraduate students who participated in different co-curricular activities during the fall semesters between 2014 and 2017. The total number of students enrolled in those years was 78,715. The graph shows that the majority (58,701) of the students were involved in Physical Education activities with a dropout rate of 7.10%, followed by cultural diffusion with 40,768 students enrolled and a dropout rate of 7.10%; while a smaller number of students (25,115), participated in some student society activity with a dropout rate of 6.31%.

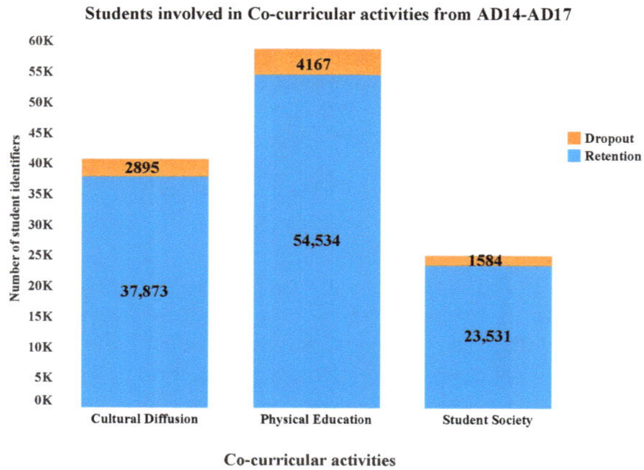

Figure 4. Number of High School and Undergraduate students who were enrolled in co-curricular activities during the fall semesters from AD14 to AD17.

Figure 5 shows the information on the co-curricular activities that belong specifically to the Tecnologico de Monterrey's LiFE program implemented since AD18. The number of students enrolled in these three generations was 64,611. According to the graph, more than half of the students (36,908) participated in Athletic Sports with a dropout rate of 6.09%. The Student Society Leadership was the second activity with a participation of 21,429 students and a dropout rate of 6.10%, followed by Art Culture with 20,849 students and a dropout rate of 6.02%. Compared to this last activity, slightly fewer students participated in the Wellness activities (20,052) with a dropout rate of 5.91%. Participation in activities related to Life-Work Mentoring was the least preferred by students with a participation of 12,863 but with the highest percentage of dropouts of 7.40%.

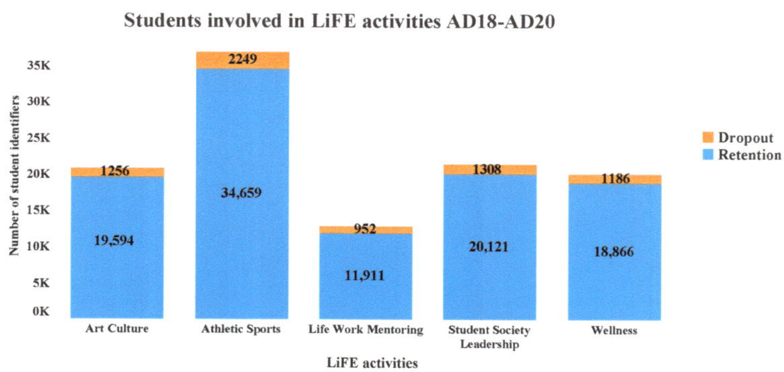

Figure 5. Number of High School and Undergraduate students who were enrolled in LiFE activities during the fall semesters from AD18 to AD20.

It is worth mentioning that a student could have participated in one or more activities at the same time.

4. Conclusions

Through this data descriptor, a non-identifiable dataset of 121,584 High School and Undergraduate students from Tecnologico de Monterrey was provided in order to contribute to the scientific community with data that will allow it to generate more accurate models to predict student dropout in higher education institutions. The generation of an appropriate model based on this dataset would benefit the students, by having timely and personalized strategies from their institution that support their permanence in their career, as well as the institution, by improving their statistics of student degree completion and their student investment costs.

The dataset is made up of variables reported in the literature as good predictors of school dropout as well as variables of the institution that are part of the student life. The contribution of more data related to the variables found in the literature from an institution other than their own could allow testing models already developed in their own institution to find new findings or improve those models.

On the other hand, the new variables (student life) could provide new relationships between the factors already studied that could enhance the development of new or improved models to predict student performance and identify at-risk students. Most papers use traditional Machine Learning algorithms (e.g., logistic regression, k-nearest neighbors, and decision tree-based ensemble models) [13,34]. However, only 5% of the studies have applied unsupervised learning algorithms [16]. Furthermore, the emergence of Explainable Artificial Intelligence (XAI) tools has made it possible to use advanced Machine Learning algorithms for interpretable dropout prediction [35–37].

Author Contributions: Conceptualization, J.A.-U. and R.A.R.D.; methodology, J.A.-U., P.M.-A., A.L.M.H. and H.G.C.; software, A.L.M.H. and P.M.-A.; validation, J.A.-U., A.L.M.H., J.E.M.G., R.A.R.D., R.M., I.H. and V.H.; formal analysis, J.A.-U., P.M.-A. and A.L.M.H.; investigation, J.A.-U., P.M.-A., R.M., I.H., V.H. and J.E.M.G.; resources, H.G.C. and R.A.R.D.; data curation, A.L.M.H. and P.M.-A.; writing—original draft preparation, J.A.-U. and P.M.-A.; writing—review and editing, R.M., I.H., A.L.M.H., J.E.M.G., V.H., R.A.R.D. and H.G.C.; visualization, P.M.-A. and J.A.-U.; supervision, H.G.C. and R.A.R.D.; project administration, H.G.C.; funding acquisition, H.G.C. and R.A.R.D. All authors have read and agreed to the published version of the manuscript.

Funding: This research was funded by the Institute for the Future of Education and the APC was funded by the Tecnologico de Monterrey.

Institutional Review Board Statement: Privacy issues related to the collection, curation, and publication of student data were validated with Tecnologico de Monterrey's Data Owners and the Data Security and Information Management Departments.

Informed Consent Statement: Not applicable.

Data Availability Statement: The data presented in this descriptor is available upon request in the Institute for the Future of Education's Educational Innovation collection of the Tecnologico de Monterrey's Data Hub at https://doi.org/10.57687/FK2/PWJRSJ (accessed on 24 August 2022).

Acknowledgments: The authors would like to thank the Tecnologico de Monterrey's Analytics and Business Intelligence Department for providing the original dataset for this project. Similarly, to Yedida Betzabé López Membrila, LiFE Programs Portfolio Leader, for providing complementary data for the presented dataset. Also, to Verónica Guadalupe Barroso Sánchez, Admissions Specialist, for explaining the variables related to admissions in the dataset. Finally, to the researchers who applied for the call for their recommendations on the integration of new variables.

Conflicts of Interest: The authors declare no conflicts of interest.

Abbreviations

The following abbreviations are used in this manuscript:

AD	August–December
CC0	Creative Commons Zero
DOI	Digital Object Identifier
LiFE	Leadership and Student Education
MDPI	Multidisciplinary Digital Publishing Institute
PAA	Academic Aptitude Test (Prueba de Aptitud Académica)
PAL	Online Aptitude Test (Prueba de Aptitud en Línea)
SAP	Systemanalyse Programmentwicklung
WebI	SAP BusinessObjects Web Intelligence
XAI	Explainable Artificial Intelligence

Appendix A

Table A1. The meaning of the acronym of the program in which the student is enrolled (Part I).

Program	Meaning
ADI	Architecture and Design/Exploration
AMC	Built Environment/Exploration
ARQ	B.A. in Architecture
BIO	Bioengineering and Chemical Process/Exploration
CIS	Law, Economics and International Relations/Exploration
COM	Communication and Digital Production/Exploration
CPF	B.A. in Finance & Accounting
ESC	Creative Studies/Exploration
IA	B.S. Agronomy Engineering
IBN	B.S. Biobusiness Engineering
IBQ	Engineering-Bioengineering and Chemical Process (avenue)/Exploration
IBT	B.S. in Biotechnology Engineering
IC	B.S. Civil Engineering
ICI	Engineering-Applied Sciences (avenue)/Exploration
ICT	Engineering-Computer Science and Information Technologies (avenue)/Exploration
IDA	B.S. Automotive Engineering
IDS	B.S. Sustainable Development Engineering
IFI	B.S. in Engineering Physics
IIA	B.S. Food Industry Engineering
IID	B.S. Innovation and Development Engineering
IIN	B.S. Industrial Innovation Engineering
IIS	B.S. Industrial Engineering with minor in Systems Engineering
IIT	Engineering-Innovation and Transformation (avenue)/Exploration
IMA	B.S. Mechanical Engineering (administrator)
IMD	B.S. Biomedical Engineering
IME	B.S. Mechanical Engineering (electrician)
IMI	B.S. Digital Music Production Engineering
IMT	B.S. in Mechatronics Engineering
ING	Engineering/Exploration
INQ	B.S. Chemistry and Nanotechnology Engineering
INT	B.S. Business Informatics
IQA	B.S. Chemical Engineering (administrator)
IQP	B.S. Chemical Engineering (sustainable processes)
ISC	B.S. Computer Science and Technology
ISD	B.S. Digital Systems and Robotics Engineering
ITC	B.S. in Computer Science and Technology
ITE	B.S. Electronic and Computer Engineering

Table A1. *Cont.*

Program	Meaning
ITI	B.S. Information and Communication Technologies
ITS	B.S. Telecommunications and Electronic Systems
LAD	B.A. Animation and Digital Art
LAE	B.A. Business Administration
LAF	B.A. Financial Management
LBC	B.A. in Biosciences
LCD	B.A. Communication and Digital Media
LCMD	B.A. Communication and Digital Media
LDE	B.A. in Entrepreneurship
LDF	B.A. Law with Minor in Finance
LDI	B.A. Industrial design
LDN	B.A. Business Innovation and Management
LDP	B.A. Law with Minor in Political Science

Table A2. The meaning of the acronym of the program in which the student is enrolled (Part II).

Program	Meaning
LEC	B.A. Economics
LED	B.A. in Law
LEF	B.A. Economics and Finances
LEM	B.A. in Marketing
LIN	B.A. in International Business
LLE	B.A. Spanish Literature
LLN	B.A. International Logistics
LMC	B.A. Marketing and Communication
LMI	B.A. Journalism and Media Studies
LNB	B.A. in Nutrition and Wellness
LP	B.A. Psychology
LPL	B.A. Political Science
LPM	B.A. Advertising and Marketing Communications
LPO	B.A. Organizational Psychology
LPS	B.S. Clinical Psychology and Health
LRI	B.A. International Relations
LTS	B.A. Social Transformation
MC	Physician & Surgeon
MO	Medical and Surgical Dentist
NEG	Business/Exploration
PBB	Bicultural High School
PBI	International High School
PTB	Bilingual High School
PTM	Multicultural High School
SLD	Health Sciences/Exploration
TIE	Information Technologies and Electronics/Exploration

References

1. Latif, A.; Choudhary, A.I.; Hammayun, A.A. Economic Effects of Student Dropouts: A Comparative Study. *J. Global Econ.* **2015**, *3*, 137. [CrossRef]
2. Raisman, N. The Cost of College Attrition at Four-Year Colleges & Universities—An Analysis of 1669 US Institutions. *Policy Perspect.* **2013**, 269. Available online: https://eric.ed.gov/?q=source%3A%22Educational+Policy+Institute%22&id=ED562625 (accessed on 24 August 2022).
3. da Silva, J.J.; Roman, N.T. Predicting Dropout in Higher Education: A Systematic Review. In *Proceedings of the Anais do XXXII Simpósio Brasileiro de Informática na Educação;* SBC: Porto Alegre, Brasil, 2021; pp. 1107–1117. [CrossRef]

4. Fahd, K.; Venkatraman, S.; Miah, S.J.; Ahmed, K. Application of machine learning in higher education to assess student academic performance, at-risk, and attrition: A meta-analysis of literature. *Educ. Inf. Technol.* **2022**, *27*, 3743–3775. [CrossRef]
5. Ranjeeth, S.; Latchoumi, T.P.; Paul, P.V. A Survey on Predictive Models of Learning Analytics. *Procedia Comput. Sci.* **2020**, *167*, 37–46. [CrossRef]
6. Dutt, A.; Ismail, M.A.; Herawan, T. A Systematic Review on Educational Data Mining. *IEEE Access* **2017**, *5*, 15991–16005. [CrossRef]
7. Kumar, M.; Singh, A.J.; Handa, D. Literature Survey on Educational Dropout Prediction. *Int. J. Educ. Manag. Eng.* **2017**, *7*, 8. [CrossRef]
8. Saleem, F.; Ullah, Z.; Fakieh, B.; Kateb, F. Intelligent Decision Support System for Predicting Student's E-Learning Performance Using Ensemble Machine Learning. *Mathematics* **2021**, *9*, 2078. [CrossRef]
9. Hilliger, I.; Ortiz-Rojas, M.; Pesántez-Cabrera, P.; Scheihing, E.; Tsai, Y.S.; Muñoz-Merino, P.J.; Broos, T.; Whitelock-Wainwright, A.; Pérez-Sanagustín, M. Identifying needs for learning analytics adoption in Latin American universities: A mixed-methods approach. *Internet High. Educ.* **2020**, *45*, 100726. [CrossRef]
10. Namoun, A.; Alshanqiti, A. Predicting Student Performance Using Data Mining and Learning Analytics Techniques: A Systematic Literature Review. *Appl. Sci.* **2021**, *11*, 237. [CrossRef]
11. Cardona, T.A.; Cudney, E.A. Predicting Student Retention Using Support Vector Machines. *Procedia Manuf.* **2019**, *39*, 1827–1833. [CrossRef]
12. Lázaro Alvarez, N.; Callejas, Z.; Griol, D. Predicting computer engineering students' dropout in cuban higher education with pre-enrollment and early performance data. *J. Technol. Sci. Educ.* **2020**, *10*, 241–258. [CrossRef]
13. Nagy, M.; Molontay, R. Predicting Dropout in Higher Education Based on Secondary School Performance. In Proceedings of the 2018 IEEE 22nd International Conference on Intelligent Engineering Systems (INES), Las Palmas de Gran Canaria, Spain, 21–23 June 2018, pp. 389–394. [CrossRef]
14. Varga, E.B.; Sátán, Á. Detecting at-risk students on Computer Science bachelor programs based on pre-enrollment characteristics. *Hung. Educ. Res. J.* **2021**, *11*, 297 – 310. [CrossRef]
15. Kiss, B.; Nagy, M.; Molontay, R.; Csabay, B. Predicting Dropout Using High School and First-semester Academic Achievement Measures. In Proceedings of the 2019 17th International Conference on Emerging eLearning Technologies and Applications (ICETA), Starý Smokovec, Slovakia, 21–22 November 2019, pp. 383–389. [CrossRef]
16. Alshanqiti, A.; Namoun, A. Predicting Student Performance and Its Influential Factors Using Hybrid Regression and Multi-Label Classification. *IEEE Access* **2020**, *8*, 203827–203844. [CrossRef]
17. Hoffman, J.L.; Lowitzki, K.E. Predicting College Success with High School Grades and Test Scores: Limitations for Minority Students. *Rev. High. Educ.* **2005**, *28*, 455–474. [CrossRef]
18. Zwick, R.; Himelfarb, I. The Effect of High School Socioeconomic Status on the Predictive Validity of SAT Scores and High School Grade-Point Average. *J. Educ. Meas.* **2011**, *48*, 101–121. [CrossRef]
19. Freitas, F.A.d.S.; Vasconcelos, F.F.X.; Peixoto, S.A.; Hassan, M.M.; Dewan, M.A.A.; Albuquerque, V.H.C.D.; Filho, P.P.R. IoT System for School Dropout Prediction Using Machine Learning Techniques Based on Socioeconomic Data. *Electronics* **2020**, *9*, 1613. [CrossRef]
20. Séllei, B.; Stumphauser, N.; Molontay, R. Traits versus Grades—The Incremental Predictive Power of Positive Psychological Factors over Pre-Enrollment Achievement Measures on Academic Performance. *Appl. Sci.* **2021**, *11*, 1744. [CrossRef]
21. Terry, M. The Effects that Family Members and Peers Have on Students' Decisions to Drop out of School. *Educ. Res. Q.* **2008**, *31*, 25–38.
22. Slade, S.; Prinsloo, P. Learning Analytics: Ethical Issues and Dilemmas. *Am. Behav. Sci.* **2013**, *57*, 1510–1529. [CrossRef]
23. Ferreyra, M.M.; Avitabile, C.; Botero Álvarez, J.; Haimovich Paz, F.; Urzúa, S. *At a Crossroads: Higher Education in Latin America and the Caribbean*; The World Bank Group: Washington, DC, USA, 2017. [CrossRef]
24. Ferreira, F.H.G.; Messina, J.; Rigolini, J.; López-Calva, L.F.; Lugo, M.A.; Vakis, R. *Economic Mobility and the Rise of the Latin American Middle Class*; The World Bank Group: Washington, DC, USA, 2013. [CrossRef]
25. Lemaitre, M.J. Quality assurance in Latin America: Current situation and future challenges. *Tuning J. High. Educ.* **2017**, *5*, 21–40. [CrossRef]
26. González-Velosa, C.; Rucci, G.; Sarzosa, M.; Urzúa, S. *Returns to Higher Education in Chile and Colombia*; Technical Report, IDB Working Paper Series No. IDB-WP-587; Inter-American Development Bank: Washington, DC, USA, 2015.
27. Cobo, C.; Aguerrebere, C. Building capacity for learning analytics in Latin America. In *Learning Analytics for the Global South*; Lim, C.P., Tinio, V.L., Eds.; Foundation for Information Technology Education and Development, Inc.: Quezon City, Philippines, 2018; Volume 58, pp. 63–67.
28. Call for Proposals: Bringing New Solutions to the Challenges of Predicting and Countering Student Dropout in Higher Education. 2022. Available online: https://ifelldh.tec.mx/en/student-dropout-higher-education (accessed on 9 June 2022).
29. Tecnologico de Monterrey. Tecnologico de Monterrey. 2022. Available online: https://tec.mx/en (accessed on 11 May 2022).
30. The Tec Is Transforming Its Educational Model to Become More Flexible. 2022. Available online: https://conecta.tec.mx/en/news/national/education/tec-transforming-its-educational-model-become-more-flexible (accessed on 18 May 2022).
31. Tec de Monterrey Has Reinvented Its Student Experience, Presents LiFE. 2022. Available online: https://conecta.tec.mx/en/news/national/institution/tec-de-monterrey-has-reinvented-its-student-experience-presents-life (accessed on 18 May 2022).

32. Gestión de Datos de Investigación. 2022. Available online: https://biblioguias.cepal.org/c.php?g=495473&p=4994826 (accessed on 21 June 2022).
33. Primer on Data Management: What You Always Wanted to Know. 2022. Available online: https://old.dataone.org/sites/all/documents/DataONE_BP_Primer_020212.pdf (accessed on 21 June 2022).
34. Rastrollo-Guerrero, J.L.; Gómez-Pulido, J.A.; Durán-Domínguez, A. Analyzing and Predicting Students' Performance by Means of Machine Learning: A Review. *Appl. Sci.* **2020**, *10*, 1042. [CrossRef]
35. Baranyi, M.; Nagy, M.; Molontay, R. Interpretable Deep Learning for University Dropout Prediction. In Proceedings of the 21st Annual Conference on Information Technology Education, Virtual, 7–9 October 2020; Association for Computing Machinery: New York, NY, USA, 2020; pp. 13–19. [CrossRef]
36. Nagy, M.; Molontay, R.; Szabó, M. A Web Application for Predicting Academic Performance and Identifying the Contributing Factors. In Proceedings of the SEFI 47th Annual Conference, Budapest, Hungary, 16–19 September 2019; pp. 1794–1806.
37. Smith, B.I.; Chimedza, C.; Bührmann, J.H. Individualized help for at-risk students using model-agnostic and counterfactual explanations. *Educ. Inf. Technol.* **2022**, *27*, 1539–1558. [CrossRef]

data

MDPI

Data Descriptor

A Large-Scale Dataset of Twitter Chatter about Online Learning during the Current COVID-19 Omicron Wave

Nirmalya Thakur

Department of Electrical Engineering and Computer Science, University of Cincinnati, Cincinnati, OH 45221-0030, USA; thakurna@mail.uc.edu

Abstract: The COVID-19 Omicron variant, reported to be the most immune-evasive variant of COVID-19, is resulting in a surge of COVID-19 cases globally. This has caused schools, colleges, and universities in different parts of the world to transition to online learning. As a result, social media platforms such as Twitter are seeing an increase in conversations related to online learning in the form of tweets. Mining such tweets to develop a dataset can serve as a data resource for different applications and use-cases related to the analysis of interest, views, opinions, perspectives, attitudes, and feedback towards online learning during the current surge of COVID-19 cases caused by the Omicron variant. Therefore, this work presents a large-scale, open-access Twitter dataset of conversations about online learning from different parts of the world since the first detected case of the COVID-19 Omicron variant in November 2021. The dataset is compliant with the privacy policy, developer agreement, and guidelines for content redistribution of Twitter, as well as with the FAIR principles (Findability, Accessibility, Interoperability, and Reusability) principles for scientific data management. The paper also briefly outlines some potential applications in the fields of Big Data, Data Mining, Natural Language Processing, and their related disciplines, with a specific focus on online learning during this Omicron wave that may be studied, explored, and investigated by using this dataset.

Dataset: https://doi.org/10.5281/zenodo.6837118

Dataset License: CC-BY 4.0

Keywords: COVID-19; COVID; omicron; online learning; remote learning; online education; Twitter; dataset; tweets; social media; big data

Citation: Thakur, N. A Large-Scale Dataset of Twitter Chatter about Online Learning during the Current COVID-19 Omicron Wave. *Data* **2022**, 7, 109. https://doi.org/10.3390/data7080109

Academic Editors: Antonio Sarasa Cabezuelo and Ramón González del Campo Rodríguez Barbero

Received: 9 June 2022
Accepted: 2 August 2022
Published: 4 August 2022

1. Introduction

The first cases of the COVID-19 pandemic, caused by the SARS-CoV-2 virus, were recorded in a seafood market in Wuhan, China, in December 2019 [1]. Since then, the virus has been found in all the countries of the world. At the time of writing this paper, globally, there have been 535,342,382 cases with 6,320,324 deaths [2]. Since the initial cases in China, the SARS-CoV-2 virus has undergone multiple mutations, and as a result, multiple variants have been detected in different parts of the world. Some of these include: Alpha (B.1.1.7), Beta (B.1.351), Gamma (P.1), Delta (B.1.617.2), Epsilon (B.1.427 B.1.429), Eta (B.1.525), Iota (B.1.526), Kappa (B.1.617.1), Zeta (P.2), Mu (B.1.621, B.1.621.1), and Omicron (B.1.1.529, BA.1, BA.1.1, BA.2, BA.3, BA.4 and BA.5) [3]. Out of all these variants, the Omicron variant, first detected on 24 November 2021 from a sample collected on 9 November 2021, was classified as a Variant of Concern (VOC) by the World Health Organization (WHO) on 26 November 2021 [4]. The Omicron variant has a spike protein that contains 30 mutations [5]. It has been reported to be the most immune-evasive variant of COVID-19 and to present very strong resistance against antibody-based or plasma-based treatments [6]. According to WHO, the new cases due to this Omicron variant have been "off the charts" and are setting new records in terms of COVID-19 cases all over the world [7]. The Omicron variant currently accounts for 86% of the COVID-19 cases worldwide [8], and some of the countries that have recorded the most cases due to the SARS-CoV-2 Omicron variant include the United Kingdom (1,138,814 cases), USA (945,470 cases), Germany (245,120 cases), Denmark (218,106 cases), France (110,959 cases), Canada (92,341 cases), Japan (71,056 cases), India (56,125 cases), Australia (46,576 cases), Sweden (43,400 cases), Israel (39,908 cases), Poland (33,436 cases), and Brazil (32,880 cases) [9].

Since the beginning of the pandemic, many countries, such as India [10], the United States [11], the United Kingdom [12], Spain [13], Greece [14], Italy [15], Austria [16], Nigeria [17], China [18], New Zealand [19], Ireland [20], Germany [21], South Africa [22], Australia [23], France, [24], Norway [25], and several more [26], went on a complete lockdown with work from home and remote work guidelines that affected a multitude of industries and sectors. Out of all these sectors that were impacted by the nationwide lockdowns and the associated guidelines in different parts of the world, the education sector was an important one. On a global scale, universities, colleges, and schools had to switch to online education, which required its faculty, administrators, staff, and students to become familiarized with online learning and the associated tools and platforms that were necessary for this new norm of education. Due to the worldwide adoption and familiarization with various forms of tools, platforms, software, and hardware necessary for online education, the online education market is rapidly booming and is expected to reach more than USD 350 billion by 2025 [27]. Online learning may be broadly defined as *"learning experiences in synchronous or asynchronous environments using different devices (e.g., mobile phones, laptops, etc.) with internet access. In these environments, students can be anywhere (independent) to learn and interact with instructors and other students"* [28]. Online learning has a range of synonyms, and some of the most commonly used synonyms include remote education, online education, virtual education, remote learning, e-learning, distance education, virtual learning, asynchronous learning, and blended learning [28].

On a global scale, more than 43,518,726 students were affected due to in-person school closures due to COVID-19 [29]. The closing of universities, colleges, and schools was recorded in 188 countries [30], and 90% of the countries reported a switch to one or more forms of online learning [31]. Despite these promising numbers, 31% (463 million) of students in schools (in preprimary to secondary education) could not adopt online learning either due to lack of technologies, training, or accessibility, and 75% of students who belonged to the poorest households could not switch to the technologies required for online learning [31].

With the advancements in vaccine research and other forms of treatment of COVID-19 toward the later part of 2020 [32–34] and in compliance with the recommendations from various local and national policy-making bodies, different universities, colleges, and schools started to transition to hybrid (both online and in-person) learning as well as completely in-person learning [35]. However, this was associated with several challenges [36], including a surge of COVID-19 cases in students, educators, and staff members, an increase in stress and anxiety in both students and their parents, the need for allocation of funds by these educational institutions to conduct classes in a socially distant manner, and for procurement of hand sanitizers and disinfectants. Despite these challenges, education continued in both hybrid and in-person forms for a few months. However, due to the recent global surge in COVID-19 cases due to the Omicron variant [7–9], many educational institutions all over the world have transitioned back to online learning since the beginning of 2022, and several are in the process of transitioning to online learning over the next few months [37–42].

The modern-day Internet of Everything lifestyle [43] is characterized by people spending more time on the internet than ever before, with a specific focus on social media platforms. The use of social media platforms has skyrocketed in the recent past [44]. Social media usage characteristics include conversations on diverse topics such as recent issues, global challenges, emerging technologies, news, current events, politics, family, relationships, and career opportunities [45]. Twitter, one such social media platform, used by people of almost all age groups [46,47], has been rapidly gaining popularity in all parts of the world and is currently the second most visited social media platform [48]. At present, there are about 192 million daily active users on Twitter, and approximately 500 million tweets are posted on Twitter every day [49]. Mining of social media conversations, such as Tweets, to develop datasets has been of significant interest to the scientific community in the areas of Big Data, Data Mining, and Natural Language Processing, as can be seen from these recent works where relevant Tweets were mined to develop Twitter datasets on the 2020 US Presidential Election [50], 2022 Russia–Ukraine war [51], climate change [52], natural hazards [53], European Migration Crisis [54], movies [55], toxic behavior amongst adolescents [56], music [57], civil unrest [58], drug safety [59], and Inflammatory Bowel Disease [60].

In the context of the recent surge of COVID-19 cases due to the Omicron variant and its impact on the education sector, there has been a significant increase in conversations on Twitter related to online learning. Mining such conversations to develop a dataset would serve as a rich data resource for the investigation of different research questions in the fields of Big Data, Data Mining, Data Science, and Natural Language Processing, with a central focus on analyzing tweets related to online learning during this time.

Previous works [61–90] (discussed in Section 2) related to online learning since the outbreak of COVID-19 have focused on analyzing multiple factors related to online learning only in certain geographic regions, mostly by using surveys, and not on a global scale by analyzing conversations from all over the world, such as Tweets. Prior works on the development of Twitter datasets related to COVID-19 have also not focused on mining relevant tweets related to online learning during the ongoing COVID-19 Omicron wave. To address these limitations, this work proposes a dataset of more than 50,000 Tweet IDs (that correspond to the same number of Tweets) about online learning that was posted on Twitter from 9 November 2021 to 13 July 2022, which is publicly available at https://doi.org/10.5281/zenodo.6837118. The earliest date was selected as 9 November 2021, as the Omicron variant was detected for the first time in a sample that was collected on this date. The most recent date, at the time of resubmission of this journal paper after the completion of the first round of peer review and the subsequent editorial decision, was 13 July 2022.

The rest of the paper is organized as follows. Section 2 presents an overview of recent works in this field. The methodology that was followed for the development of this dataset is presented in Section 3. Section 4 provides the description of the dataset. Section 5 briefly discusses a few potential applications of this dataset. The conclusion and scope for future work are presented in Section 6, which is followed by references.

2. Literature Review

There has been a significant amount of research related to online learning since the global outbreak of COVID-19. The work by Muhammad et al. [61] was a research study that examined the attitudes of Pakistani higher education students toward compulsory digital and distance learning courses during COVID-19. In [62], Rasmitadila et al. presented a study that explored the perceptions of primary school teachers towards online learning during COVID-19. Data were collected through surveys and semi-structured interviews, and 67 teachers in primary schools participated in this study. The work by Irawan et al. [63] aimed to identify the impact of student psychology on online learning during the COVID-19 pandemic. The research method used a qualitative research type of phenomenology. The research subjects were 30 students of Mulawarman University, a university in Indonesia, who were interviewed via telephone. The work of Baticulon et al. [64] was to identify barriers to online learning from the perspective of medical students in the Philippines. The authors sent out an electronic survey to the students who participated in this study. The qualitative study presented by Hussein et al. [65] aimed to investigate the attitudes of undergraduate students towards online learning during the first few weeks of the mandatory shift to online learning caused by COVID-19. Students from two general

English courses at a university located in the United Arab Emirates were asked to write semi-guided essays and the associated data were analyzed by the authors. The work of Famularsih et al. [66] focused on studying the utilization of online learning applications in English as a Foreign Language (EFL) classrooms. The participants of this study were 35 students from a university in Salatiga, Indonesia. The data were gathered through surveys and semi-structured interviews.

The study by Sutarto et al. [67] focused on understanding the strategies used by teachers of SDIT Rabbi Radhiyya Curup, a school in Indonesia, to increase students' interest and responses to online learning during COVID-19. The data were collected by conducting semi-structured interviews, which were analyzed using the Miles and Huberman model. Almusharraf et al.'s [68] work aimed to evaluate the level of postsecondary student satisfaction with online learning platforms and learning experiences during the COVID-19 pandemic in Saudi Arabia. Quantitative research was carried out in this study by using a survey that was sent out to 283 students enrolled at a higher education institution in Saudi Arabia. These data were analyzed using SPSS. Al-Salman et al. [69] investigated the influence of digital technology, instructional and assessment quality, economic status and psychological state, and course type on Jordanian university students' attitudes towards online learning during the COVID-19 emergency transition to online learning. A total of 4037 undergraduate students from four universities participated in this study.

The aim of Bolatov et al.'s work [70] was to compare the differences between the mental state of students switching to online learning and the mental state of the students who were still using traditional learning. This study included medical students from Astana Medical University, a university in Kazakhstan. The work by Agormedah et al. [71] explored the responses of students to online learning in higher education in Ghana. The sample size of this study involved 467 students. The findings indicated that a majority of the students had a positive response to the transition to online learning. The work of Moawad et al. [72] aimed to identify the academic stressors by analyzing the worries and fears that students at the College of Education in King Saud University, a university in Saudi Arabia, experienced during the time of COVID-19. The results showed that the issue with the highest percentage of stress among students was their uncertainty over the end-of-semester exams and assessments. The work by Khan et al. [73] discussed various digital education methods, approaches, and systems that could be implemented by the education system of Bangladesh during COVID-19. The purpose of the study performed by Catalano et al. [74] was to determine teacher perceptions of students' access and participation in online learning, as well as concerns about educational outcomes among different groups of learners. The work of Kapasia et al. [75] aimed to assess the impact of the nationwide lockdown on account of COVID-19 on undergraduate and postgraduate students in West Bengal, a state in India. The authors conducted an online survey that included 232 students. In [76], Burns et al. performed a conceptual analysis on student wellbeing at universities in the United Kingdom with a specific focus on the psychosocial impact the pandemic had on students. Küsel et al. [77] performed a study to evaluate German university students' readiness for using digital media and online learning in their tertiary education and compared the findings with the results from the same study performed on students in the United States. A total of 72 students from universities in Germany and 176 students from universities in the United States were a part of this study. Darayseh et al. [78] analyzed the impact of COVID-19 on modes of teaching, with a specific focus on science education in schools in the United Arab Emirates. Questionnaires were deployed through an online platform, and a total of 62 science teachers participated in this study. Tsekhmister et al. [79] conducted a study to evaluate the effectiveness of virtual reality technology and online teaching systems among medical students of Bogomolets National Medical University, a university in Ukraine. The study was performed using a questionnaire that contained 15 questions with five options to comprehensively evaluate these technologies.

Arsaliev et al.'s work [80] aimed to investigate whether an online format was effective in providing education for ethnocultural competence development. A combination of digital surveys, tests, questionnaires, and online class interviews were used in this study that involved 120 students at Southern Federal University, a university in Russia. Cárdenas-Cruz et al.'s [81] work aimed to facilitate the acquisition of specific transversal skills of undergraduate students at the University of Granada in Spain during the outbreak by means of an integrated online working system. Papouli et al. [82] aimed to explore Greek social-work students' views on the use of digital technology during their stay at home due to the coronavirus lockdown. A total of 550 students from different universities in Greece participated in this study. In [83], Parmigiani et al. designed a qualitative study aimed at investigating the factors affecting e-inclusion during COVID-19. A total of 785 teachers at the University of Genoa, a university in Italy, participated in this study. Resch et al. [84] focused on analyzing the effects of COVID-19 on university students' social and academic integration, based on Tinto's integration theory. A total of 640 university students in Austria completed an online survey pertaining to academic and social integration in this study. The purpose of the study by Noah et al. [85] was to examine the impacts of Google classroom as an online learning delivery platform in a secondary school during the COVID-19 pandemic in Nigeria. The study included 140 participants. Chen et al. [86] studied user satisfaction in the context of using online education platforms in China during COVID-19. The work used a combination of questionnaires and a back propagation neural network.

Drane et al. [87] performed a comprehensive review of existing works to present the impact of 'learning at home' on the educational outcomes of vulnerable children in Australia during the COVID-19 pandemic. The work of Mukuna et al. [88] explored the perceived challenges of online teaching encountered by educators in a school in the Thabo Mofutsanyana District in South Africa. A total of six educators participated in this study. In [89], Hsiao presented the results of a study to explore the influences of course type and gender on distance learning performance. A total of 18,085 students from a university in Taiwan comprised the sample size of this study. Nafrees et al. [90] performed an analysis to determine the factors of awareness of students about online learning among undergraduate students at Southeastern University, a university in Sri Lanka. The study comprised about 400 questionnaires, and a total of 310 responses from students were analyzed by the authors. The findings showed that most students preferred to use WebEx over other platforms for their online education due to the user-friendliness of WebEx.

In terms of mining relevant conversations related to a specific topic on Twitter since the outbreak of COVID-19, the prior works in this field have focused on the development of datasets for healthcare misinformation [91], misleading information [92], vaccine misinformation [93], patient identification [94], updates related to vaccine development [95], and rumors related to COVID-19 [96].

Despite these emerging works in the fields of online learning and the development of Twitter datasets, there exist multiple limitations. First, these works in the field of online learning have been confined to studying or analyzing the success or failure, degrees of acceptance, and associated factors related to online learning in specific geographic regions in countries such as Pakistan [61], Indonesia [62,63,66,67], Philippines [64], UAE [65], Saudi Arabia [68,72], Jordan [69], Kazakhstan [70], Ghana [71], Bangladesh [73], the United States [74,77], India [75], the United Kingdom [76], Germany [77], the UAE [78], Ukraine [79], Russia [80], Spain [81], Greece [82], Italy [83], Austria [84], Nigeria [85], China [86], Australia [87], South Africa [88], Taiwan [89], and Sri Lanka [90], and not on a global level. Second, due to the lack of datasets such as Twitter conversations related to online learning from global users, the data that were analyzed in these studies were mostly in the form of surveys that were conducted in these respective geographic regions. Third, the Twitter datasets related to COVID-19 [91–96] do not focus on online learning and the ongoing chatter on Twitter about the same amidst the global rise of COVID-19 cases due to the Omicron variant. The dataset proposed in this paper aims to address all these limitations.

3. Methodology

This section describes the methodology that was followed for the development of this dataset, which is available at https://doi.org/10.5281/zenodo.6837118. The dataset contains a total of 52,984 Tweet IDs that correspond to the same number of tweets about online learning, which were publicly posted on Twitter from 9 November 2021 to 13 July 2022. This section also outlines how this work and the associated dataset development is in compliance with the privacy policy, developer agreement, and guidelines for content redistribution of Twitter, as well as follows the FAIR (Findability, Accessibility, Interoperability, and Reusability) principles for scientific data management. These are discussed in Sections 3.1–3.3, respectively.

3.1. Process for Dataset Development

As this work focuses on developing a Twitter dataset, the privacy policy, developer agreement, and guidelines for content redistribution of Twitter [97,98] were thoroughly studied, and after studying the same, it was concluded that mining relevant tweets from Twitter to develop a dataset (comprising only Tweet IDs) is in compliance with all these policies of Twitter. Therefore, this dataset contains only Tweet IDs and does not contain any other information related to the respective Tweets that were mined. A detailed explanation of this compliance is mentioned in Section 3.2.

The tweets were collected by using the Search Twitter "operator" [99] available in RapidMiner studio [100] and the Advanced Search feature of the Twitter API. RapidMiner is a data science platform that allows the development, implementation, and testing of various algorithms, processes, and applications in the fields of Big Data, Data Mining, Data Science, Artificial Intelligence, Machine Learning, and their related areas. There are various RapidMiner products available such as RapidMiner Studio, RapidMiner AI Hub, and RapidMiner Radoop. For this work, the RapidMiner studio, version 9.10, was downloaded and installed on a laptop with the Microsoft Windows 10 Home operating system with Intel (R) Pentium (R) Silver N5030 CPU @ 1.10GHz, 1101 Mhz, 4 Core (s), and 4 Logical Processor (s). In the RapidMiner platform, "process" and "operator" are two commonly used terminologies. An "operator" represents a specific function or operation, for instance, fetch data from a social media platform such as Twitter based on a specific set of guidelines or to perform a specific operation on a dataset. RapidMiner has a number of in-built "operators". It also allows users to develop "operators" from scratch. A collection of "operators" that are connected in a logical and executable sequence to achieve a desired purpose is called a "process". A "process" may also contain just one "operator" if the complete functionality of the "process" can be found in one in-built or user-defined "operator". The Search Twitter "operator", an in-built "operator" of RapidMiner, works by connecting with the Twitter API and by complying with the Twitter API standard search policies [101] to fetch tweets between two given dates that contain one or more keywords or phrases which are provided as input to this "operator". As there are different keywords that Twitter users can use to refer to both COVID-19, the Omicron variant, and online learning; therefore, a bag of words was developed based on studying commonly used synonyms, phrases, and terms used to refer to online learning [102], COVID-19 and the Omicron variant [103]. These synonyms, terms, and phrases, all of which were included in the data collection process, are shown in Table 1.

Table 1. List of commonly used synonyms, terms, and phrases for online learning and COVID-19.

Terminology	List of Synonyms and Terms
COVID-19	Omicron, COVID, COVID19, coronavirus, coronavirus pandemic, COVID-19, corona, corona outbreak, omicron variant, SARS-CoV-2, corona virus
online learning	online education, online learning, remote education, remote learning, e-learning, elearning, distance learning, distance education, virtual learning, virtual education, online teaching, remote teaching, virtual teaching, online class, online classes, remote class, remote classes, distance class, distance classes, virtual class, virtual classes, online course, online courses, remote course, remote courses, distance course, distance courses, virtual course, virtual courses, online school, virtual school, remote school, online college, online university, virtual college, virtual university, remote college, remote university, online lecture, virtual lecture, remote lecture, online lectures, virtual lectures, remote lectures

There are various forms of educational structures and educational systems followed by different countries all over the world. For instance, in the United States, early childhood education is followed by primary school (also called elementary school), middle school, secondary school (also called high school), and then postsecondary (tertiary) education. Postsecondary education includes nondegree programs that lead to certificates and diplomas plus six degree levels: associate, bachelor, first professional, master, advanced intermediate, and research doctorate. The US system does not offer a second or higher doctorate but does offer postdoctorate research programs [104]. A different educational structure is followed in India [105]. The school system in India has four levels: lower primary school (age 6 to 10), upper primary school (age 11 and 12), high school (age 13 to 15), and higher secondary school (age 17 and 18). The lower primary school is divided into five "standards", upper primary school into two, high school into three, and higher secondary school into two. Another different educational structure can be seen in the United Kingdom (UK). The education system in the UK is divided into four main parts, primary education, secondary education, further education, and higher education. Children in the UK have to legally attend primary and secondary education, which runs from about five years old until the student is 16 years old. The education system in the UK is also split into "key stages": Key Stage 1 (age 5 to 7), Key Stage 2 (age 7 to 11), Key Stage 3 (age 11 to 14), and Key Stage 4 (age 14 to 16) [106]. This study focuses on collecting tweets about online education or online learning on a global scale (and not tweets originating from any specific country specific to its educational structure or educational system). So, a comprehensive list of keywords (as shown in Table 1) was developed that would most commonly be used to refer to online education or online learning in different parts of the world, irrespective of the educational structure followed in that specific geographic region. The effectiveness of this approach can be seen from the different worldwide educational systems that are the subject matters of the tweets present in the dataset proposed as a result of this work. For instance, in this dataset, Tweet ID: 1458685065152450565 refers to online education in India; Tweet ID: 1462489169079513090 refers to online education in the United States; Tweet ID: 1462475208644874242 refers to online education in Pakistan; Tweet ID: 1462373712389238787 refers to online education in Indonesia; Tweet ID: refers to online education in the UK; Tweet ID: 1462357217479434241 refers to online education in Ukraine; Tweet ID: 1462512737402109952 refers to online education in Nigeria; Tweet ID: 1462315144411856897 refers to online education in Spain; Tweet ID: 1462411445035941891 refers to online education in Malaysia, and so on.

Tweets were searched using this "process" that comprised the Search Twitter "operator" in a way that it consisted of at least one synonym, term, or phrase used to refer to COVID-19 and at least one synonym, term, or phrase used to refer to online learning. The Search Twitter "operator" is not case-sensitive, so it returned the tweets based on keyword matching by ignoring the case (uppercase or lowercase).

The output of this RapidMiner "process" comprised multiple attributes such as the Tweet ID, Tweet Source (the source used to post the Tweet such as Twitter for Android, Twitter for IOS, etc.), Text of the Tweet, Retweet count, and the username of the Twitter user who posted the Tweet, all of which is public information that can be mined in compliance with the guidelines set forth in the Twitter API standard search policies. However, as per the developer policy, privacy policy, and content redistribution guidelines of Twitter, all the attributes other than the Tweet IDs were deleted by using data filters. Therefore, the dataset consists of only Tweet IDs. These Tweet IDs were grouped into different .txt files based on the timeline of the associated tweets. The description and details of these dataset files are presented in Section 4.

The complete information associated with a tweet, such as the text of a tweet, username, user ID, timestamp, retweet count, etc., can be obtained from a Tweet ID by following a process known as hydration of Tweet ID [107]. Researchers in the field of Big Data, Data Mining, and Natural Language Processing, with a specific focus on Twitter research, have developed multiple tools for the hydration of Tweet IDs. Some of the most commonly used tools include the Hydrator app [108], Social Media Mining Toolkit [109], and Twarc [110], all of which work by complying with the policies of accessing the Twitter API. Any of these tools can be used on this dataset to obtain the associated information, such as the text of a tweet, username, user ID, timestamp, and retweet count for all the Tweet IDs. A step-by-step process on how to use one of these tools, the Hydrator app, for hydrating all the Tweet IDs in this dataset is mentioned in Appendix A.

A couple of things are worth mentioning here. First, Twitter allows users the option to delete a tweet, which would mean that there would be no retrievable Tweet text and other related information (upon hydration) for a Tweet ID of a deleted tweet. All the Tweet IDs available in this dataset correspond to tweets that have not been deleted at the time of writing this paper. Second, the Twitter API's search feature does not return an exhaustive list of tweets that were posted in a specific date range. So, it is possible that multiple tweets that might have been posted in between this date range were not returned by the Twitter API's search feature when the data collection was performed and are thus not a part of this dataset.

3.2. Compliance with Twitter Policies

The privacy policy of Twitter [97] states *"Twitter is public and Tweets are immediately viewable and searchable by anyone around the world"*. To add, the Twitter developer agreement [98] defines tweets as *"public data"*. The guidelines for Twitter content redistribution [98] state *"If you provide Twitter Content to third parties, including downloadable datasets or via an API, you may only distribute Tweet IDs, Direct Message IDs, and/or User IDs (except as described below)"*. It also states *"We also grant special permissions to academic researchers sharing Tweet IDs and User IDs for non-commercial research purposes. Academic researchers are permitted to distribute an unlimited number of Tweet IDs and/or User IDs if they are doing so on behalf of an academic institution and for the sole purpose of non-commercial research"*. Therefore, it may be concluded that mining relevant tweets from Twitter to develop a dataset (comprising only Tweet IDs) and to share the same is in compliance with the privacy policy, developer agreement, and content redistribution guidelines of Twitter.

3.3. Compliance with FAIR

This section outlines how this dataset is compliant with the FAIR (Findability, Accessibility, Interoperability, and Reusability) principles for scientific data management [111]. The dataset is findable, as it has a unique and permanent DOI, which was assigned by Zenodo. The dataset is accessible online. It is interoperable due to the use of .txt files for data representation that can be downloaded, read, and analyzed across different computer systems and applications. The dataset is reusable as the associated tweets and related information, such as user ID, username, retweet count, etc., for all the Tweet IDs, can be obtained by the process of hydration in compliance with Twitter policies (Appendix A) for data analysis and interpretation.

4. Data Description

This section provides a detailed description of this dataset. The raw version of the dataset comprised 67,319 tweets. This included multiple duplicate tweets. The duplicate tweets were recorded mostly because several Twitter users used a list of different hashtags referring to either online learning and/or the Omicron variant of COVID-19 in the same tweet, probably for increased audience engagement. For instance, as per the methodology described in Section 3, Tweet ID: 1464533235367510019 was captured twice, as it contains two synonyms ("omicron" and "covid") from the list of synonyms presented in Table 1. Therefore, after the data collection process was completed as described in Section 3, data preprocessing and data cleaning were performed using RapidMiner to remove duplicate tweets. After the removal of duplicate tweets, the dataset comprised 52,984 Tweet IDs corresponding to the same number of tweets about online learning posted on Twitter between 9 November 2021 (the sample collected on this date was the first case of Omicron) to 13 July 2022 (the most recent date at the time of resubmission of this paper to this journal after the completion of the first round of peer review and the subsequent editorial decision). The dataset is available at https://doi.org/10.5281/zenodo.6837118. The dataset comprises nine .txt files. Table 2 presents the description of each of these dataset files along with the number of Tweet IDs present in each of them. As can be seen from Table 2, the greatest number of Tweets were posted in January 2022. The fact that the tweets of only 13 days in July 2022 were mined is the likely reason why July 2022 accounts for the least number of tweets as per this table.

Table 2. Description of all the files present in this dataset that comprises tweets about online learning during the current COVID-19 Omicron Wave.

Filename	No. of Tweet IDs	Date Range of the Associated Tweets
TweetIDs_November_2021.txt	1283	1 November 2021 to 30 November 2021
TweetIDs_December_2021.txt	10,545	1 December 2021 to 31 December 2021
TweetIDs_January_2022.txt	23,078	1 January 2022 to 31 January 2022
TweetIDs_February_2022.txt	4751	1 February 2022 to 28 February 2022
TweetIDs_March_2022.txt	3434	1 March 2022 to 31 March 2022
TweetIDs_April_2022.txt	3355	1 April 2022 to 30 April 2022
TweetIDs_May_2022.txt	3120	1 May 2022 to 31 May 2022
TweetIDs_June_2022.txt	2361	1 June 2022 to 30 June 2022
TweetIDs_July_2022.txt	1057	1 June 2022 to 13 July 2022

Table 3 presents some characteristic features of this dataset. As can be seen from Table 3, the tweets are present in 34 different languages in this dataset. The most common language is English (50,539 Tweets), which is followed by Indonesian (527 Tweets), Tagalog (525 Tweets), Estonian (364 Tweets), Spanish (236 Tweets), Hindi (179 Tweets), and 28 other languages. All these tweets were posted on 237 different days between 9 November 2021 and 13 July 2022. The highest number of Tweets was recorded on 5 January 2022 (2067 Tweets), which is followed by 6 January 2022 (1592 Tweets), 3 January 2022 (1465 Tweets), 4 January 2022 (1355 Tweets), and the other dates. A total of 17,950 distinct Twitter users posted these tweets, who have a total follower count of 4,345,192,697. The combined favorite count and retweet count of all the tweets present in this dataset are 3,273,263 and 556,980, respectively. A total of 5722 Tweets present in this dataset were posted by Twitter users with a verified Twitter account, and the remaining Tweets came from an unverified Twitter account. The number of distinct URLs that can be found embedded in these Tweets is 7869. The URL that occurs the greatest number of times (30 times) in the Tweets points to a list of online courses for COVID-19 safety at work [112]. The URL that occurs the second greatest number of times (29 times) is a YouTube video that is also an online course on COVID-19 [113].

Table 3. Characteristic features of this dataset that comprises tweets about online learning during the current COVID-19 Omicron Wave

Characteristic Feature	Count
Languages in which the Tweets are available	34
Distinct days when the Tweets were posted	237
Distinct users who posted the Tweets	17,950
Total follower count of all the Twitter users who posted the Tweets	4,345,192,697
Number of Tweets from a verified Twitter account	5722
Number of Tweets from an unverified Twitter account	47,262
Total favorite count of all the Tweets	3,273,263
Total retweet count of all the Tweets	556,980
Distinct URLs embedded in the Tweets	7869

5. Potential Applications: Brief Overview

This dataset of more than 50,000 Tweet IDs is expected to help advance interdisciplinary research in different fields such as Big Data, Data Science, Data Mining, Natural Language Processing, Healthcare, and their related disciplines. A few potential applications and use-case scenarios that may be investigated using this dataset include performing sentiment analysis [114], performing aspect-based sentiment analysis [115], predicting popular tweets [116], detecting sarcasm [117], developing topic modeling [118], tracking retweeting patterns [119], ranking tweets [120], performing content value analysis [121], tracking credibility of information [122], detecting conspiracy theories [123], predicting emoji usage patterns [124], studying the relevance of information [125], detecting satire [126], detecting deception [127], extracting categorical topics and emerging issues [128], characterizing Twitter users [129], and detection of Twitter user demographics [130] in the context of Twitter chatter related to online learning during the current Omicron wave of COVID-19.

6. Conclusions

The outbreak of COVID-19 led to schools, colleges, and universities in almost all parts of the world closing and transitioning to online learning. The development of vaccines and other forms of treatment towards the end of 2020 led to some of these educational institutions reopening and starting to function in a hybrid as well as in a completely in-person manner. The recent surge of COVID-19 cases globally due to the Omicron variant, the most immune-evasive variant of COVID-19 that presents very strong resistance against antibody-based or plasma-based treatments, has resulted in several such educational institutions switching to online learning once again. This has led to an increase in the number of online conversations, specifically on Twitter, related to online learning since the first detected case of the Omicron variant in November 2021. Mining such tweets to develop a dataset would serve as a data resource for interdisciplinary research related to the analysis of interest, views, opinions, perspectives, attitudes, and feedback towards online learning during the current surge of COVID-19 cases caused due to this variant. The prior works in this field did not focus on the development of a similar data resource. Therefore, this work presents an open-access dataset of more than 50,000 Tweet IDs (that correspond to the same number of tweets) about online learning posted on Twitter between 9 November 2021 (the sample collected on this date was the first case of Omicron) and 13 July 2022 (the most recent date at the time of resubmission of this journal paper after the completion of the first round of peer review and the subsequent editorial decision). The dataset is compliant with the privacy policy, developer agreement, and guidelines for content redistribution of Twitter, as well as with the FAIR principles (Findability, Accessibility, Interoperability, and Reusability) principles for scientific data management. The paper also briefly outlines a few potential research directions that may be investigated using this dataset. Future work on this project would involve updating the dataset with more recent tweets to ensure that the scientific community has access to the recent data in this regard.

Funding: This research received no external funding.

Institutional Review Board Statement: Not applicable.

Informed Consent Statement: Not applicable.

Data Availability Statement: The data presented in this study are publicly available at https://doi.org/10.5281/zenodo.6837118.

Conflicts of Interest: The author declares no conflict of interest.

Appendix A

The following is the step-by-step process for using the Hydrator app [105] to hydrate this dataset or, in other words, to obtain the text of the tweet, user ID, username, retweet count, language, tweet URL, source, and other public information related to all the Tweet IDs present in this dataset. The Hydrator app works in compliance with the policies for accessing and calling the Twitter API.

1. Download and install the desktop version of the Hydrator app [131].
2. Click on the "Link Twitter Account" button on the Hydrator app to connect the app to an active Twitter account.
3. Click on the "Add" button to upload one of the dataset files (in .txt format, such as TweetIDs_June_2022.txt). This process adds the dataset file to the Hydrator app.
4. If the file upload is successful, the Hydrator app will show the total number of Tweet IDs present in the file. For instance, for the file, "TweetIDs_June_2022.txt", the app would show the Number of Tweet IDs as 2361.
5. Provide details for the respective fields: Title, Creator, Publisher, and URL in the app, and click on "Add Dataset" to add this dataset to the app.
6. The app would automatically redirect to the "Datasets" tab. Click on the "Start" button to start hydrating the Tweet IDs. During the hydration process, the progress indicator would increase, indicating the number of Tweet IDs that have been successfully hydrated and the number of Tweet IDs that are pending hydration.
7. After the hydration process ends, a .jsonl file would be generated by the app that the user can choose to save on the local storage.
8. The app would also display a "CSV" button in place of the "Start" button. Clicking on this "CSV" button would generate a .csv file with detailed information about the tweets, which would include the text of the tweet, user ID, username, retweet count, language, tweet URL, source, and other public information related to the tweet.
9. Repeat steps 3–8 for hydrating all the files of this dataset.

References

1. Wu, Y.-C.; Chen, C.-S.; Chan, Y.-J. Overview of the 2019 Novel Coronavirus (2019-NCoV): The Pathogen of Severe Specific Contagious Pneumonia (SSCP): The Pathogen of Severe Specific Contagious Pneumonia (SSCP). *J. Chin. Med. Assoc.* **2020**, *83*, 217–220. [CrossRef]
2. COVID Live. Coronavirus Statistics—Worldometer. Available online: https://www.worldometers.info/coronavirus/ (accessed on 6 June 2022).
3. CDC. SARS-CoV-2 Variant Classifications and Definitions. Available online: https://www.cdc.gov/coronavirus/2019-ncov/variants/variant-classifications.html (accessed on 6 June 2022).
4. Classification of Omicron (B.1.1.529): SARS-CoV-2 Variant of Concern. Available online: https://www.who.int/news/item/26-11-2021-classification-of-omicron-(b.1.1.529)-sars-cov-2-variant-of-concern (accessed on 6 June 2022).
5. Gobeil, S.M.-C.; Henderson, R.; Stalls, V.; Janowska, K.; Huang, X.; May, A.; Speakman, M.; Beaudoin, E.; Manne, K.; Li, D.; et al. Structural diversity of the SARS-CoV-2 Omicron spike. *Mol. Cell* **2022**, *82*, 2050–2068.e6. [CrossRef] [PubMed]
6. Schmidt, F.; Muecksch, F.; Weisblum, Y.; Da Silva, J.; Bednarski, E.; Cho, A.; Wang, Z.; Gaebler, C.; Caskey, M.; Nussenzweig, M.C.; et al. Plasma Neutralization of the SARS-CoV-2 Omicron Variant. *N. Engl. J. Med.* **2022**, *386*, 599–601. [CrossRef] [PubMed]

7. Feiner, L. WHO Says Omicron Cases Are "off the Charts" as Global Infections Set New Records. Available online: https://www.cnbc.com/2022/01/12/who-says-omicron-cases-are-off-the-charts-as-global-infections-set-new-records.html (accessed on 6 June 2022).
8. Weekly Epidemiological Update on COVID-19—22 March 2022. Available online: https://www.who.int/publications/m/item/weekly-epidemiological-update-on-covid-19---22-march-2022 (accessed on 6 June 2022).
9. SARS-CoV-2 Omicron Variant Cases Worldwide 2022. Available online: https://www.statista.com/statistics/1279100/number-omicron-variant-worldwide-by-country/ (accessed on 6 June 2022).
10. Lancet, T. India under COVID-19 lockdown. *Lancet* **2020**, *395*, 1315. [CrossRef]
11. Surano, F.V.; Porfiri, M.; Rizzo, A. Analysis of lockdown perception in the United States during the COVID-19 pandemic. *Eur. Phys. J. Spéc. Top.* **2021**, *231*, 1625–1633. [CrossRef]
12. Jallow, H.; Renukappa, S.; Suresh, S. The impact of COVID-19 outbreak on United Kingdom infrastructure sector. *Smart Sustain. Built Environ.* **2020**, *10*, 581–593. [CrossRef]
13. Tejedor, S.; Cervi, L.; Pérez-Escoda, A.; Jumbo, F.T. Digital Literacy and Higher Education during COVID-19 Lockdown: Spain, Italy, and Ecuador. *Publications* **2020**, *8*, 48. [CrossRef]
14. Fountoulakis, K.N.; Apostolidou, M.K.; Atsiova, M.B.; Filippidou, A.K.; Florou, A.K.; Gousiou, D.S.; Katsara, A.R.; Mantzari, S.N.; Padouva-Markoulaki, M.; Papatriantafyllou, E.I.; et al. Self-reported changes in anxiety, depression and suicidality during the COVID-19 lockdown in Greece. *J. Affect. Disord.* **2021**, *279*, 624–629. [CrossRef]
15. Guzzetta, G.; Riccardo, F.; Marziano, V.; Poletti, P.; Trentini, F.; Bella, A.; Andrianou, X.; Del Manso, M.; Fabiani, M.; Bellino, S.; et al. The Impact of a Nation-Wide Lockdown on COVID-19 Transmissibility in Italy. *arXiv* **2020**, arXiv:2004.12338.
16. Probst, T.; Stippl, P.; Pieh, C. Changes in Provision of Psychotherapy in the Early Weeks of the COVID-19 Lockdown in Austria. *Int. J. Environ. Res. Public Health* **2020**, *17*, 3815. [CrossRef]
17. Oyediran, W.O.; Omoare, A.M.; Owoyemi, M.A.; Adejobi, A.O.; Fasasi, R.B. Prospects and limitations of e-learning application in private tertiary institutions amidst COVID-19 lockdown in Nigeria. *Heliyon* **2020**, *6*, e05457. [CrossRef] [PubMed]
18. Lau, H.; Khosrawipour, V.; Kocbach, P.; Mikolajczyk, A.; Schubert, J.; Bania, J.; Khosrawipour, T. The positive impact of lockdown in Wuhan on containing the COVID-19 outbreak in China. *J. Travel Med.* **2020**, *27*, taaa037. [CrossRef]
19. Chan, D.Z.; Stewart, R.A.; Kerr, A.J.; Dicker, B.; Kyle, C.V.; Adamson, P.D.; Devlin, G.; Edmond, J.; El-Jack, S.; Elliott, J.M.; et al. The impact of a national COVID-19 lockdown on acute coronary syndrome hospitalisations in New Zealand (ANZACS-QI 55). *Lancet Reg. Health West. Pac.* **2020**, *5*, 100056. [CrossRef]
20. Fahy, S.; Moore, J.; Kelly, M.; Flannery, O.; Kenny, P. Analysing the variation in volume and nature of trauma presentations during COVID-19 lockdown in Ireland. *Bone Jt. Open* **2020**, *1*, 261–266. [CrossRef] [PubMed]
21. LeMenager, T.; Neissner, M.; Koopmann, A.; Reinhard, I.; Georgiadou, E.; Müller, A.; Kiefer, F.; Hillemacher, T. COVID-19 Lockdown Restrictions and Online Media Consumption in Germany. *Int. J. Environ. Res. Public Health* **2020**, *18*, 14. [CrossRef] [PubMed]
22. Stiegler, N.; Bouchard, J.-P. South Africa: Challenges and successes of the COVID-19 lockdown. *Ann. Med. Psychol.* **2020**, *178*, 695–698. [CrossRef] [PubMed]
23. Matheson, A.; McGannon, C.J.; Malhotra, A.; Palmer, K.R.; Stewart, A.E.; Wallace, E.M.; Mol, B.W.; Hodges, R.J.; Rolnik, D.L. Prematurity Rates During the Coronavirus Disease 2019 (COVID-19) Pandemic Lockdown in Melbourne, Australia. *Obstet. Gynecol.* **2021**, *137*, 405–407. [CrossRef]
24. Di Domenico, L.; Pullano, G.; Sabbatini, C.E.; Boëlle, P.-Y.; Colizza, V. Impact of lockdown on COVID-19 epidemic in Île-de-France and possible exit strategies. *BMC Med.* **2020**, *18*, 240. [CrossRef]
25. Lehmann, S.; Skogen, J.C.; Haug, E.; Mæland, S.; Fadnes, L.T.; Sandal, G.M.; Hysing, M.; Bjørknes, R. Perceived consequences and worries among youth in Norway during the COVID-19 pandemic lockdown. *Scand. J. Public Health* **2021**, *49*, 755–765. [CrossRef]
26. Onyeaka, H.; Anumudu, C.K.; Al-Sharify, Z.T.; Egele-Godswill, E.; Mbaegbu, P. COVID-19 pandemic: A review of the global lockdown and its far-reaching effects. *Sci. Prog.* **2021**, *104*, 368504211019854. [CrossRef]
27. Research and Markets Ltd. Online Education Market & Global Forecast, by End User, Learning Mode (Self-Paced, Instructor Led), Technology, Country, Company. Available online: https://www.researchandmarkets.com/reports/4876815/ (accessed on 15 July 2022).
28. Singh, V.; Thurman, A. How Many Ways Can We Define Online Learning? A Systematic Literature Review of Definitions of Online Learning (1988–2018). *Am. J. Distance Educ.* **2019**, *33*, 289–306. [CrossRef]
29. Education: From Disruption to Recovery, UNSECO Report. Available online: https://en.unesco.org/covid19/educationresponse (accessed on 6 June 2022).
30. Education and COVID-19. Available online: https://data.unicef.org/topic/education/covid-19/ (accessed on 6 June 2022).
31. COVID-19: Are Children Able to Continue Learning during School Closures? Available online: https://data.unicef.org/resources/remote-learning-reachability-factsheet/ (accessed on 6 June 2022).
32. Stasi, C.; Fallani, S.; Voller, F.; Silvestri, C. Treatment for COVID-19: An overview. *Eur. J. Pharmacol.* **2020**, *889*, 173644. [CrossRef] [PubMed]
33. Peng, Y.; Tao, H.; Satyanarayanan, S.K.; Jin, K.; Su, H. A Comprehensive Summary of the Knowledge on COVID-19 Treatment. *Aging Dis.* **2021**, *12*, 155–191. [CrossRef] [PubMed]

34. Bartoli, A.; Gabrielli, F.; Alicandro, T.; Nascimbeni, F.; Andreone, P. COVID-19 treatment options: A difficult journey between failed attempts and experimental drugs. *Intern. Emerg. Med.* **2021**, *16*, 281–308. [CrossRef] [PubMed]
35. Reopening Schools after COVID-19 Closures Considerations for States. Available online: http://files.eric.ed.gov/fulltext/ED609236.pdf (accessed on 6 June 2022).
36. Gunawan, M.; Setiawan, A.A.; Leonita, I. Neville School Reopening during COVID-19 Pandemic: Is It Safe? A Systematic Review. Available online: https://jamsa.amsa-international.org/index.php/main/article/view/380 (accessed on 1 August 2022).
37. Lockdowns, School Closures Return to Mainland China. Available online: https://www.usnews.com/news/education-news/articles/2022-03-14/lockdowns-school-closures-return-to-mainland-china (accessed on 6 June 2022).
38. Sachdev, C. India Postpones In-School Learning as Omicron Surges. Available online: https://theworld.org/stories/2022-01-07/india-postpones-school-learning-omicron-surges (accessed on 6 June 2022).
39. School Systems around the World Debate New Closures as Omicron Spreads. Available online: https://www.washingtonpost.com/world/2022/01/07/global-school-closures-omicron/ (accessed on 6 June 2022).
40. Nearly 6000 Public Schools in Japan at Least Partially Closed Amid Omicron Wave. Available online: https://www.japantimes.co.jp/news/2022/02/04/national/school-closures-omicron/ (accessed on 8 June 2022).
41. Khan, N. Hong Kong to Shut Schools to Fight Omicron; Foreigners Rush to Leave. Available online: https://www.wsj.com/articles/hong-kong-sets-all-schools-for-covid-19-response-centers-11645530595 (accessed on 1 August 2022).
42. Collin Binkley (Associated Press). Dozens of US Colleges Starting Semester Online. Available online: https://www.10tv.com/article/news/nation-world/colleges-online-omicron-covid-remote-learning/507-63ea4bd0-9ccf-40cd-a373-e54da40e2fdb (accessed on 6 June 2022).
43. Snyder, T.; Byrd, G. The Internet of Everything. *Computer* **2017**, *50*, 8–9. [CrossRef]
44. Boulianne, S. Social media use and participation: A meta-analysis of current research. *Inf. Commun. Soc.* **2015**, *18*, 524–538. [CrossRef]
45. Kavada, A. Social Media as Conversation: A Manifesto. *Soc. Media Soc.* **2015**, *1*, 205630511558079. [CrossRef]
46. Liu, Y.; Singh, L.; Mneimneh, Z. A Comparative Analysis of Classic and Deep Learning Models for Inferring Gender and Age of Twitter Users. In Proceedings of the 2nd International Conference on Deep Learning Theory and Applications, Online, 7–9 July 2021; SciTePress–Science and Technology Publications: Setúbal, Portugal, 2021.
47. Özbaş-Anbarlı, Z. Living in digital space: Everyday life on Twitter. *Commun. Soc.* **2021**, *34*, 31–47. [CrossRef]
48. Gruzd, A.; Wellman, B.; Takhteyev, Y. Imagining Twitter as an Imagined Community. *Am. Behav. Sci.* **2011**, *55*, 1294–1318. [CrossRef]
49. Aslam, S. Twitter by the Numbers (2022): Stats, Demographics & Fun Facts. Available online: https://www.Omnicoreagency.com (accessed on 13 July 2022).
50. Chen, E.; Deb, A.; Ferrara, E. #Election2020: The first public Twitter dataset on the 2020 US Presidential election. *J. Comput. Soc. Sci.* **2021**, *5*, 1–18. [CrossRef]
51. Haq, E.-U.; Tyson, G.; Lee, L.-H.; Braud, T.; Hui, P. Twitter Dataset for 2022 Russo-Ukrainian Crisis. *arXiv* **2022**, arXiv:2203.02955. [CrossRef]
52. Effrosynidis, D.; Karasakalidis, A.I.; Sylaios, G.; Arampatzis, A. The climate change Twitter dataset. *Expert Syst. Appl.* **2022**, *204*, 117541. [CrossRef]
53. Meng, L.; Dong, Z.S. Natural Hazards Twitter Dataset. *arXiv* **2020**, arXiv:2004.14456.
54. Urchs, S.; Wendlinger, L.; Mitrovic, J.; Granitzer, M. MMoveT15: A Twitter Dataset for Extracting and Analysing Migration-Movement Data of the European Migration Crisis 2015. In Proceedings of the 2019 IEEE 28th International Conference on Enabling Technologies: Infrastructure for Collaborative Enterprises (WETICE), Napoli, Italy, 12–14 June 2019; IEEE: New York, NY, USA, 2019; pp. 146–149.
55. Dooms, S.; De Pessemier, T.; Martens, L. MovieTweetings: A Movie Rating Dataset Collected from Twitter. In Proceedings of the Workshop on Crowdsourcing and Human Computation for Recommender Systems (CrowdRec 2013), Held in Conjunction with the 7th ACM Conference on Recommender Systems (RecSys 2013), Hong Kong, China, 12 October 2013.
56. Wijesiriwardene, T.; Inan, H.; Kursuncu, U.; Gaur, M.; Shalin, V.L.; Thirunarayan, K.; Sheth, A.; Arpinar, I.B. ALONE: A Dataset for Toxic Behavior among Adolescents on Twitter. In *Social Informatics*; Lecture Notes in Computer Science; Springer International Publishing: Cham, Switzerland, 2020; pp. 427–439. ISBN 9783030609740.
57. Zangerle, E.; Pichl, M.; Gassler, W.; Specht, G. #nowplaying Music Dataset: Extracting Listening Behavior from Twitter. In Proceedings of the First International Workshop on Internet-Scale Multimedia Management—WISMM'14, Orlando, FL, USA, 7 November 2014; ACM Press: New York, NY, USA, 2014.
58. Sech, J.; DeLucia, A.; Buczak, A.L.; Dredze, M. Civil Unrest on Twitter (CUT): A Dataset of Tweets to Support Research on Civil Unrest. In Proceedings of the Sixth Workshop on Noisy User-Generated Text (W-NUT 2020), Online, 19 November 2020; Association for Computational Linguistics: Stroudsburg, PA, USA, 2020; pp. 215–221.
59. Tekumalla, R.; Banda, J.M. A Large-Scale Twitter Dataset for Drug Safety Applications Mined from Publicly Existing Resources. *arXiv* **2020**, arXiv:2003.13900.
60. Stemmer, M.; Parmet, Y.; Ravid, G. What Are IBD Patients Talking about on Twitter? In *ICT for Health, Accessibility and Wellbeing*; Springer International Publishing: Cham, Switzerland, 2021; pp. 206–220. ISBN 9783030942083.

61. Adnan, M.; Anwar, K. Online Learning amid the COVID-19 Pandemic: Students' Perspectives. *J. Pedagog. Sociol. Psychol.* **2020**, *2*, 45–51. [CrossRef]
62. Rasmitadila, R.; Aliyyah, R.R.; Rachmadtullah, R.; Samsudin, A.; Syaodih, E.; Nurtanto, M.; Tambunan, A.R.S. The Perceptions of Primary School Teachers of Online Learning during the COVID-19 Pandemic Period: A Case Study in Indonesia. *J. Ethn. Cult. Stud.* **2020**, *7*, 90–109. [CrossRef]
63. Irawan, A.W.; Dwisona, D.; Lestari, M. Psychological Impacts of Students on Online Learning during the Pandemic COVID-19. *KONSELI J. Bimbing. Konseling (E-J.)* **2020**, *7*, 53–60. [CrossRef]
64. Baticulon, R.E.; Sy, J.J.; Alberto, N.R.I.; Baron, M.B.C.; Mabulay, R.E.C.; Rizada, L.G.T.; Tiu, C.J.S.; Clarion, C.A.; Reyes, J.C.B. Barriers to Online Learning in the Time of COVID-19: A National Survey of Medical Students in the Philippines. *Med. Sci. Educ.* **2021**, *31*, 615–626. [CrossRef]
65. Hussein, E.; Daoud, S.; Alrabaiah, H.; Badawi, R. Exploring undergraduate students' attitudes towards emergency online learning during COVID-19: A case from the UAE. *Child. Youth Serv. Rev.* **2020**, *119*, 105699. [CrossRef]
66. Famularsih, S. Students' Experiences in Using Online Learning Applications Due to COVID-19 in English Classroom. *Stud. Learn. Teach.* **2020**, *1*, 112–121. [CrossRef]
67. Sutarto, S.; Sari, D.P.; Fathurrochman, I. Teacher strategies in online learning to increase students' interest in learning during COVID-19 pandemic. *J. Konseling Pendidik.* **2020**, *8*, 129. [CrossRef]
68. Almusharraf, N.; Khahro, S. Students Satisfaction with Online Learning Experiences during the COVID-19 Pandemic. *Int. J. Emerg. Technol. Learn. (iJET)* **2020**, *15*, 246. [CrossRef]
69. Al-Salman, S.; Haider, A.S. Jordanian University Students' Views on Emergency Online Learning during COVID-19. *Online Learn.* **2021**, *25*, 286–302. [CrossRef]
70. Bolatov, A.K.; Seisembekov, T.Z.; Askarova, A.Z.; Baikanova, R.K.; Smailova, D.S.; Fabbro, E. Online-Learning due to COVID-19 Improved Mental Health Among Medical Students. *Med. Sci. Educ.* **2020**, *31*, 183–192. [CrossRef] [PubMed]
71. Agormedah, E.K.; Henaku, E.A.; Ayite, D.M.K.; Ansah, E.A. Online Learning in Higher Education during COVID-19 Pandemic: A case of Ghana. *J. Educ. Technol. Online Learn.* **2020**, *3*, 183–210. [CrossRef]
72. Moawad, R.A. Online Learning during the COVID-19 Pandemic and Academic Stress in University Students. *Rev. Rom. Pentru Educ. Multidimens.* **2020**, *12*, 100–107. [CrossRef]
73. Khan, M.M.; Rahman, S.M.T.; Islam, S.T.A. Online Education System in Bangladesh during COVID-19 Pandemic. *Creat. Educ.* **2021**, *12*, 441–452. [CrossRef]
74. Catalano, A.J.; Torff, B.; Anderson, K.S. Transitioning to online learning during the COVID-19 pandemic: Differences in access and participation among students in disadvantaged school districts. *Int. J. Inf. Learn. Technol.* **2021**, *38*, 258–270. [CrossRef]
75. Kapasia, N.; Paul, P.; Roy, A.; Saha, J.; Zaveri, A.; Mallick, R.; Barman, B.; Das, P.; Chouhan, P. Impact of lockdown on learning status of undergraduate and postgraduate students during COVID-19 pandemic in West Bengal, India. *Child. Youth Serv. Rev.* **2020**, *116*, 105194. [CrossRef]
76. Burns, D.; Dagnall, N.; Holt, M. Assessing the Impact of the COVID-19 Pandemic on Student Wellbeing at Universities in the United Kingdom: A Conceptual Analysis. *Front. Educ.* **2020**, *5*, 582882. [CrossRef]
77. Küsel, J.; Martin, F.; Markic, S. University Students' Readiness for Using Digital Media and Online Learning—Comparison between Germany and the USA. *Educ. Sci.* **2020**, *10*, 313. [CrossRef]
78. Al Darayseh, A.S. The Impact of COVID-19 Pandemic on Modes of Teaching Science in UAE Schools. *J. Educ. Pract.* **2020**, *11*, 110–115. [CrossRef]
79. Tsekhmister, Y.V.; Konovalova, T.; Tsekhmister, B.Y.; Agrawal, A.; Ghosh, D. Evaluation of Virtual Reality Technology and Online Teaching System for Medical Students in Ukraine During COVID-19 Pandemic. *Int. J. Emerg. Technol. Learn.* **2021**, *16*, 127–139. [CrossRef]
80. Arsaliev, S.M.-K.; Andrienko, A.S. The Development of Ethnocultural Competence of University Students during COVID-19 Pandemic in Russia. In Proceedings of the 2020 3rd International Seminar on Education Research and Social Science (ISERSS 2020), Kuala Lumpur, Malaysia, 24–26 December 2021; Atlantis Press: Paris, France, 2021.
81. Cárdenas-Cruz, A.; Gómez-Moreno, G.; Matas-Lara, A.; Romero-Palacios, P.J.; Parrilla-Ruiz, F.M. An example of adaptation: Experience of virtual clinical skills circuits of internal medicine students at the Faculty of Medicine, University of Granada (Spain) during the COVID-19 pandemic. *Med. Educ. Online* **2022**, *27*, 2040191. [CrossRef] [PubMed]
82. Papouli, E.; Chatzifotiou, S.; Tsairidis, C. The use of digital technology at home during the COVID-19 outbreak:Views of social work students in Greece. *Soc. Work Educ.* **2020**, *39*, 1107–1115. [CrossRef]
83. Parmigiani, D.; Benigno, V.; Giusto, M.; Silvaggio, C.; Sperandio, S. E-inclusion: Online special education in Italy during the COVID-19 pandemic. *Technol. Pedagog. Educ.* **2020**, *30*, 111–124. [CrossRef]
84. Resch, K.; Alnahdi, G.; Schwab, S. Exploring the effects of the COVID-19 emergency remote education on students' social and academic integration in higher education in Austria. *High. Educ. Res. Dev.* **2022**, 1–15. [CrossRef]
85. Oyarinde, O.N.; Komolafe, O.G. Impact of Google Classroom as an Online Learning Delivery during COVID-19 Pandemic: The Case of a Secondary School in Nigeria. *J. Educ. Soc. Behav. Sci.* **2020**, *33*, 53–61. [CrossRef]
86. Chen, T.; Peng, L.; Yin, X.; Rong, J.; Yang, J.; Cong, G. Analysis of User Satisfaction with Online Education Platforms in China during the COVID-19 Pandemic. *Healthcare* **2020**, *8*, 200. [CrossRef]

87. Drane, C.; Vernon, L.; O'shea, S. The Impact of "Learning at Home" on the Educational Outcomes of Vulnerable Children in Australia during the COVID-19 Pandemic. Available online: https://www.ncsehe.edu.au/wp-content/uploads/2020/04/NCSEHE_V2_Final_literaturereview-learningathome-covid19-final_30042020.pdf (accessed on 6 June 2022).
88. Mukuna, K.R.; Aloka, P.J.O. Exploring Educators' Challenges of Online Learning in COVID-19 at a Rural School, South Africa. *Int. J. Learn. Teach. Educ. Res.* **2020**, *19*, 134–149. [CrossRef]
89. Hsiao, Y.-C. Impacts of course type and student gender on distance learning performance: A case study in Taiwan. *Educ. Inf. Technol.* **2021**, *26*, 6807–6822. [CrossRef] [PubMed]
90. Nafrees, A.; Roshan, A.; Baanu, A.N.; Nihma, M.F.; Shibly, F. Awareness of Online Learning of Undergraduates during COVID-19 with special reference to South Eastern University of Sri Lanka. *J. Physics Conf. Ser.* **2020**, *1712*, 012010. [CrossRef]
91. Cui, L.; Lee, D. CoAID: COVID-19 Healthcare Misinformation Dataset. *arXiv* **2020**, arXiv:2006.00885.
92. Elhadad, M.K.; Li, K.F.; Gebali, F. COVID-19-FAKES: A Twitter (Arabic/English) Dataset for Detecting Misleading Information on COVID-19. In *Advances in Intelligent Networking and Collaborative Systems*; Springer International Publishing: Cham, Switzerland, 2021; pp. 256–268. ISBN 9783030577957.
93. Hayawi, K.; Shahriar, S.; Serhani, M.; Taleb, I.; Mathew, S. ANTi-Vax: A novel Twitter dataset for COVID-19 vaccine misinformation detection. *Public Health* **2021**, *203*, 23–30. [CrossRef] [PubMed]
94. Nasser, N.; Karim, L.; El Ouadrhiri, A.; Ali, A.; Khan, N. n-Gram based language processing using Twitter dataset to identify COVID-19 patients. *Sustain. Cities Soc.* **2021**, *72*, 103048. [CrossRef]
95. DeVerna, M.R.; Pierri, F.; Truong, B.T.; Bollenbacher, J.; Axelrod, D.; Loynes, N.; Torres-Lugo, C.; Yang, K.-C.; Menczer, F.; Bryden, J. CoVaxxy: A Collection of English-Language Twitter Posts about COVID-19 Vaccines. *arXiv* **2021**, arXiv:2101.07694.
96. Cheng, M.; Wang, S.; Yan, X.; Yang, T.; Wang, W.; Huang, Z.; Xiao, X.; Nazarian, S.; Bogdan, P. A COVID-19 Rumor Dataset. *Front. Psychol.* **2021**, *12*, 644801. [CrossRef]
97. Privacy Policy. Available online: https://twitter.com/en/privacy/previous/version_15 (accessed on 6 June 2022).
98. Developer Agreement and Policy. Available online: https://developer.twitter.com/en/developer-terms/agreement-and-policy (accessed on 6 June 2022).
99. RapidMiner GmbH Search Twitter—RapidMiner Documentation. Available online: https://docs.rapidminer.com/latest/studio/operators/data_access/applications/twitter/search_twitter.html (accessed on 6 June 2022).
100. Mierswa, I.; Wurst, M.; Klinkenberg, R.; Scholz, M.; Euler, T. YALE: Rapid Prototyping for Complex Data Mining Tasks. In Proceedings of the 12th ACM SIGKDD International Conference on Knowledge Discovery and Data Mining—KDD'06, Philadelphia, PA, USA, 20–23 August 2006; ACM Press: New York, NY, USA, 2006.
101. Using Standard Search. Available online: https://developer.twitter.com/en/docs/twitter-api/v1/tweets/search/guides/standard-operators (accessed on 6 June 2022).
102. Anohina, A. Analysis of the Terminology Used in the Field of Virtual Learning. *J. Educ. Technol. Soc.* **2005**, *8*, 91–102.
103. Ma, H.; Shen, L.; Sun, H.; Xu, Z.; Hou, L.; Wu, S.; Fang, A.; Li, J.; Qian, Q. COVID term: A bilingual terminology for COVID-19. *BMC Med. Inform. Decis. Mak.* **2021**, *21*, 231. [CrossRef]
104. Structure of U.S. Education. Available online: https://www2.ed.gov/about/offices/list/ous/international/usnei/us/edlite-structure-us.html (accessed on 15 July 2022).
105. The Education System in India. Available online: https://www.gnu.org/education/edu-system-india.en.html (accessed on 15 July 2022).
106. British Education System. Available online: https://www.brightworldguardianships.com/en/guardianship/british-education-system/ (accessed on 15 July 2022).
107. Lamsal, R. Hydrating Tweet IDs. Available online: https://theneuralblog.com/hydrating-tweet-ids/ (accessed on 6 June 2022).
108. Hydrator: Turn Tweet IDs into Twitter JSON & CSV from Your Desktop! Available online: https://github.com/DocNow/hydrator (accessed on 6 June 2022).
109. Tekumalla, R.; Banda, J.M. Social Media Mining Toolkit (SMMT). *Genom. Inform.* **2020**, *18*, e16. [CrossRef]
110. Twarc: A Command Line Tool (and Python Library) for Archiving Twitter JSON. Available online: https://github.com/DocNow/twarc (accessed on 6 June 2022).
111. Wilkinson, M.D.; Dumontier, M.; Aalbersberg, I.J.; Appleton, G.; Axton, M.; Baak, A.; Blomberg, N.; Boiten, J.W.; da Silva Santos, L.B.; Bourne, P.E.; et al. The FAIR Guiding Principles for scientific data management and stewardship. *Sci. Data* **2016**, *3*, 160018. [CrossRef] [PubMed]
112. Coronavirus (COVID 19) Online Training Certificate Courses for Workplace, Employees, Workers Australia. Available online: https://www.sentrient.com.au/covid-19-coronavirus-courses (accessed on 15 July 2022).
113. Chew, P. LearnT-SMArET Online Course (18-11-2021). COVID-19: Peter Chew Pandemic to Endemic Strategy. Available online: https://www.youtube.com/watch?v=zLkUPY5Kt6c (accessed on 15 July 2022).
114. Carvalho, J.; Plastino, A. On the evaluation and combination of state-of-the-art features in Twitter sentiment analysis. *Artif. Intell. Rev.* **2020**, *54*, 1887–1936. [CrossRef]
115. Wang, J.; Xu, B.; Zu, Y. Deep Learning for Aspect-Based Sentiment Analysis. In Proceedings of the 2021 International Conference on Machine Learning and Intelligent Systems Engineering (MLISE), Chongqing, China, 9–11 July 2021; IEEE: New York, NY, USA, 2021; pp. 267–271.

116. Hong, L.; Dan, O.; Davison, B.D. Predicting Popular Messages in Twitter. In Proceedings of the 20th international conference companion on World Wide Web—WWW'11, Hyderabad, India, 28 March–1 April 2011. ACM Press: New York, NY, USA, 2011.
117. Bouazizi, M.; Ohtsuki, T.O. A Pattern-Based Approach for Sarcasm Detection on Twitter. *IEEE Access* **2016**, *4*, 5477–5488. [CrossRef]
118. Alvarez-Melis, D.; Saveski, M. Topic Modeling in Twitter: Aggregating Tweets by Conversations. In Proceedings of the Tenth International AAAI Conference on Web and Social Media, Cologne, Germany, 17–20 May 2016.
119. Boyd, D.; Golder, S.; Lotan, G. Tweet, Tweet, Retweet: Conversational Aspects of Retweeting on Twitter. In Proceedings of the 2010 43rd Hawaii International Conference on System Sciences, Honolulu, HI, USA 5–8 January 2010; IEEE: New York, NY, USA, 2010; pp. 1–10.
120. Uysal, I.; Croft, W.B. User Oriented Tweet Ranking: A Filtering Approach to Microblogs. In Proceedings of the 20th ACM International Conference on Information and Knowledge Management—CIKM'11, Glasgow, UK, 24–28 October 2011; ACM Press: New York, NY, USA, 2011.
121. André, P.; Bernstein, M.; Luther, K. Who Gives a Tweet?: Evaluating Microblog Content Value. In Proceedings of the ACM 2012 Conference on Computer Supported Cooperative Work—CSCW'12, Seattle, WA, USA, 11–15 February 2012; ACM Press: New York, NY, USA, 2012.
122. Ito, J.; Song, J.; Toda, H.; Koike, Y.; Oyama, S. Assessment of Tweet Credibility with LDA Features. In Proceedings of the 24th International Conference on World Wide Web—WWW'15 Companion, Florence, Italy, 18–22 May 2015; ACM Press: New York, NY, USA, 2015.
123. Stephens, M. A geospatial infodemic: Mapping Twitter conspiracy theories of COVID-19. *Dialogues Hum. Geogr.* **2020**, *10*, 276–281. [CrossRef]
124. Wu, C.; Wu, F.; Wu, S.; Huang, Y.; Xie, X. Tweet Emoji Prediction Using Hierarchical Model with Attention. In Proceedings of the 2018 ACM International Joint Conference and 2018 International Symposium on Pervasive and Ubiquitous Computing and Wearable Computers, Singapore, 8–12 October 2018; ACM: New York, NY, USA, 2018.
125. Mccreadie, R.; Macdonald, C. Relevance in Microblogs: Enhancing Tweet Retrieval Using Hyperlinked Documents. Available online: http://terrierteam.dcs.gla.ac.uk/publications/oair2013_McCreadie.pdf (accessed on 7 June 2022).
126. Salas-Zárate, M.D.P.; Paredes-Valverde, M.A.; Rodríguez-García, M.; Valencia-Garcia, R.; Alor-Hernández, G. Automatic detection of satire in Twitter: A psycholinguistic-based approach. *Knowl.-Based Syst.* **2017**, *128*, 20–33. [CrossRef]
127. Alowibdi, J.S.; Buy, U.A.; Yu, P.S.; Ghani, S.; Mokbel, M. Deception detection in Twitter. *Soc. Netw. Anal. Min.* **2015**, *5*, 32. [CrossRef]
128. Zheng, L.; Han, K. Extracting Categorical Topics from Tweets Using Topic Model. In *Information Retrieval Technology*; Springer: Berlin/Heidelberg, Germany, 2013; pp. 86–96. ISBN 9783642450679.
129. Zahra, K.; Azam, F.; Butt, W.H.; Ilyas, F. A Framework for User Characterization Based on Tweets Using Machine Learning Algorithms. In Proceedings of the 2018 VII International Conference on Network, Communication and Computing—ICNCC 2018, Taipei City, Taiwan, 14–16 December 2018; ACM Press: New York, NY, USA, 2018.
130. Sloan, L.; Morgan, J.; Housley, W.; Williams, M.; Edwards, A.; Burnap, P.; Rana, O. Knowing the Tweeters: Deriving Sociologically Relevant Demographics from Twitter. *Sociol. Res. Online* **2013**, *18*, 74–84. [CrossRef]
131. Hydrator. Available online: https://github.com/DocNow/hydrator/releases (accessed on 1 August 2022).

data

MDPI

Article

A Cross-Sectional Study on Mental Health of School Students during the COVID-19 Pandemic in India

Sibnath Deb [1], Samarjit Kar [2,*], Shayana Deb [3], Sanjib Biswas [4], Aehsan Ahmad Dar [5] and Tusharika Mukherjee [6]

1 Rajiv Gandhi National Institute of Youth Development, Post Box No. 6, Sriperumbudur Post, Sriperumbudur 602105, India; sibnath@rgniyd.gov.in or sibnath23@gmail.com
2 National Institute of Technology Durgapur, Mahatma Gandhi Rd, A-Zone, Durgapur 713209, India
3 Department of Psychology, CHRIST (Deemed to Be University), Hosur Rd, Bhavani Nagar, S.G. Palya, Bengaluru 560029, India; shayana.deb@arts.christuniversity.in
4 Calcutta Business School, Bishnupur 743503, India; sanjibb@acm.org
5 Department of Psychology, School of Liberal Arts & Social Sciences, SRM University Neeru Konda, Amaravati 522502, India; aehsan.a@srmap.edu.in
6 Department of Psychology, Justus-Liebig University Giessen, Ludwigstraße 23, 35390 Gießen, Germany; tmukherjee1@kol.amity.edu
* Correspondence: samarjit.kar@maths.nitdgp.ac.in

Abstract: The broad objective of the present study is to assess the levels of anxiety and depression of school students during the COVID-19 lockdown phase and their association with students' background, stress, concerns and social support. In this regard, the present study follows a novel two stage approach. In the first phase, an empirical survey was carried out, based on multivariate statistical analysis, wherein a group of 273 school students participated in the study voluntarily. In the second phase, a novel Picture Fuzzy FFA (PF-FFA) method was applied for understanding the dynamics of facilitating and prohibiting factors for three categories of focus groups (FG), formulated on the basis of attendance in online classes. Findings revealed a significant impact of anxiety and depression on mental health. Further, PF-FFA examinedthe impact of the driving forces that steered children to attend class as contrasted to the the impact of the restricting forces.

Keywords: school students; COVID-19; mental health; social support; picture fuzzy force field analysis (PF-FFA); level based weight assessment (LBWA)

Citation: Deb, S.; Kar, S.; Deb, S.; Biswas, S.; Dar, A.A.; Mukherjee, T. A Cross-Sectional Study on Mental Health of School Students during the COVID-19 Pandemic in India. *Data* **2022**, 7, 99. https://doi.org/10.3390/data7070099

Academic Editors: Antonio Sarasa Cabezuelo and Ramón González del Campo Rodríguez Barbero

Received: 15 June 2022
Accepted: 13 July 2022
Published: 18 July 2022

Publisher's Note: MDPI stays neutral with regard to jurisdictional claims in published maps and institutional affiliations.

1. Introduction

Childhood is a golden phase in every individual's life. During this phase, children attend schools, befriend other children and enjoy their association, develop good habits, imbibe values from their teachers and move ahead in life for further career growth and development. The lessons which children learn in the school build the foundation for their future career. Cognitive, social and personality development of children take shape through school education. Sadly, education at every level has been adversely affected globally since the outbreak of COVID-19 in January 2020. The impact of continuous lockdown in a phased manner, to arrest the rapid spread of the pandemic, caused psychological distress for school students, such as depression and anxiety, and affected their quality of life [1–6]. Recent evidence highlights that women are more vulnerable to depression and anxiety during adversities [4,6]. It has also been observed that young children, hailing from poor income families, manifest greater risks of mental health challenges [7,8].

Rural students, in developing countries like India, were the worst victims of the situation, with lack of online education, due to lack of internet facilities and/or poor internet connectivity. The Remote Learning Reachability report [9] indicated the widening learning gap owing to the digital divide. Poor economic conditions did not allow a majority of the rural children to buy smartphones and/or laptops, and data cards, for online education. Less than 25 percent of households were reported to be equipped with access to the internet [9], among which more than 80 percent of the students in government schools had no access to any educational material in Odisha, Jharkhand, Bihar, Chattisgarh, and Uttar Pradesh [10]. As a result, cognitive development and nutrition [11] of rural school students in India was more affected, when compared to urban students. Further, social isolation and confinement at home, without much physical activity, caused mental distress for the school students [12].

1.1. Magnitude of School Student Enrolment in India, Computer Facilities and Impact

About 1.49 million schools in India offer education at different levels [13]. Available data highlights that almost 265 million students were taught in the schools during 2019–2020 and only 22% of schools in India had internet facilities [14]. Unfortunately, among the government schools, less than 12% had internet in 2019–2020, while less than 30% had functional computer facilities, which adversely affected online education during the pandemic, especially for students in government schools and in the rural areas. Therefore, it is estimated that closure of educational institutions on account of COVID-19 has affected the education of over 320 million children from pre-primary to tertiary levels [15].

The challenges for urban students during the pandemic were slightly different from rural students, although most urban students got the opportunity for online education. Most of the students in the urban areas of India live in two or three room flats. Therefore, physical mobility was restricted. Attending continuous online classes without much break in between the classes was psychologically distressful and tended to aggravate their miseries. Available evidence indicates that COVID-19 has had an adverse impact on the mental and physical health of people beyond geographical boundaries [16–21], with a high incidence of distress, depression, and anxiety among adolescents and the youth [7,22].

Regarding the efficacy of the online mode of teaching and learning process, a mixed image has been reported. A few studies reported positive outcomes of online teaching and learning processes [23–29], while a few reported the opposite picture i.e., negative outcomes such as poor communication between teachers and students, poor internet connectivity, lack of concentration and so on [30–32].

1.2. Research Objectives

There has been limited research addressing the mental health of school children vis-à-vis COVID-19. Therefore, the present study attempted to examine the status of mental health of Indian school students on account of COVID-19 lockdown, in terms of anxiety and depression. Further, the study assessed the association between the status of mental health, their background, stress, worries and support facilities. The following hypotheses were developed for verification.

Hypothesis 1. *Anxiety and depression of school students differ significantly in terms of gender and grade.*

Hypothesis 2. *There exists an association between feeling stressed for staying at home for a long period during the COVID-19 pandemic, and anxiety and depression.*

Hypothesis 3. *There exists an association between perception about the online teaching mode, and anxiety and depression.*

Hypothesis 4. *There exists an association between social support from family and friends, and anxiety and depression.*

Hypothesis 5. *There exists an association between worries about catching COVID-19 and future career of the students with anxiety and depression.*

Moving further, we were also inquisitive about the dynamics between the facilitating and prohibiting factors that influence the level of participation of the children during online classes. For this purpose, we aimed to carry out an FFA on three focused groups (FG) classified as "Always Attend" (FG-1), "Sometimes Attend" (FG-2) and "Very Rarely and Rarely Attend" (FG-3).

1.3. Contributions of the Paper

The present paper adds value to the extant literature in the following ways.

(a) Within our best possible search, we noticed scanty work related to the mental health of school children. In this regard, our study puts forth a new perspective for the educational leaders, parents and policymakers.
(b) The present paper provides a first of its kind integrated framework of empirical multivariate analysis and PFS based FFA, grounded on a psychological perspective.
(c) A new framework of PF-FFA is proposed in the broad domain of change management.

The rest of the paper is structured as follows. Section 2 describes the research methodology. In Section 3 we present the findings and in Section 4 we include necessary discussions. The concluding remarks and recommendations are provided in Section 5.

2. Materials and Methods

The present research is designed in two stages, i.e., stage 1 (empirical multivariate analysis) and stage 2 (FG opinion based PF-FFA). The stages are interconnected and provide a comprehensive framework for assessment.

2.1. Stage 1

2.1.1. Study Design

An online cross-sectional survey was carried out among Indian school students between 3 June 2020, and 3 August 2020, the period of outbreak of COVID-19.

2.1.2. Sample

A group of 237 school students from Grade IX to XII, aged between 14 to 18 years, participated in the online survey voluntarily.

2.1.3. Study Tools

Semi-structured questionnaire (developed by Deb, 2020): this was developed to understand the school students' perception of the online mode of teaching and their issues and concerns during the COVID-19 pandemic. The questionnaire consisted of five sections, viz.:

Section I: Background Information
Section II: Online Mode of Teaching, Learning and Examinations
Section III: Health
Section IV: Perceived Stress and Coping Strategies
Section V: Mental Health of School Students

Section I consisted of 10 questions pertaining to socio-demographic details, including gender, age, grade, type of family, number of siblings, family monthly income, number of rooms in the house, educational background and occupation of parents, and history of chronic health problems of any family member, while Section II consisted of 20 questions related to school students' perception of online classes, challenges faced by them in attending the online classes, students' perception about online and face to face teaching methods, experience in writing online examinations and so on. Some of the questions related to Section II were as follows:

- Did you attend online classes offered by your school?
- Did you face an internet connectivity problem?
- How did you find the online mode of teaching?
- Could you clarify your doubts, ask questions and get the answers?

Section III comprised seven items related to health concerns of school students, arising due to attending continuous classes, fear of catching COVID-19, physical and leisure time activities. Section IV included six questions related to perceived stress and coping strategies.

The face validity of the interview schedule was ascertained by two experts. A five-point scale was used to capture the response of the subjects to most of the questions while for some questions, a dichotomous mode of response was sought. For example, in a question like 'did you face an internet connectivity problem?', the mode of response was captured by using a 5-point scale (1 = Always: 5 = Never).

(a) Depression Scale:

This brief scale consisted of two items of the Reynolds Adolescent Depression Scale-2nd Edn. (RADS-2), and the items included: (i) I feel that no one cares about me; and (ii) I feel worried. The mode of responses include 'almost never', 'sometimes', 'a lot of the time' and 'all the time'. Score "0" is assigned to "almost never" while score "3" is assigned to "all the time". A high score indicates high depression. The Cronbach's Alpha of RADS-2 short version with the present sample was 0.66.

(b) Anxiety Scale:

The brief Anxiety Scale consisted of six items of the Multidimensional Anxiety Scale for Children (MASC). Some of the items included: (i) I get scared when my parents go away; (ii) I avoid going to places without my family; (iii) I feel restless and on the edge. The mode of responses includes 'never true about me', 'rarely true about me', 'sometimes true about me' and 'often true about me'. Score "0" is assigned to "never true about me" while score "3" is assigned to "often true about me". A high score indicates high anxiety. The Cronbach's Alpha of RADS-2 short version with the present sample was 0.68.

2.1.4. Data Collection and Analysis

Data was drawn via an online survey. Data was primarily reported using descriptive statistics. Differences in the prevalence of depression and anxiety across each demographic variable were tested by using independent samples t-test and one-way ANOVA. All the analysis was done by using IBM SPSS version 23.0.

2.2. *Stage 2*

In the next stage, a FG study was conducted and a PF-FFA was carried out.

2.2.1. Description of the FGs

In the study, the respondents were classified into three categories, such as "Always Attend (AA)", "Sometimes Attend (SA)" and "Very Rarely and Rare Attend (VRA)", based on their attendance during online classes. It was observed that a higher number of students fall under category 1 (AA). According to the size of the three categories, three FGs were formed, following convenient sampling.

- FG 1: The representative sample consists of 20 students belonging to AA category.
- FG 2: In this group 10 students from SA category are included.
- FG 3: 05 (five) students from the VRA category.

2.2.2. Identification of the Factor

An exploratory discussion was carried out with the FGs separately and after accumulating the views, six facilitating factors and six prohibiting factors were finalized, as given in Table 1.

Table 1. Facilitating and Prohibiting Factors.

Facilitating Factors		Prohibiting Factors	
S/L	Description of the Factor	S/L	Description of the Factor
P1	Less travelling	N1	Lack of infrastructure
P2	Access to distant courses	N2	Physical health issue
P3	Staying together with family	N3	Difficulty in online learning
P4	Enjoy online class	N4	Worry about Covid-19
P5	Free time	N5	Worry about future
P6	Enjoy online exam	N6	Movement restriction

2.2.3. Force Field Analysis (FFA)

The concept of FFA has its genesis in the seminal work on the three step model of planned organizational change management, proposed by Lewin. Later, Lewin [33] proposed the framework of FFA, which is based on two types of forces, namely, Driving Forces (DF), which favour the change and act as the enablers, and Restraining Forces (RF) which tend to restrict the change from taking place. Given a situation or requirement, FFA analyses the interplay among various RFs and DFs while making a transition from the "As is" state to the "To Be" stage, by embracing the change [34]. FFA helps in formulating dynamic business strategies to withstand strategic regression and market competition, as utilized by Paquin and Koplyay [35].

In the context of psychological analysis and organizational change management, FFA has been applied extensively by the researchers. For instance, Hlalele [36] conducted a vulnerability analysis of drought conditions from the phenomenological perspective, supported by FFA. Youssef and Mostafa [37] extended the application focus of FFA to the area of cloud computing adoption in organizations and utilized a combined model of FFA, along with pairwise comparison and the Delphi method. To understand the DFs and RFs supporting the adoption of environmental strategy by firms involved in the hotel industry of Taiwan, Mak and Chang [38] applied FFA.

In a recent study [39], the authors explored the factors that influenced students and their families regarding online learning and reported that safe home environment, leisure time, food, family bonding, economical aspects and flexibility, are some of the supporting factors, while technical issues such as network glitches, distractions, stress and anxiety, lack of real-life experiences, and social distancing, were the adverse factors. The authors finally advocated for a "blended or hybrid" mode of learning. In a different scenario, Ramos et al. [40] conducted a field study on Small and Medium Enterprises (SME) in the Philippines, to assess their readiness toward adopting the Internet of Things (IoT) and related technologies, based on FFA and causal analysis, using Structural Equation Modelling (SEM).

In this context, it may be noted that FFA has been used for solving various issues related to engineering, management and social sciences, and has been applied in conjunction with other multivariate techniques such as SEM. However, the application of FFA with imprecise information and analysis in an uncertain environment seems to be rare.

2.2.4. PFS

The concept of PFS was developed as an extension of the intuitionistic fuzzy sets (IFS). Unlike IFS, PFS indicates the degree of refusal and brings about better accuracy and granularity in the analysis, which involves a considerable amount of subjectivity and impreciseness in the available information [41]. Due to its potential for superior analysis under uncertainty, PFS has been utilized by various researchers (e.g., [42–48]) in distinct situations, for multi-criteria decision making (MCDM) related problems. In the following section, we mention some of the basic definitions, operations and properties of PFS [49,50]. The preliminary definitions, operations and properties of PFS are given in Appendix A.

2.2.5. LBWA Method

LBWA is an algorithm designed by Žižović & Pamučar [51], to decide criteria weights. LBWA offers a lesser number of pairwise comparisons (only $(n-1)$ number of criteria comparisons) and thus enables operation with a lesser computational complexity. It works efficiently with a large criteria set, to provide better consistency and robustness of results, and equally operates with subjective and objective information. LBWA finds its applications in many complex real-life problems, such as facility location selection [52]; selection of airport ground access mode [53]; military operations [54–56]; healthcare management during crisis [57]; location selection for offshore wind farms [58]; sustainable energy management [59,60]; and social entrepreneurship [42]. The computational steps of the algorithm are described in Appendix B.

2.2.6. The Proposed PF-FFA Method

The algorithm of the proposed PF-FFA method is described below.

Suppose,

C_j, where $j = 1, 2, \ldots n$ (n is finite and ≥ 2): The number of criteria. In our paper, the criteria represent a list of six facilitating/prohibiting factors.

E_t, where $t = 1, 2, \ldots m$ (m is finite and ≥ 2): The number of respondents who have provided their opinions during the study.

Then, for each of the facilitating and prohibiting factors separately, the following steps are followed for computation

Step 1. Formulation of the linguistic rating matrix

$$\varphi^t = \left[\varphi_1^t \ \varphi_2^t \ldots \varphi_n^t\right]$$

Here, φ_j^t is the rating of the factor C_j by the respondent E_t based on the relative importance of the corresponding factor. We use the linguistic expression in terms of Yes (Y) (if the challenging factor is perceived as impactful, i.e., positive membership), No (N) (if the challenging factor is perceived to have very little or no impact, i.e., negative membership), and Can't say (A) (if it is not possible to precisely assess the impact, i.e., neutral view). We do not include the option of refusal as the factors are derived through discussion with the respondents.

Step 2. Formulation of the PF factor weight matrix

The factor weight matrix is represented as $= [_j]_{n\times 1}$. Here, $_j = \left\langle \mu_j, \eta_j, \upsilon_j \right\rangle$ is a PFN representing the importance of the factor C_j considering the responses of all respondents. We follow the demonstration of Jovčić et al. [61] to derive the PFNs.

Step 3. Calculation of the actual scores

The actual scores are calculated using the steps followed in Si et al. (2019) (refer the Equations (A26)–(A31) as given in Appendix A)

The actual scores of all PFNs corresponding to the factors (facilitating/prohibiting factors) are calculated separately.

Step 4. Determination of weights of the factors

We use the computational steps of the LBWA algorithm (as given in the Appendix B) to derive the weights of the factors.

Step 5. Finding out the aggregated scores of the facilitating and prohibiting factors

We use PFWA operator (see Expression (A32) in Appendix A) to calculate the aggregated scores of the facilitating and prohibiting factors separately.

Step 6. Comparison of aggregated scores of the facilitating and prohibiting factors

The aggregate score is obtained through defuzzification (see Expressions (A21) to (A23) in Appendix A).

The decision rule:

If $Score_{Facilitating} > Score_{Prohibiting}$, we conclude that the change is supported;
If $Score_{Facilitating} < Score_{Prohibiting}$, we conclude that the change is dominated and prevented;
If $Score_{Facilitating} = Score_{Prohibiting}$, no adequate evidence to support the movement

3. Results

In this section, we summarize the results of the data analysis using the methodological steps as described in Section 2. The results are exhibited stage-wise.

3.1. *Stage 1*

3.1.1. Description of the Sample

Table 2 depicts the description of the sample. Of the 273 participants, 54.9% (150/273) were male and 45.1% (123/273) were female. The majority of the respondents came from single families (74.7%), less than a quarter of the participants (23.4%) were from joint families and 1.8% were living with their relatives. Close to a quarter of the respondents were studying in the 9th grade (24.2%), 18.7% were in their 10th grade, 22.0% were studying in the 11th grade and more than one third of the students were studying in the 12th grade (35.2%).

Table 2. Description of the Sample (*N* = 273).

Variable	*N*	(%)
Gender		
Male	150	54.9
Female	123	45.1
Family Type		
Joint	64	23.4
Single	204	74.7
Staying with relative family	05	1.8
Grade		
9th class	66	24.2
10th class	51	18.7
11th class	60	22.0
12th class	96	35.2
Age		
14 to 15 years	134	49.1
16 to 18 years	139	50.9
Siblings		
Only child	88	32.2
1 sibling	159	58.2
2 siblings and above	26	9.5
Monthly income		
Less than 20,000 INR	48	17.6
20,001 to 50,000 INR	86	31.5
50,001 to 100,000 INR	72	26.4
100,001 to 150,000 INR	42	15.4
150,001 and above INR	25	9.2
How did you find the online mode of teaching?		
Most effective	09	3.3
Effective	59	21.6
Moderately effective	104	38.1
Not so effective	67	24.5
Not at all effective	34	12.5

Table 2. *Cont.*

Variable	N	(%)
Are you worried that you will catch COVID-19?		
Highly worried	41	15.0
Worried	50	18.3
Worried to some extent	75	27.5
Not so worried	54	19.8
Not at all worried	53	19.4
Do you feel stressed from staying at home for a long period during COVID-19 pandemic?		
Highly stressed	118	43.2
Stressed	51	18.7
Moderately stressed	46	16.8
Rarely stressed	27	9.9
Not so stressed	31	11.4
Are you worried about your future career?		
Highly worried	119	43.6
Worried	77	28.2
Worried to some extent	39	14.3
Not so worried	19	7.0
Not at all worried	19	7.0
Do you get emotional support from your family when you need it?		
Always	105	38.5
Most of the time	66	24.2
Sometimes	58	21.2
Rarely	29	10.6
Never	15	5.5
Do you have friends who extend support at times of any crisis or challenge?		
Always	108	39.6
Most of the time	52	19.0
Sometimes	63	23.1
Rarely	23	8.4
Never	27	9.9

Half of the participants were aged 18–20 years (49.1%) and another half were aged 16 to 20 years (50.9%). More than half of the participants had 1 sibling (58.2%), one third of the respondents were the only children (32.2%) and 9.5% had two or more siblings. The monthly income of one third of the sample was 20,001 to 50,000 INR (31.5%), more than a quarter had 50,001 to 100,000 INR (26.4%), less than a quarter had less than 20,000 INR (17.6%), 15.4% had 100,001 to 150,000 INR and 9.2% had 150,001 INR and above monthly familial income.

With respect to online mode of teaching during lockdown, more than one-third of the respondents found it moderately effective (38.1%), a quarter of the participants found it to be not so effective (24.5%), less than a quarter found it effective (21.6%) and more than one-tenth did not find it effective (12.5%). Only 3.3% found it most effective. Less than a quarter of the students were worried about catching COVID-19 (18.3%) while 15% and 27.5% reported being highly worried and worried to some extent.

Regarding stress from staying at home for a long period during COVID-19, nearly half of the participants felt highly stressed (43.2%), less than a quarter felt stressed (18.7%) or moderately stressed (16.8%), 9.9% felt rarely stressed and 11.4% did not feel much stressed. About half of the participants were highly worried about their future career (43.6%) and over a quarter were in the worried category (28.2%).

More than one-third of the participants received emotional support (38.5%) from the family during the lockdown period, while a quarter of the respondents received emotional support most of the time (24.2%). As far as emotional support from friends is concerned, 39.6%, 19% and 23.1% received it always, most of the time and sometimes, respectively (Table 2).

3.1.2. Description of Anxiety and Depression

Data pertaining to description of anxiety and depression is provided in Table 3, indicating the mean anxiety (mean = 8.05; *SD* = 4.10) and depression (mean= 3.10; *SD* = 1.86) levels reported by the students during COVID-19 lockdown, in the range of 0 to 18 and 0 to 6, respectively.

Table 3. Description of anxiety and depression (N = 273).

	Mean	*SD*	Actual Score Range	Possible Score Range
Anxiety	8.05	4.10	0–18	0–18
Depression	3.10	1.86	0–6	0–6

Note: SD = Standard deviation.

3.1.3. Levels of Anxiety among Students during COVID-19 Lockdown

Table 4 highlights the levels of anxiety of school students during COVID-19 lockdown. More than one-third of the participants (37.7%) reported having low levels of anxiety, nearly half of the participants (46.9%) reported moderate levels of anxiety, and over one-tenth of the sample (13.2%) reported high levels of anxiety, while 2.2% of the participants did not report any anxiety.

Table 4. Levels of anxiety (N = 273).

Level	Score Range	n	%
No anxiety	0	6	2.2
Low anxiety	1 to 6	103	37.7
Moderate anxiety	7 to 12	128	46.9
High anxiety	13 to 18	36	13.2
Total		273	100

3.1.4. Levels of Depression among Students during COVID-19 Lockdown

Table 5 reflects the levels of depression of students during COVID-19 lockdown. A quarter of the participants (25.3%) reported a low level of depression. Over one-third of the respondents (34.8%) reported a moderate level of depression. More than a quarter of the participants (27.5%) reported high levels of depression and over one-tenth of the samples (12.5%) did not report depression.

Table 5. Levels of depression (N = 273).

Level	Score Range	n	%
No depression	0	34	12.5
Low depression	1 to 2	69	25.3
Moderate depression	3 to 4	95	34.8
High depression	5 to 6	75	27.5
Total		273	100

3.1.5. Association of Anxiety and Depression with Demographic Variables

Data provided in Table 6 indicates a significant association between grade and depression [F (3, 269) = 4.15, $p < 0.01$]. Depression was found to be higher among the students of the 11th grade, followed by students of the 12th grade and the 9th grade. Therefore, Hypothesis 1 that is "*Anxiety and depression of school students differ significantly in terms of gender and grade*" is partially accepted, as there was no significant gender difference with respect to anxiety and depression, although grade-wise significant difference was observed.

Regarding the stress resulting from long stay at home during COVID-19 lockdown, among adolescent students, a significant association was found with anxiety [F (4, 268) = 2.48, $p < 0.05$] and depression [F (4, 268) = 8.10, $p < 0.001$]. Students who felt stressed during the lockdown reported significantly higher rates of anxiety, as compared to students who did not feel stressed. The levels of depression were found to be higher among students who felt stressed, followed by those who felt highly stressed, rarely stressed and not so stressed. Hence Hypothesis 2 i.e., "*There exists an association between feeling stressed from staying at home for a long period during COVID-19 pandemic and anxiety and depression*" has been accepted.

A significant association was found between an online mode of teaching during COVID-19 lockdown, and anxiety [F (4, 268) = 4.20, $p < 0.01$] and depression [F (4, 268) = 8.44, $p < 0.001$]. Students who did not find online teaching effective at all reported greater levels of anxiety, followed by those who found online teaching moderately effective and effective. Students who did not find online teaching effective at all reported higher rates of depression, followed by those who did not find online teaching so effective, moderately effective, effective and most effective, respectively. As for Hypothesis 2, Hypothesis 3 i.e., "*There exists an association between perception about the online mode of teaching and anxiety and depression*" has been accepted.

With respect to emotional support from family, a significant association was found with depression only [F (4, 268) = 12.13, $p < 0.001$]. Students who did not receive emotional support from their family demonstrated higher levels of depression, followed by those who received emotional support rarely, sometimes, most of the time and always.

Further analysis of data revealed a significant association between peer group support during COVID-19 lockdown and depression [F (4, 268) = 2.83, $p < 0.05$]. Students who always received support from friends were found to be less depressed than students who never received any support from their friends during times of crisis. Therefore, Hypothesis 4 i.e., "*There exists an association between social support from family and friends and anxiety and depression*" is partially tenable as significant association was found between poor social support and depression only.

Worries of school students about infection with COVID-19 was found to be associated with high levels of anxiety [F(4, 268) = 2.83, $p < 0.05$]. That is, the students who demonstrated worries about catching COVID-19 showed higher anxiety levels than those who were not at all worried.

Similarly, worries of school students about their future career caused higher levels of anxiety [F (4, 268) = 5.66, $p < 0.001$] and depression [F(4, 268) = 6.62, $p < 0.001$]. Students who felt worried about their future reported higher anxiety, followed by those who felt rarely worried and not so worried during the lockdown. Similarly, students who felt worried about their future career demonstrated higher rates of depression, followed by those who felt moderately worried, rarely worried and not so worried.

Looking at the analysis of data, it has been observed that Hypothesis 5 (There exists an association between worries about catching COVID-19 and future career of the students with anxiety and depression) has been tenable in case of association with fears of catching COVID-19 and anxiety, while in the case of worries of the school students about future career, a significant number of school students reported to be suffering from high levels of anxiety and depression.

Table 6. Means, Standard Deviation and t/F Scores of Anxiety and Depression (N = 273).

	Anxiety	Depression
Gender		
Male	7.82 (4.25) 1.04	2.96 (1.90) 1.34
Female	8.32 (3.92)	3.26 (1.80)
Grade		
9th class	7.55 (4.23) 0.66	2.48 (1.73) 4.15 **
10th class	8.53 (4.64)	2.94 (2.0)
11th class	8.31 (3.74)	3.50 (1.84)
12th class	7.96 (3.95)	3.34 (1.79)
Do you feel stressed from staying at home for along period during COVID-19 pandemic?		
Highly stressed	8.34 (4.56) 2.48 *	3.52 (1.73) 8.10 ***
Stressed	9.06 (3.30)	3.57 (1.81)
Moderately stressed	7.37 (3.64)	2.83 (1.83)
Rarely stressed	7.81 (4.19)	2.11 (1.76)
Not so stressed	6.45 (3.62)	1.97 (1.74)
How did you find the online mode of teaching?		
Most effective	6.33 (5.31) 4.20 **	1.11 (1.26) 8.44 ***
Effective	6.92 (4.16)	2.37 (1.82)
Moderately effective	8.04 (3.63)	3.15 (1.77)
Not so effective	8.15 (3.53)	3.42 (1.77)
Not at all effective	10.26 (5.25)	4.06 (1.76)
Do you get emotional support from your family when you need it?		
Always	7.86 (4.51) 0.69	2.43 (1.78) 12.13 ***
Most of the time	7.76 (3.77)	2.92 (1.74)
Sometimes	8.07 (3.39)	3.53 (1.67)
Rarely	8.62 (4.35)	3.97 (1.72)
Never	9.40 (4.73)	5.13 (1.41)
Do you have friends who extend support at times of any crisis or challenge?		
Always	7.90 (4.42) 1.93	2.74 (1.91) 2.83 *
Most of the time	8.52 (3.66)	3.44 (1.78)
Sometimes	7.49 (3.59)	3.24 (1.77)
Rarely	7.17 (4.03)	2.74 (1.79)
Never	9.74 (4.47)	3.81 (1.82)
Are you worried that you will catch COVID-19?		
Highly worried	8.61 (4.21) 3.10 *	3.22 (1.96) 1.36
Worried	9.06 (4.47)	3.30 (1.90)
Worried to some extent	7.88 (4.13)	3.35 (1.66)
Not so worried	8.43 (3.69)	2.72 (1.87)
Not at all worried	6.49 (3.68)	2.83 (1.98)
Are you worried about your future career?		
Highly worried	9.09 (4.31) 5.66 ***	3.62 (1.85) 6.62 ***
Worried	7.95 (3.61)	3.06 (1.79)
Moderately worried	7.21 (3.99)	2.54 (1.59)
Rarely worried	6.32 (3.11)	2.11 (1.70)
Not so worried	5.32 (3.27)	2.05 (1.87)

Note: Figures out of the brackets are the means and those within the brackets are standard deviations; figures after the brackets are t values and F values. Differences in the prevalence across each demographic variable were tested by using independent samples t-test and one-way ANOVA. * $p < 0.05$, ** $p < 0.01$ and *** $p < 0.001$.

3.2. Stage 2

We present the findings of the data analysis for the FGs separately. We follow the steps as described in Section 2.

3.2.1. Analysis of the Responses of the FG-1 (i.e., AA Category)

Tables 7 and 8 provide the responses of the respondents of FG 1, related to facilitating and prohibiting factors respectively. Tables 9 and 10 indicate the actual score values and weights (obtained using LBWA) of the factors (facilitating and prohibiting). Tables 11 and 12 show the intermediate calculations of LBWA. However, the decision of PF-FFA largely depends on the weights of the factor and the PFWA calculation. Therefore, sensitivity analysis is of paramount importance here. Sensitivity analysis is carried out to check the stability of the result, obtained by using a MCDM algorithm under the influence of changes in the given conditions, for example, calculation of the criteria weights, changes in the criteria and alternative sets, interplay among the alternatives and criteria among the others [62–68]. In our paper, we change the values of the coefficient of elasticity and examine the changes in the criteria weights. Figures 1 and 2 confirm that the weights obtained using LBWA method are stable with respect to the varying values of the coefficient of elasticity, for the facilitating and prohibiting factors.

Table 7. Rating of the facilitating factors by the respondents of FG-1.

Respondent	Facilitating Factors					
	P1	P2	P3	P4	P5	P6
R1	Y	A	Y	A	Y	N
R2	Y	A	Y	N	Y	N
R3	Y	A	Y	Y	Y	A
R4	Y	A	Y	Y	Y	Y
R5	A	A	Y	Y	Y	A
R6	A	A	N	A	Y	Y
R7	Y	Y	N	Y	Y	A
R8	N	N	Y	Y	Y	A
R9	N	A	Y	Y	A	Y
R10	A	Y	N	A	N	Y
R11	Y	Y	A	N	N	Y
R12	Y	A	N	Y	Y	Y
R13	N	A	Y	Y	N	N
R14	N	A	A	A	Y	Y
R15	A	A	Y	Y	A	Y
R16	N	Y	N	Y	Y	A
R17	N	A	Y	Y	A	A
R18	N	Y	A	A	N	N
R19	Y	Y	Y	Y	Y	A
R20	N	Y	Y	Y	Y	N
μ	0.4	0.35	0.6	0.65	0.65	0.4
η	0.2	0.6	0.15	0.25	0.15	0.35
ν	0.4	0.05	0.25	0.1	0.2	0.25

Table 8. Rating of the prohibiting factors by the respondents of FG-1.

Respondent	Prohibiting Factors					
	N1	N2	N3	N4	N5	N6
R1	Y	Y	Y	N	Y	Y
R2	Y	Y	Y	N	N	Y
R3	N	Y	Y	A	A	Y
R4	N	Y	N	N	A	A
R5	A	N	Y	A	A	Y
R6	Y	A	A	N	N	A
R7	Y	N	Y	Y	A	A
R8	N	N	Y	Y	Y	N
R9	N	N	Y	Y	A	N
R10	N	N	N	Y	N	A
R11	N	N	N	N	N	Y
R12	Y	N	Y	A	Y	N
R13	Y	N	Y	N	A	A
R14	Y	Y	Y	N	N	A
R15	N	Y	Y	A	A	N
R16	A	Y	N	A	A	N
R17	N	Y	N	N	N	A
R18	N	Y	N	Y	N	A
R19	N	N	N	N	A	N
R20	Y	Y	A	Y	N	Y
μ	0.4	0.5	0.55	0.3	0.15	0.3
η	0.1	0.05	0.1	0.25	0.45	0.4
ν	0.5	0.45	0.35	0.45	0.4	0.3

Table 9. Actual score values of the facilitating factors.

Factors	PGD	NGD	Abs_Score	Act_Score	Weight	Rank
P1	0.25	0.35	0.40	0.4364	0.1065	6
P2	0.30	0.00	0.70	0.5316	0.1538	4
P3	0.05	0.20	0.75	0.8654	0.1730	3
P4	0.00	0.05	0.95	0.9828	0.2307	1
P5	0.00	0.15	0.85	0.9808	0.1977	2
P6	0.25	0.20	0.55	0.5156	0.1384	5
					Σ = 1.000	

(PIS: <0.65, 0.15, 0.05>; Avg _η = 0.283).

Table 10. Actual score values of the prohibiting factors.

Factors	PGD	NGD	Abs_Score	Act_Score	Weight	Rank
N1	0.15	0.20	0.65	0.7429	0.17300	3
N2	0.05	0.15	0.80	0.9697	0.19771	2
N3	0.00	0.05	0.95	1.0857	0.23066	1
N4	0.25	0.15	0.60	0.5854	0.13840	5
N5	0.40	0.10	0.50	0.4082	0.10646	6
N6	0.25	0.00	0.75	0.6383	0.15377	4
				Σ	1.00000	

(PIS: <0.55, 0.05, 0.03>; Avg _η = 0.225).

Table 11. Weight calculation for facilitating factors (LBWA).

Criteria	C4	C5	C3	C2	C6	C1	
Level	1	1	1	1	1	2	
Integer value	0	1	2	3	4	1	
Function	1.000	0.857	0.750	0.667	0.600	0.462	Σ
Criteria weights	0.2307	0.1977	0.1730	0.1538	0.1384	0.1065	1.00

Table 12. Weight calculation for prohibiting factors (LBWA).

Criteria	C3	C2	C1	C6	C4	C5	
Level	1	1	1	1	1	2	
Integer Value	0	1	2	3	4	1	
Function	1.000	0.857	0.750	0.667	0.600	0.462	Σ
Criteria weights	0.2307	0.1977	0.1730	0.1538	0.1334	0.1065	1.00

Moving ahead, we calculate the aggregate scores of facilitating and prohibiting factors. Tables 13 and 14 show the aggregate scores for facilitating and prohibiting factors.

Table 13. Aggregate score of facilitating factors.

PFWA	μ	η	ν	π
	0.55047	**0.24214**	**0.15894**	**0.04845**
$Score_{Facilitating}$	0.70525			

Table 14. Aggregate score of prohibiting factors.

PFWA	μ	η	ν	Π
	0.41205	**0.14377**	**0.40126**	**0.04292**
$Score_{Prohibiting}$	0.50562			

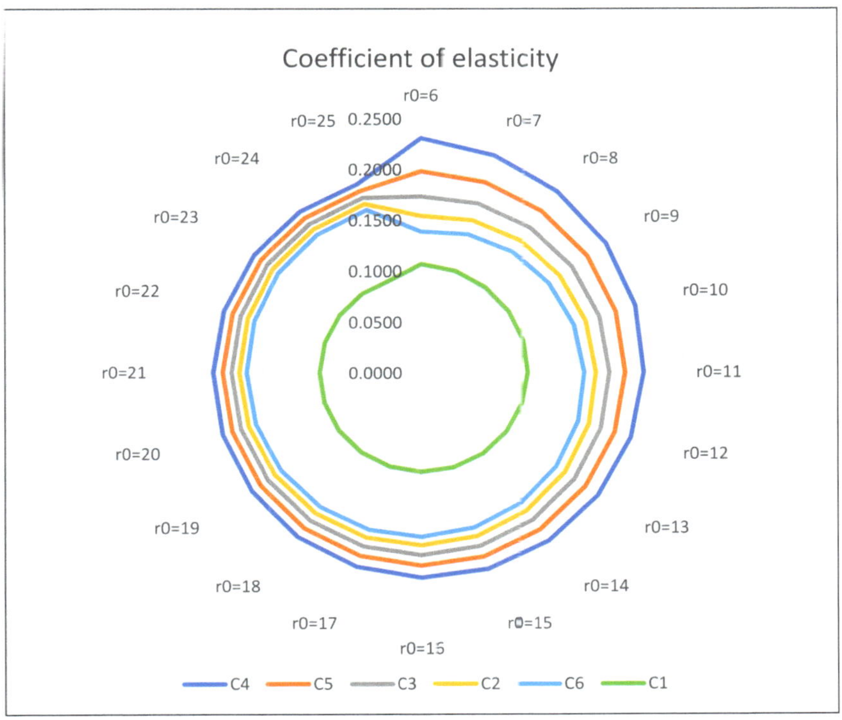

Figure 1. Sensitivity analysis (facilitating actors for FG 1).

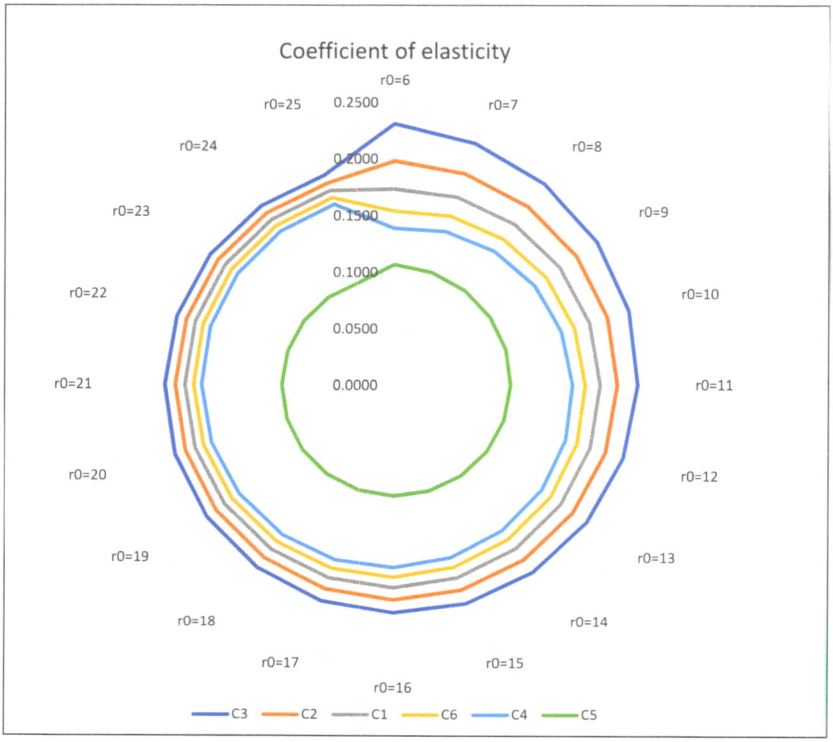

Figure 2. Sensitivity analysis (prohibiting factors for FG 1).

It is observed that $Score_{Facilitating} > Score_{Prohibiting}$ which implies that, for the category of AA, the respondents of FG 1 indicate that there was a stronger driving force as compared to the obstacles. As a result, children found it useful to attend classes regularly.

In a similar way, we carried out the calculations for two other categories, such as SA (represented by FG 2) and VRA (represented by FG 3).

3.2.2. Analysis of the Responses of the FG-2 (i.e., SA Category)

In the similar way (like the previous Section 3.2.1) we carry out the analysis (see Tables 15–20).

Table 15. Rating of the facilitating factors by the respondents of FG-2.

Respondent	Facilitating Factors					
	P1	**P2**	**P3**	**P4**	**P5**	**P6**
R1	Y	A	Y	Y	Y	A
R2	Y	N	Y	N	A	A
R3	N	A	Y	N	N	Y
R4	A	A	Y	Y	Y	A
R5	Y	Y	N	A	Y	N
R6	A	Y	A	Y	N	Y
R7	A	N	A	Y	A	Y
R8	Y	Y	Y	A	Y	Y
R9	N	A	Y	A	N	Y
R10	Y	Y	N	Y	Y	A
μ	0.5	0.4	0.6	0.5	0.5	0.5
η	0.3	0.4	0.2	0.3	0.2	0.4
ν	0.2	0.2	0.2	0.2	0.3	0.1

Table 16. Rating of the prohibiting factors by the respondents of FG-2.

Respondent	Prohibiting Factors					
	N1	N2	N3	N4	N5	N6
R1	N	Y	Y	Y	A	Y
R2	N	Y	Y	A	A	A
R3	A	Y	N	N	A	Y
R4	Y	N	Y	N	A	N
R5	Y	N	Y	N	A	Y
R6	Y	A	A	Y	N	A
R7	Y	N	A	A	Y	N
R8	A	Y	N	N	A	A
R9	Y	Y	Y	Y	Y	N
R10	N	Y	Y	N	Y	Y
μ	0.5	0.6	0.6	0.3	0.3	0.4
η	0.2	0.1	0.2	0.2	0.5	0.3
ν	0.3	0.3	0.2	0.5	0.2	0.3

Table 17. Actual score values of the facilitating factors (FG 2).

Factors	PGD	NGD	Abs_Score	Act_Score	Weight	Rank
P1	0.10	0.10	0.80	0.8000	0.1701	3
P2	0.20	0.10	0.70	0.6364	0.1276	6
P3	0.00	0.10	0.90	1.0000	0.2187	1
P4	0.10	0.10	0.80	0.8000	0.1531	4
P5	0.10	0.20	0.70	0.7778	0.1392	4
P6	0.10	0.00	0.90	0.8182	0.1914	2

(PIS: <0.6, 0.2, 0.1>; Avg _η = 0.3).

Table 18. Actual score values of the prohibiting factors (FG 2).

Factors	PGD	NGD	Abs_Score	Act_Score	Weight	Rank
N1	0.10	0.10	0.80	0.8421	0.17300	3
N2	0.00	0.10	0.90	1.0588	0.23066	1
N3	0.00	0.00	1.00	1.0526	0.19771	2
N4	0.30	0.30	0.40	0.4211	0.10646	6
N5	0.30	0.00	0.70	0.5600	0.13840	5
N6	0.20	0.10	0.70	0.6667	0.15377	4

(PIS: <0.6, 0.1, 0.2>; Avg _η = 0.25).

Table 19. Aggregate score of facilitating factors (FG 2).

PFWA	μ	H	ν	π
	0.51261	**0.28441**	**0.18532**	**0.01765**
$Score_{Facilitating}$	0.66653			

Table 20. Aggregate score of prohibiting factors (FG 2).

PFWA	μ	η	ν	π
	0.49253	**0.20594**	**0.27641**	**0.02512**
$Score_{Prohibiting}$	0.61077			

It may be noted that the criteria weights are decided using the LBWA method, following the usual calculations. One such type of calculation is already shown in detail for FG 1. Like FG 1, here also we observe stability in the results obtained by using the LBWA approach.

In case of FG 2 (i.e., SA category) we observe that $Score_{Facilitating}$ is marginally greater than $Score_{Prohibiting}$ was reflected in their participation level during online classes. Now, we move towards the FG 3 (i.e., VRA category).

3.2.3. Analysis of the Responses of the FG-3 (i.e., VRA Category)

Here also we use the LBWA method, following the usual calculations (see Tables 21–26).

Table 21. Rating of the facilitating factors by the respondents of FG-3.

Respondent	Facilitating Factors					
	P1	P2	P3	P4	P5	P6
R1	A	N	A	N	Y	N
R2	N	N	A	A	Y	N
R3	A	N	N	N	A	N
R4	Y	A	A	Y	N	A
R5	Y	Y	Y	N	Y	Y
μ	0.4	0.2	0.2	0.2	0.6	0.2
η	0.4	0.2	0.6	0.2	0.2	0.2
ν	0.2	0.6	0.2	0.6	0.2	0.6

Table 22. Rating of the prohibiting factors by the respondents of FG-3.

Respondent	Prohibiting Factors					
	N1	N2	N3	N4	N5	N6
R1	Y	Y	Y	Y	A	Y
R2	Y	A	N	N	Y	N
R3	N	Y	A	N	N	A
R4	A	N	N	A	Y	A
R5	Y	Y	Y	Y	Y	Y
μ	0.6	0.6	0.4	0.4	0.6	0.4
η	0.2	0.2	0.2	0.2	0.2	0.4
ν	0.2	0.2	0.4	0.4	0.2	0.2

Table 23. Actual score values of the facilitating factors (FG 3).

Factors	PGD	NGD	Abs_Score	Act_Score	Weight	Rank
P1	0.20	0.00	0.80	0.7273	0.2867	2
P2	0.40	0.40	0.20	0.2222	0.0683	4
P3	0.40	0.00	0.60	0.4615	0.1593	3
P4	0.40	0.40	0.20	0.2222	0.0652	5
P5	0.00	0.00	1.00	1.1111	0.3583	1
P6	0.40	0.40	0.20	0.2222	0.0623	6

(PIS: <0.6, 0.2, 0.2>; Avg _η = 0.3).

Table 24. Actual score values of the prohibiting factors (FG 3).

Factors	PGD	NGD	Abs_Score	Act_Score	Weight	Rank
N1	0.00	0.00	1.00	1.0345	0.21870	1
N2	0.00	0.00	1.00	1.0345	0.19136	2
N3	0.20	0.20	0.60	0.6207	0.13917	5
N4	0.20	0.20	0.60	0.6207	0.12757	6
N5	0.00	0.00	1.00	1.0345	0.17010	3
N6	0.20	0.00	0.80	0.6857	0.15309	4

(PIS: <0.6, 0.2, 0.2>; Avg _η = 0.2333).

Table 25. Aggregate score of facilitating factors (FG 3).

PFWA	μ	η	ν	π
	0.42535	**0.29061**	**0.24798**	**0.03606**
$Score_{Facilitating}$	0.59188			

Table 26. Aggregate score of prohibiting factors (FG 3).

PFWA	μ	η	ν	π
	0.52577	**0.22239**	**0.24062**	**0.01122**
$Score_{Prohibiting}$	0.64418			

Now, here (for FG 3 i.e., VRA category) we notice that $Score_{Facilitating} < Score_{Prohibiting}$. The level of participation in online classes is also rare, which clearly justifies the findings of the PF-FFA for FG 3.

4. Discussion

The present online study is successful in understanding the level of anxiety and depression among school students in India during the COVID-19 lockdown phase and its association with their background, stress, worries, and social support facilities.

A total of 273 school students participated voluntarily in the online study. Almost equal numbers of male and female school students (male 54.9% vs. female 45.1%) participated in the study, and they were in grades IX to XII; there was similar interest among students of both genders to participate in the study and share their views and concerns. Attending online classes was enjoyable initially but gradually, students perceived it to be challenging because of various reasons, like internet connectivity problems, power problems and sitting in front of a computer for an extended period. Therefore, students' perception of the online mode of teaching was examined. About one-fifth of the students perceived the online teaching mode as most effective and effective while about-two-fifths perceived it to be moderately effective. The rest found it not so effective and not at all effective. This might be due to various factors, such as not being able to follow classes properly, a hectic time table without any break between the classes, internet connectivity problems, ineffective methods of teaching and also boredom on the part of teachers, thereby leading to a casual method of disseminating knowledge. A study on school students in Romania revealed that the availability of equipment for accessing the internet and the ability of the teaching staff were crucial in effectiveness of online education [69]. Some of the previous studies corroborate our findings, i.e., internet connectivity problems and lack of physical interactions for clarification of academic queries caused a lot of anxiety among students [2,23,27].

Further, through our newly proposed PF-FFA methodology, we ascertain that the dynamics between the facilitating and prohibiting factors determine the intentions of the children regarding attending online classes.

Human beings prefer to remain connected with others, to and share their personal feelings and thinking, which enhances their subjective experience of happiness [70]. School children get maximum happiness while interacting with their classmates in the school. Therefore, social isolation causes high stress for school students [71], as demonstrated by our findings. The findings of the present study indicate that more than half of the school students viewed school suspension as highly stressful.

Uncertainty about the situation, i.e., when the school will reopen, caused worries for a large number of students (about 70%) especially for grade X and XII students, as the final board examinations are very important for every child, for their future growth and prospects. Jung, Horta, and Postiglione [72] showed that unexpected occurrences during the pandemic led to unprepared decisions and psychological disruptions in the education sector. Emotional support at times of crisis or pandemic is very helpful to cope with the situation [73]. In the present study, more than three-fifths of school students were reported to have received emotional support from their family members and friends which can positively impact an individual's personal resilience [74].

As far as anxiety of school students in a pandemic like COVID-19 is concerned, the findings of the study disclosed that half of the students reported experiencing moderate levels of anxiety while 13.2% suffered from high levels of anxiety. Under normal circumstances, students go to school and gain knowledge by attending regular classes and clarifying their academic queries. Attending regular classes is essential to complete the syllabus, so that they can write the examinations with maximum effectiveness. In a country like India, where students' performance is assessed based on the results of written examinations, suspension of classes because of the pandemic-induced lockdown, caused a lot of anxiety, although online classes were arranged. Online teaching does not help to clarify all the queries and sometimes teaching is also not clear, due to poor connectivity. Monotony and/or stress on the part of the teachers, while taking continuous online classes, affect the quality of teaching. However, individual coping capacity plays an important role in dealing with various crisis situations. The study also revealed that more than one-third (34.8%) and one-fourth (27.5%) of students were suffering from moderate and high levels of depression, respectively. Continuous social isolation and worrying about the future, fear of getting COVID-19 and exposure to media information on deaths caused by COVID-19, were depressing for the school students. Interaction with classmates in the school campus helps students share their personal feelings and issues with their peers. Since students were effectively under house arrest and unable to meet their friends, they were emotionally upset.

Cross analysis of data highlights an association between grade and depression, i.e., grade XI students were more victims of depression, followed by grade XII students. The school students who perceived social isolation stress were more vulnerable to anxiety and depression. Perception about the online mode of teaching is also associated with anxiety, i.e., the students who reported that the online mode of teaching was not effective have been suffering from more anxiety and depression. Similarly, social support from family and friends were found to be beneficial when dealing with crisis situations. In fact, emotional support helped school students to remain emotionally stable and happy, despite prolonged lockdown, and they utilized their time effectively in studies, and with their family members. The findings of the present study with respect to the benefits of social support are similar to the outcomes of some of the previous studies [17,18,25].

5. Conclusions and Recommendations

In conclusion, it might be stated that COVID-19 caused great anxiety and depression to a large number of school students, mostly because of social isolation and discontinuation of the physical mode of the teaching–learning process. The study revealed that 13.2%, 46.9% and 37.7% of the students were suffering from high, moderate and low levels of anxiety, while 27.5%, 34.8% and 25.3% have been assessed to be suffering from high, medium and low levels of depression. Female school students were suffering from more depression and anxiety, as compared to their male counterparts. Further, a significant association was found between grade, social isolation, feeling of stress, ineffective mode of teaching, worries about catching COVID-19, worry about future career, lack of social support, anxiety and depression. In addition, the PF-FFA analysis provides a visible understanding of the interplay of the facilitating and prohibiting factors (i.e., DFs and RFs) that steer the children and determine their behavioural intentions in response to the changing scenario of learning, as imposed by COVID-19.

It is recommended for school administration to arrange online mental health support services urgently for school students who exhibit high anxiety and depression, in addition to organizing online parent and teacher interaction meetings, for offering school-based family counseling for parents, to deal sensitively with children's emotions.

The questionnaire was distributed online to students via the school administration in different states of India. Since the study took place relatively early during the lockdown months in India and students were still getting accustomed to the online education system, the percentage of response has been low. We recommend using longitudinal research to understand the trajectory of mental health issues among students through the pandemic, especially as the pandemic continues to surge in waves globally. Larger sample size is likely to enhance the generalizability of the studies. Though online education can itself have varied impacts on students, being infected by the virus once or multiple times during the course of the pandemic can hinder a student's academic progress and impact wider aspects of students' education. However, understanding that the pandemic is common for all can assist in coping with the associated distress. Controlling for these factors would provide additional confirmatory information.

Further, in this paper we have used PFS based analysis for FFA. The calculation of weights plays a central role in determining the outcome of FFA. The present paper uses the LBWA method with which we have carried out the sensitivity analysis. The result of sensitivity analysis shows that there is a stability in the weight calculation process. However, a further study may check the consistency aspect for validation purposes. An algorithm such as the Full Consistency Method (FUCOM) may be used to find the weights using PFNs. A comparison of the weights calculated by using both FUCOM and LBWA may be made for further validation. There may a further study using Spherical Fuzzy Sets (SFS), a generalization of PFS, to conduct the FFA. Subsequently, the outcomes (by using PFS and SFS) may be compared. Nevertheless, the present paper has its own usefulness and we believe that this paper may add value to the growing literature in the stated field.

Author Contributions: Conceptualization, S.D. (Sibnath Deb), S.K., S.D. (Shayana Deb), S.B., A.A.D. and T.M.; Data curation, S.D. (Shayana Deb); Formal analysis, S.D. (Sibnath Deb), S.D. (Shayana Deb), and S.B.; Funding acquisition, S.K. and S.B.; Investigation, S.D. (Sibnath Deb), S.D. (Shayana Deb), A.A.D. and T.M.; Methodology, S.K. and S.B.; Project administration, S.D. (Sibnath Deb) and A.A.D.; Resources, S.D. (Sibnath Deb); Supervision, S.K. and T.M.; Validation, S.D. (Sibnath Deb) and T.M.; Writing—original draft, S.D. (Sibnath Deb), S.D. (Shayana Deb), S.B. and A.A.D.; Writing—review & editing, S.K. and T.M. All authors have read and agreed to the published version of the manuscript.

Funding: This was a syndicated study. No external funding is involved.

Institutional Review Board Statement: The authors declare that the present study was subjected to ethical approval and obtained the clearance (Ref.No.RGNIYD/ADMIN/20-21/SEC/001). All the procedures performed while collecting data from the participants were in accordance with the ethical standards of the institutional and/or national research committee and with the 1964 Helsinki declaration. Participation in the study was voluntary and participants were assured of the confidentiality of information. Prior consent was taken from the participants and/or their legal guardian(s) after explaining the purpose and modality of the study.

Informed Consent Statement: Prior consent was taken from the participants and/or their legal guardian(s) after explaining the purpose and modality of the study.

Data Availability Statement: The data that support the findings of this study are available on request from the corresponding author for maintaining confidentiality of the respondents. However, necessary information for carrying out the analysis is provided.

Acknowledgments: We are thankful to the schools that consented to share the survey with their students. We are also thankful to all the school students for participating in the study and providing information.

Conflicts of Interest: Authors declare no conflict of interest in the publication of this research.

Appendix A. Preliminaries of PFS

Appendix A.1. Definition

Let $\widetilde{A}$ denote a PFS on a universe of discourse U. Then, $\widetilde{A}$ is defined as

$$\widetilde{A} = \langle x, \mu_{\widetilde{A}}(x), \eta_{\widetilde{A}}(x), v_{\widetilde{A}}(x) \rangle \tag{A1}$$

where, $x \in U$; $\mu_{\widetilde{A}}(x), \eta_{\widetilde{A}}(x), v_{\widetilde{A}}(x) \in [0, 1]$ are the degrees of positive, neutral and negative membership of x in $\widetilde{A}$ respectively such that

$$0 \leq \mu_{\widetilde{A}}(x) + \eta_{\widetilde{A}}(x) + v_{\widetilde{A}}(x) \leq 1 \; \forall \, x \in U \tag{A2}$$

Here, if $\eta_{\widetilde{A}}(x) = 0$ then it becomes the intuitionistic fuzzy set (IFS) and if both $\eta_{\widetilde{A}}(x) = v_{\widetilde{A}}(x) = 0$, $\widetilde{A}$ becomes a traditional fuzzy set.

The degree of refusal ($\pi_{\widetilde{A}}(x)$) is defined as

$$\pi_{\widetilde{A}}(x) = 1 - \left(\mu_{\widetilde{A}}(x) + \eta_{\widetilde{A}}(x) + v_{\widetilde{A}}(x)\right) \forall\, x \in U \tag{A3}$$

For a given element x in U, a PFN is represented as

$$A = \{\{\mu_A, \eta_A, v_A \in [0,1] and\ 0 \leq \mu_A + \eta_A + v_A \leq 1\ \}\} \tag{A4}$$

Appendix A.2. Properties

Let $\widetilde{A} = \langle x,\ \mu_{\widetilde{A}}(x), \eta_{\widetilde{A}}(x), v_{\widetilde{A}}(x)\ \rangle$ and $\widetilde{B} = \langle x,\ \mu_{\widetilde{B}}(x), \eta_{\widetilde{B}}(x), v_{\widetilde{B}}(x)\ \rangle$ be two PFS $\forall\, x \in U$, and then the following properties are defined

$$\widetilde{A} \cup \widetilde{B} = \{x \in U\} \tag{A5}$$

$$\widetilde{A} \cap \widetilde{B} = \{x \in U\} \tag{A6}$$

$$\widetilde{A}^{\,c} = \{x \in U\} \tag{A7}$$

$$\widetilde{A} \subseteq \widetilde{B}\, if\ \left(\mu_{\widetilde{A}}(x) \leq \mu_{\widetilde{B}}(x),\ \eta_{\widetilde{A}}(x) \leq \eta_{\widetilde{B}}(x),\ v_{\widetilde{A}}(x) \geq v_{\widetilde{B}}(x)\ \forall\, x \in U\right) \tag{A8}$$

$$\widetilde{A} = \widetilde{B}\, if\ \widetilde{A} \subseteq \widetilde{B}\ and\ \widetilde{B} \subseteq \widetilde{A} \tag{A9}$$

$$\widetilde{A} \subseteq \widetilde{B}\ and\ \widetilde{B} \subseteq \widetilde{C} \Rightarrow \widetilde{A} \subseteq \widetilde{C} \tag{A10}$$

$$\left(\widetilde{A}^{\,c}\right)^c = \widetilde{A} \tag{A11}$$

Appendix A.3. Basic Operations

Let $A = (\mu_A, \eta_A, v_A\ )$ and $B = (\mu_B, \eta_B, v_B\ )$ be any two PFNs. The following are some of the basic operations.

$$A \oplus B = (\mu_A + \mu_B - \mu_A\mu_B,\ \eta_A\eta_B,\ v_A v_B) \tag{A12}$$

$$A \otimes B = (\mu_A\mu_B,\ \eta_A + \eta_B - \eta_A\eta_B,\ v_A + v_B - v_A v_B) \tag{A13}$$

$$\lambda A = \left(1 - (1-\mu_A)^{\lambda},\ \eta_A{}^{\lambda},\ v_A{}^{\lambda}\right);\ \lambda > 0 \tag{A14}$$

$$A^{\lambda} = (\mu_A{}^{\lambda},\ 1 - (1-\eta_A)^{\lambda},\ 1 - (1-v_A)^{\lambda};\ \lambda > 0 \tag{A15}$$

$$A \oplus B = B \oplus A \tag{A16}$$

$$A \otimes B = B \otimes A \tag{A17}$$

$$\left(A^{\lambda 1}\right)^{\lambda 2} = A^{\lambda 1 \lambda 2} \tag{A18}$$

$$\lambda\,(A \oplus B) = \lambda A \oplus \lambda B \tag{A19}$$

$$(A \otimes B)^{\lambda} = A^{\lambda} \otimes\ B^{\lambda} \tag{A20}$$

Appendix A.4. Defuzzification

The defuzzification of a PFN A is done in the following steps [75,76]:

Step 1. Defining new positive and negative memberships

$$\mu_{\acute{A}} = \mu_A + \frac{\eta_A}{2} \tag{A21}$$

$$v_{\acute{A}} = v_A + \frac{\eta_A}{2} \tag{A22}$$

Step 2. Calculation of defuzzication value

$$\gamma_A = \mu_{\acute{A}} + \pi_A \left(\frac{1 + \mu_{\acute{A}} - \upsilon_{\acute{A}}}{2} \right) \quad \text{(A23)}$$

Appendix A.5. Score and Accuracy Functions

The score function of any PFN is calculated as

$$S_A = \mu_A - \upsilon_A \quad \text{(A24)}$$

The accuracy function is defined as

$$H_A = \mu_A + \eta_A + \upsilon_A \quad \text{(A25)}$$

Rule for comparison

$$If\ S_A \prec S_B,\ then\ A \prec B$$
$$If\ S_A \succ S_B,\ then\ A \succ B$$
$$If\ S_A = S_B,\ H_A \prec H_B,\ then\ A \prec B$$
$$If\ S_A = S_B,\ H_A \succ H_B,\ then\ A \succ B$$
$$If\ S_A = S_B,\ H_A = H_B,\ then\ A = B$$

Appendix A.6. Absolute and Actual Score

Computational steps [77] are described below.
Step 1. Identification of the positive ideal solution (PIS)
For a set of n number of PFNs, PIS is given as

$$Z^+ = (\mu^+, \eta^+, \upsilon^+) = \mu_i, \eta_i, \upsilon_i), \text{ where } i = 1, 2, \ldots n \quad \text{(A26)}$$

Step 2. Find out goal differences for each PFN
Positive goal difference (PGD):

$$\mu_{i+} = \mu^- - \mu_i \quad \text{(A27)}$$

Negative goal difference (NGD):

$$\upsilon_{i-} = \upsilon_i - \upsilon^+ \quad \text{(A28)}$$

Step 3. Find out the average neutral degree (Avg_ η)

$$\underline{\eta} = \frac{1}{n} \sum_{i=1}^{n} \eta_i \quad \text{(A29)}$$

Step 4. Calculation of the absolute score for each PFN

$$S_{i(abs)} = (1 - \mu_{i+}) - \upsilon_{i-} \quad \text{(A30)}$$

Step 5. Derive the actual score for each PFN

$$S_{i(act)} = \frac{S_{i(abs)}}{1 - (\underline{\eta} - \eta_i)} \quad \text{(A31)}$$

Here, the following rules are applicable

$$\text{If } S_{A(act)} \succ S_{B(act)} \text{ then } A \succ B$$

$$\text{If } S_{A(act)} = S_{B(act)} \text{ then if } \mu_A > \mu_B \text{ and } \eta_A \geq \eta_B \text{ then } A \succ B$$

$$\text{If } S_{A(act)} = S_{B(act)} \text{ and } \mu_A \geq \mu_B \text{ and } \eta_A < \eta_B \text{ then if } \upsilon_A \leq \upsilon_B \text{ then } A \succ B, \text{ otherwise } A \prec B$$

As $(\underline{\eta} - \eta_i) \neq 1$, $S_{i(act)}$ is always finite.

Appendix A.7. Aggregation Operator

Let $A_j = (\mu_j, \eta_j, v_j)$ $(j = 1, 2, \ldots n)$ be a collection of PFNs. Then the Picture Fuzzy Weighted Average (PFWA) is defined as [78]

$$PFWA_w(A_1, A_2, A_3 \ldots A_n) = \oplus_{j=1}^{n} (w_j A_j) == \left(1 - \prod_{j=1}^{n}\left(1 - \mu_{A_j}\right)^{w_j}, \prod_{j=1}^{n}\left(\eta_{A_j}\right)^{w_j}, \prod_{j=1}^{n}\left(\nu_{A_j}\right)^{w_j}\right) \quad \text{(A32)}$$

Here, w_j is the corresponding weight of A_j $(j = 1, 2, \ldots n)$ with the conditions that

$$w_j > 0; \sum_{j=1}^{n} w_j = 1$$

In this paper, w_j is derived using the LBWA method based on actual scores as used by Biswas et al. [42].

Appendix B. Computational Steps of LBWA Algorithm

Let the criteria set be given by $\mathbb{C} = \{C_1, C_2, C_3 \ldots \ldots C_n\}$. Let the ith criterion $(C_i \in \mathbb{C})$ be the most important one as opined by the respondents.

Step 1: Formation of subsets of criteria by grouping, based on level of significance.

The grouping process is described below.

Level L_1: Group the criteria and form the subset with the criteria that are having equal to or up to twice as less as the significance of the criterion C_i

Level L_2: Group the criteria and form the subset with the criteria having exactly twice as less as the significance of the criterion C_i or up to three times as less as the significance of the criterion C_i

Level L_3: Group the criteria and form the subset with the criteria having exactly three times as less as the significance of the criterion C_i or up to four times as less as the significance of the criterion C_i

———————————————————————————

Level L_k: Group the criteria and form the subset with the criteria having exactly 'k' times as less as the significance of the criterion C_i or up to 'k + 1' times as less as the significance of the criterion C_i

Hence,

$$L = L_1 \cup L_2 \cup L_3 \ldots \ldots \cup L_k \quad \text{(A33)}$$

If $s(C_j)$ is the significance of the jth criterion, we note that

$$L_k = \{C_j \in L : k \leq s(C_j) \leq k+1\} \quad \text{(A34)}$$

Also, the following condition holds good to appropriately define the grouping,

$$L_p \cap L_q = \varnothing; \text{ where } p, q \in \{1, 2, \ldots k\} \text{ and } p \neq q \quad \text{(A35)}$$

Step 2: Comparison of factors according to the significance within the subsets

Based on the comparison, each criterion $C_j \in L_k$ is assigned with an integer value $I_{C_j} \in \{0, 1, 2 \ldots \ldots r\}$; where, r is the maximum value on the scale for comparison and is given by:

$$r = \max\{|L_1|, |L_2|, |L_3| \ldots \ldots |L_k|\} \quad \text{(A36)}$$

Conditions used in this context are as follows.

The most important criterion is assigned with an integer value of zero. In other words,

$$I_{C_i} = 0 \quad \text{(A37)}$$

If C_p is more significant than C_q then

$$I_{C_p} < I_{C_q} \quad \text{(A38)}$$

If C_p is equally significant with C_q then

$$I_{C_p} = I_{C_q} \quad \text{(A39)}$$

Step 3: Find out the elasticity coefficient

The elasticity coefficient r_0 is defined as any real number with the condition $r_0 > r$ and $\tau \in \mathbb{R}$; Where $\mathbb{R}$ represents a set of real numbers

Step 4: Calculate the influence function of the criteria

For a particular criterion, $C_j \in L_k$; the influence function can be defined as $f : L \rightarrow R$

It is calculated as

$$f(C_j) = \frac{r_0}{\delta\, r_0 + I_{C_j}} \quad \text{(A40)}$$

Here, ∂ is the number of level or subset to which C_j belongs and $I_{C_j} \in \{0, 1, 2 \ldots\ldots r\}$ is the value assigned to the criterion C_j within that level

Step 5: Calculation of the optimum values of the priority weights of the criteria

For most significant criterion:

$$w_i = \frac{1}{1 + f(C_1) + f(C_2) + \ldots + f(C_n)} \quad \text{(A41)}$$

where, $i \in j; j = 1, 2, \ldots, n$, the number of criteria

For other factors:

$$w_{j \neq i} = f(C_j) w_i \quad \text{(A42)}$$

Decision rule: rank the criteria in descending order of criticality based on the weight values.

References

1. Duggal, D.; Sacks-Zimmerman, A.; Liberta, T. The Impact of Hope and Resilience on Multiple Factors in Neurosurgical Patients. *Cureus* **2016**, *8*, e849. [CrossRef] [PubMed]
2. Lee, J. Mental health effects of school closures during COVID-19. *Lancet Child Adolesc. Health* **2020**, *4*, 421. [CrossRef]
3. Qin, Z.; Shi, L.; Xue, Y.; Lin, H.; Zhang, J.; Liang, P.; Lu, Z.; Wu, M.; Chen, Y.; Zheng, X.; et al. Prevalence and Risk Factors Associated with Self-reported Psychological Distress Among Children and Adolescents During the COVID-19 Pandemic in China. *JAMA Netw. Open* **2021**, *4*, e2035487. [CrossRef] [PubMed]
4. Qiu, J.; Shen, B.; Zhao, M.; Wang, Z.; Xie, B.; Xu, Y. A nationwide survey of psychological distress among Chinese people in the COVID-19 epidemic: Implications and policy recommendations. *Gen. Psychiatry* **2020**, *33*, e100213. [CrossRef] [PubMed]
5. Zhou, X. Managing psychological distress in children and adolescents following the COVID-19 epidemic: A cooperative approach. *Psychol. Trauma Theory Res. Pract. Policy* **2020** 12, S76–S78. [CrossRef]
6. Zhou, S.-J.; Zhang, L.-G.; Wang, L.-L.; Guo, Z.-C.; Wang, J.-Q.; Chen, J.-C.; Liu, M.; Chen, X.; Chen, J.-X. Prevalence and socio-demographic correlates of psychological health problems in Chinese adolescents during the outbreak of COVID-19. *Eur. Child Adolesc. Psychiatry* **2020**, *29*, 749–758 [CrossRef]
7. Dangi, R.R.; George, M. Psychological Perception of Students during COVID-19 Outbreak in India. *High Technol. Lett.* **2020**, *26*, 142–144.
8. Smith, L.; Jacob, L.; Trott, M.; Yakkundi A.; Butler, L.; Barnett, Y.; Armstrong, N.C.; McDermott, D.; Schuch, F.; Meyer, J.; et al. The association between screen time and mental health during COVID-19: A cross sectional study. *Psychiatry Res.* **2020**, *292*, 113333. [CrossRef]
9. UNICEF. India Case Study: Situation Analysis on the Effects of and Responses to COVID-19 on the Education Sector in Asia. 2021. Available online: https://www.unicef.org/rosa/media/16511/file/India%20Case%20Study.pdf (accessed on 10 December 2021).
10. Vyas, A. Status Report: Government and Private Schools During COVID-19. *Oxfam India* **2020**, *4*, 1.
11. Jena, P.K. Impact of pandemic COVID-19 on education in India. *Int. J. Curr. Res.* **2020**, *12*, 12582–12586.

12. Mahapatra, A.; Sharma, P. Education in times of COVID-19 pandemic: Academic stress and its psychosocial impact on children and adolescents in India. *Int. J. Soc. Psychiatry* **2020**, *67*, 397–399. [CrossRef] [PubMed]
13. UDISE. Unified District Information System for Education (UDISE) Report 2019–2020, Provisional. 2021. Available online: https://udiseplus.gov.in/#/home (accessed on 22 November 2021).
14. The Hindu. In Academic Year 2019–2020, only 22% Indian Schools Had Internet. The Hindu, Chennai, 2 July 2021. Available online: https://www.thehindu.com/news/national/in-academic-year-2019-20-only-22-indian-schools-had-internet/article35082011.ece (accessed on 22 December 2021).
15. Samuel, S.; Saksena, K. The Effect of COVID-19 Second Wave on Primary Education, Education Times, Chennai India. 2021, p. 2. Available online: https://www.educationtimes.com/article/school-guide/84117353/portal-exclusive-all-you-need-to-know-about-the-impact-of-covid-19-second-wave-on-childrens-education (accessed on 2 January 2022).
16. Dangi, R.R.; Dewett, P.; Joshi, P. Stress level and coping strategies among youth during coronavirus disease lockdown in INDIA. *SSRN Electron. J.* **2020**, *8*, 605–617. [CrossRef]
17. Gaudin, J.M., Jr.; Pollane, L. Social networks, stress and child abuse. *Child. Youth Serv. Rev.* **1983**, *5*, 91–102. [CrossRef]
18. Hefner, J.; Eisenberg, D. Social support and mental health among college students. *Am. J. Orthopsychiatry* **2009**, *79*, 491–499. [CrossRef]
19. Grover, S.; Sahoo, S.; Mehra, A.; Avasthi, A.; Tripathi, A.; Subramanyan, A.; Pattojoshi, A.; Rao, G.P.; Saha, G.; Mishra, K.; et al. Psychological impact of COVID-19 lockdown: An online survey from India. *Indian J. Psychiatry* **2020**, *62*, 354–362. [CrossRef] [PubMed]
20. Liang, L.; Ren, H.; Cao, R.; Hu, Y.; Qin, Z.; Li, C.; Mei, S. The Effect of COVID-19 on Youth Mental Health. *Psychiatr. Q.* **2020**, *91*, 841–852. [CrossRef]
21. Torales, J.; O'Higgins, M.; Castaldelli-Maia, J.M.; Ventriglio, A. The outbreak of COVID-19 coronavirus and its impact on global mental health. *Int. J. Soc. Psychiatry* **2020**, *66*, 317–320. [CrossRef]
22. Hawes, M.T.; Szenczy, A.K.; Klein, D.N.; Hajcak, G.; Nelson, B.D. Increases in depression and anxiety symptoms in adolescents and young adults during the COVID-19 pandemic. *Psychol. Med.* **2021**, 1–9. [CrossRef]
23. Keskin, S.; Yurdugül, H. Factors affecting students' preferences for online and blended learning: Motivational vs. cognitive. *Eur. J. Open Distance E-Learn.* **2019**, *22*, 72–86. [CrossRef]
24. Neuhauser, C. Learning Style and Effectiveness of Online and Face-to-Face Instruction. *Am. J. Distance Educ.* **2002**, *16*, 99–113. [CrossRef]
25. Reblin, M.; Uchino, B.N. Social and emotional support and its implication for health. *Curr. Opin. Psychiatry* **2008**, *21*, 201–205. [CrossRef] [PubMed]
26. Smart, K.L.; Cappel, J.J. Students' Perceptions of Online Learning: A Comparative Study. *J. Inf. Technol. Educ. Res.* **2006**, *5*, 201–219. [CrossRef]
27. Thongsri, N.; Shen, L.; Bao, Y. Investigating factors affecting learner's perception toward online learning: Evidence from ClassStart application in Thailand. *Behav. Inf. Technol.* **2019**, *38*, 1243–1258. [CrossRef]
28. Demuyakor, J. Coronavirus (COVID-19) and Online Learning in Higher Institutions of Education: A Survey of the Perceptions of Ghanaian International Students in China. *Online J. Commun. Media Technol.* **2020**, *10*, e202018. [CrossRef]
29. Wang, C.; Chudzicka-Czupała, A.; Grabowski, D.; Pan, R.; Adamus, K.; Wan, X.; Hetnał, M.; Tan, Y.; Olszewska-Guizzo, A.; Xu, L.; et al. The Association Between Physical and Mental Health and Face Mask Use During the COVID-19 Pandemic: A Comparison of Two Countries with Different Views and Practices. *Front. Psychiatry* **2020**, *11*, 569981. [CrossRef]
30. Adnan, M.; Anwar, K. Online learning amid the COVID-19 pandemic: Students' perspectives. *Online Submiss.* **2020**, *2*, 45–51. [CrossRef]
31. Agustina, P.Z.R.; Cheng, T.H. How students' perspectives about online learning amid the COVID-19 pandemic? *Stud. Learn. Teach.* **2020**, *1*, 133–139.
32. Alawamleh, M.; Al-Twait, L.M.; Al-Saht, G.R. The effect of online learning on communication between instructors and students during COVID-19 pandemic. *Asian Educ. Dev. Stud.* **2020**, *11*, 380–400. [CrossRef]
33. Lewin, K. *Field Theory in Social Science*; Harper Row: London, UK, 1951.
34. Baulcomb, J.S. Management of change through force field analysis. *J. Nurs. Manag.* **2003**, *11*, 275–280. [CrossRef]
35. Paquin, J.-P.; Koplyay, T. Force Field Analysis and Strategic Management: A Dynamic Approach. *Eng. Manag. J.* **2007**, *19*, 28–37. [CrossRef]
36. Hlalele, B.M. Application of the force-field technique to drought vulnerability analysis: A phenomenological approach. *Jàmbá J. Disaster Risk Stud.* **2019**, *11*, 589. [CrossRef] [PubMed]
37. Youssef, A.; Mostafa, A.M. Critical Decision-Making on Cloud Computing Adoption in Organizations Based on Augmented Force Field Analysis. *IEEE Access* **2019**, *7*, 167229–167239. [CrossRef]
38. Mak, A.H.; Chang, R.C. The driving and restraining forces for environmental strategy adoption in the hotel Industry: A force field analysis approach. *Tour. Manag.* **2019**, *73*, 48–60. [CrossRef]
39. Shamsher, S.; Praba, T.; Sethuraman, K.R. The force field analysis of online learning. *Asian J. Med. Health Sci.* **2021**, *4*, 154–163.
40. Ramos, P.N.; Enteria, M.L.B.; Norona, M.I. Readiness Model Development in the Adoption of Internet of Things (IoT) among Philippine Manufacturing SMEs Using Force Field Analysis Approach and Structural Equation Modelling. In Proceedings of

the Second Asia Pacific International Conference on Industrial Engineering and Operations Management, Surakarta, Indonesia, 14–16 September 2021.

41. Wang, C.; Zhou, X.; Tu, H.; Tao, S. Some geometric aggregation operators based on picture fuzzy sets and their application in multiple attribute decision making. *Ital. J. Pure Appl. Math* **2017**, *37*, 477–492.
42. Biswas, S.; Majumder, S.; Pamucar, D.; Dawn, S.K. An Extended LBWA Framework in Picture Fuzzy Environment Using Actual Score Measures Application in Social Enterprise Systems. *Int. J. Enterp. Inf. Syst.* **2021**, *17*, 37–68. [CrossRef]
43. Biswas, S.; Pamucar, D.; Chowdhury, P.; Kar, S. A New Decision Support Framework with Picture Fuzzy Information: Comparison of Video Conferencing Platforms for Higher Education in India. *Discret. Dyn. Nat. Soc.* **2021**, *2021*, 2046097. [CrossRef]
44. Duong, T.T.T.; Thao, N.X. A novel dissimilarity measure on picture fuzzy sets and its application in multi-criteria decision making. *Soft Comput.* **2021**, *25*, 15–25. [CrossRef]
45. Gocer, F. A Novel Interval Value Extension of Picture Fuzzy Sets into Group Decision Making: An Approach to Support Supply Chain Sustainability in Catastrophic Disruptions. *IEEE Access* **2021**, *9*, 117080–117096. [CrossRef]
46. Khan, M.J.; Kumam, P.; Deebani, W.; Kumam, W.; Shah, Z. Bi-parametric distance and similarity measures of picture fuzzy sets and their applications in medical diagnosis. *Egypt. Inform. J.* **2021**, *22*, 201–212. [CrossRef]
47. Mahmood, T.; Waqas, H.M.; Ali, Z.; Ullah, K.; Pamucar, D. Frank aggregation operators and analytic hierarchy process based on interval-valued picture fuzzy sets and their applications. *Int. J. Intell. Syst.* **2021**, *36*, 7925–7962. [CrossRef]
48. Tchier, F.; Ali, G.; Gulzar, M.; Pamučar, D.; Ghorai, G. A New Group Decision-Making Technique under Picture Fuzzy Soft Expert Information. *Entropy* **2021**, *23*, 1176. [CrossRef]
49. Cuong, B.C.; Kreinovich, V. Picture fuzzy sets. *J. Comput. Sci. Cybern.* **2014**, *30*, 409–420.
50. Cuong, B.C.; Kreinovich, V. Picture Fuzzy Sets-a new concept for computational intelligence problems. In Proceedings of the 2013 Third World Congress on Information and Communication Technologies (WICT 2013), Hanoi, Vietnam, 15–18 December 2013; pp. 1–6.
51. Žižović, M.; Pamucar, D. New model for determining criteria weights: Level Based Weight Assessment (LBWA) model. *Decis. Mak. Appl. Manag. Eng.* **2019**, *2*, 126–137. [CrossRef]
52. Biswas, S.; Pamucar, D. Facility Location Selection for B-Schools in Indian Context: A Multi-Criteria Group Decision Based Analysis. *Axioms* **2020**, *9*, 77. [CrossRef]
53. Pamucar, D.; Deveci, M.; Canıtez, F.; Lukovac, V. Selecting an airport ground access mode using novel fuzzy LBWA-WASPAS-H decision making model. *Eng. Appl. Artif. Intell.* **2020**, *93*, 103703. [CrossRef]
54. Hristov, N.; Pamucar, D.; Amine, M.E. Application of a D Number based LBWA Model and an Interval MABAC Model in Selection of an Automatic Cannon for Integration into Combat Vehicles. *Def. Sci. J* **2021**, *71*, 34–45. [CrossRef]
55. Božanić, D.; Jurišić, D.; Erkić, D. LBWA–Z-MAIRCA model supporting decision making in the army. *Oper. Res. Eng. Sci. Theory Appl.* **2020**, *3*, 87–110. [CrossRef]
56. Božanić, D.; Ranđelović, A.; Radovanović, M.; Tešić, D. A hybrid lbwa-ir-mairca multi-criteria decision-making model for determination of constructive elements of weapons. *Facta Univ. Ser. Mech. Eng.* **2020**, *18*, 399–418. [CrossRef]
57. Pamučar, D.; Žižović, M.; Marinković, D.; Doljanica, D.; Jovanović, S.; Brzaković, P. Development of a Multi-Criteria Model for Sustainable Reorganization of a Healthcare System in an Emergency Situation Caused by the COVID-19 Pandemic. *Sustainability* **2020**, *12*, 7504. [CrossRef]
58. Deveci, M.; Özcan, E.; John, R.; Covrig, C.-F.; Pamucar, D. A study on offshore wind farm siting criteria using a novel interval-valued fuzzy-rough based Delphi method. *J. Environ. Manag.* **2020**, *270*, 110916. [CrossRef] [PubMed]
59. Ecer, F.; Pamucar, D.; Mardani, A.; Alrasheedi, M. Assessment of renewable energy resources using new interval rough number extension of the level based weight assessment and combinative distance-based assessment. *Renew. Energy* **2021**, *170*, 1156–1177. [CrossRef]
60. Pamučar, D.; Behzad, M.; Božanić, D.; Behzad, M. Decision making to support sustainable energy policies corresponding to agriculture sector: Case study in Iran's Caspian Sea coastline. *J. Clean. Prod.* **2020**, *292*, 125302. [CrossRef]
61. Jovčić, S.; Simić, V.; Průša, P.; Dobrodolac, M. Picture Fuzzy ARAS Method for Freight Distribution Concept Selection. *Symmetry* **2020**, *12*, 1062. [CrossRef]
62. Pamucar, D.; Torkayesh, A.E.; Biswas, S. Supplier selection in healthcare supply chain management during the COVID-19 pandemic: A novel fuzzy rough decision-making approach. *Ann. Oper. Res.* **2022**, 1–43. [CrossRef]
63. Pamučar, D.; Žižović, M.; Biswas, S.; Božanić, D. A new logarithm methodology of additive weights (lmaw) for multi-criteria decision-making: Application in logistics. *Facta Univ. Series Mech. Eng.* **2021**, *19*, 361–380. [CrossRef]
64. Biswas, S.; Majumder, S.; Dawn, S.K. Comparing the Socioeconomic Development of G7 and BRICS Countries and Resilience to COVID-19: An Entropy–MARCOS Framework. *Bus. Perspect. Res.* **2021**, *10*, 286–303. [CrossRef]
65. Biswas, S.; Pamucar, D.; Kar, S.; Sana, S.S. A New Integrated FUCOM–CODAS Framework with Fermatean Fuzzy Information for Multi-Criteria Group Decision-Making. *Symmetry* **2021**, *13*, 2430. [CrossRef]
66. Pramanik, P.K.D.; Biswas, S.; Pal, S.; Marinković, D.; Choudhury, P. A Comparative Analysis of Multi-Criteria Decision-Making Methods for Resource Selection in Mobile Crowd Computing. *Symmetry* **2021**, *13*, 1713. [CrossRef]
67. Biswas, S.; Anand, O.P. Logistics Competitiveness Index-Based Comparison of BRICS and G7 Countries: An Integrated PSI-PIV Approach. *IUP J. Supply Chain. Manag.* **2020**, *17*, 32–57.

68. Biswas, S. Measuring performance of healthcare supply chains in India: A comparative analysis of multi-criteria decision making methods. *Decis. Making Appl. Manag. Eng.* **2020**, *3*, 162–189. [CrossRef]
69. Butnaru, G.; Niță, V.; Anichiti, A.; Brînză, G. The Effectiveness of Online Education during COVID 19 Pandemic—A Comparative Analysis between the Perceptions of Academic Students and High School Students from Romania. *Sustainability* **2021**, *13*, 5311. [CrossRef]
70. Kroll, C. Towards a Sociology of Happiness: The Case of an Age Perspective on the Social Context of Well-Being. *Sociol. Res. Online* **2014**, *19*, 1–18. [CrossRef]
71. Larsen, L.; Helland, M.S.; Holt, T. The impact of school closure and social isolation on children in vulnerable families during COVID-19: A focus on children's reactions. *Eur. Child Adolesc. Psychiatry* **2021**, 1–11. [CrossRef]
72. Jung, J.; Horta, H.; Postiglione, G.A. Living in uncertainty: The COVID-19 pandemic and higher education in Hong Kong. *Stud. High. Educ.* **2021**, *46*, 107–120. [CrossRef]
73. Ferren, M. Social and Emotional Supports for Educators during and after the Pandemic. 2021. Available online: https://www.americanprogress.org/article/social-emotional-supports-educators-pandemic/ (accessed on 14 June 2022).
74. Labrague, L.J.; De los Santos, J.A.A.; Falguera, C. Social and Emotional Loneliness among College Students during the COVID-19 Pandemic: The Predictive Role of Coping Behaviours, Social Support, and Personal Resilience. *Perspect. Psychiatr. Care* **2021**. [CrossRef]
75. Xu, X.-G.; Shi, H.; Xu, D.-H.; Liu, H.-C. Picture Fuzzy Petri Nets for Knowledge Representation and Acquisition in Considering Conflicting Opinions. *Appl. Sci.* **2019**, *9*, 983. [CrossRef]
76. Son, L.H. Measuring analogousness in picture fuzzy sets: From picture distance measures to picture association measures. *Fuzzy Optim. Decis. Mak.* **2016**, *16*, 359–378. [CrossRef]
77. Si, A.; Das, S.; Kar, S. An Approach to Rank Picture Fuzzy Numbers for Decision Making Problems. *Decis. Making Appl. Manag. Eng.* **2019**, *2*, 54–64. [CrossRef]
78. Wei, G. Picture fuzzy aggregation operators and their application to multiple attribute decision making. *J. Intell. Fuzzy Syst.* **2017**, *33*, 713–724. [CrossRef]

data

MDPI

Article

Development of a Model Using Data Mining Technique to Test, Predict and Obtain Knowledge from the Academics Results of Information Technology Students

Wisam Ibrahim [1], Sanjar Abdullaev [1], Hussein Alkattan [1,*], Oluwaseun A. Adelaja [2,*] and Alhumaima Ali Subhi [1,3]

1. Department of System Programming, South Ural State University, Chelyabinsk 454080, Russia; wsamkreem5@gmail.com (W.I.); abdullaevsm@susu.ru (S.A.); alhumaimaali@uodiyala.edu.iq (A.A.S.)
2. Information Communication and Technology Department, Lagos State University, Lagos 102101, Nigeria
3. Electronic and Computer Center, University of Diyala, Baqubah 32010, Iraq

* Correspondence: alkattan.hussein92@gmail.com (H.A.); oluwaseun.adelaja@lasu.edu.ng (O.A.A.)

Abstract: Due to the huge amount of data obtained from students' academic results in most tertiary institutions such as the colleges, polytechnics and universities, data mining has become one of the most effective tools for discovering vital knowledge from students' dataset. The discovered knowledge can be productive in understanding numerous challenges in the scope of education and providing possible solutions to these challenges. The main objective of this research is to utilize the J48 decision algorithm model to test, classify and predict the students' dataset by identifying some important attributes and instances. The analysis was conducted on the final year students' academic results in C# programming amongst five universities which was imported in csv excel file dataset in WEKA environment. These training datasets contained the scores obtained in the examinations, grade remarks, grades, gender, and department. The knowledge extracted for the prediction model will help both the tutors and students to determine the success grade performance in the future. Flow lines, J48 decision trees, confusion matrices and a program flowchart were generated from the students' dataset. The KAPPA value obtained from the prediction in this research ranges from 0.9070–0.9582 which perfectly agrees with the standard for an ideal analysis on datasets.

Keywords: data mining tools; WEKA; J48 algorithm; KAPPA value; predict; confusion matrix; csv

Citation: Ibrahim, W.; Abdullaev, S.; Alkattan, H.; Adelaja, O.A.; Subhi, A.A. Development of a Model Using Data Mining Technique to Test, Predict and Obtain Knowledge from the Academics Results of Information Technology Students. *Data* **2022**, *7*, 67. https://doi.org/10.3390/data7050067

Academic Editors: Antonio Sarasa Cabezuelo and Ramón González del Campo Rodríguez Barbero

Received: 9 May 2022
Accepted: 22 May 2022
Published: 23 May 2022

1. Introduction

The students' academic performance is an important aspect in most tertiary educational system, particularly the higher learning institutions. The excellent records achieved amongst students' academic performances in examinations have become one of the key factors in considering tertiary institutions on the highly ranked Q.S world university rating system [1]. In the world today, a huge amount of students' data increases daily which makes it very critical to perform analysis on data to discover and retrieve useful information likewise knowledge from this data. There are numerous techniques that have been proposed in the evaluation (which involves testing, prediction and knowledge discovery of dataset) of students' academic performance. Data mining is one of the most common techniques utilized to analyze the academic performance of students and it has been recently applied in a vast approach regarding the educational sectors [2]. Data mining, also known as Knowledge discovery from data (KDD), can be defined a process of discovering interesting patterns and knowledge from stored data. Data Mining has various methods for used analyzing which include classification, clustering, and association rules [3]. Data mining could also be referred to as data dredging, which is a multidisciplinary field that obtains relevant information from large amount of data at the confluence among other specializations which includes artificial intelligence, statistics, databases, and information science [4]. In the educational sectors, one of the major objectives is to provide learning processes that allow for understanding students and their learning paths, termed as Educational Data Mining and Learning Analytics (EDM/LA). Educational Data Mining (EDM) is a discipline that focuses on extraction of useful information and knowledge from huge educational database, thereby utilizing this useful information and knowledge dredged to predict students' academic performance [5]. Apart from extracting and analyzing educational data, Educational Data Mining can enhance and develop students' performance in the teaching and learning domain [6]. There are several works in Educational Data Mining and Learning Analytics (EDM/LA) which has been devoted to prediction methods of student performance. According to [7], the authors compared different decision trees based on the students' academic performance for prediction. The decision trees were able to reveal the total number of students with excellent grades and those with failed grades, as this prediction effectively improved both the teaching/learning process in the institution and mitigated the failure rate amongst the students.

WEKA is a Data Mining tool used for managing the experimental analysis for data mining process such as (predictions, classification, clustering, association rule and evaluation); it also provides a flexible support for machine learning research and serves as a tool for introducing people to machine learning in the educational environment [8]. This research work focuses on using the J48 decision tree Classification model in WEKA to analyze the students' academic performance of Information Technology (I.T) department in five universities across five countries which includes Iraq, Sudan, Nigeria, South Africa, and India. The data was obtained from the records of the undergraduate students in the final year study of the five countries in the second semester of examinations. The authors in [9] revealed the taxonomy for Data mining approaches and this was illustrated pictorially, see Figure 1.

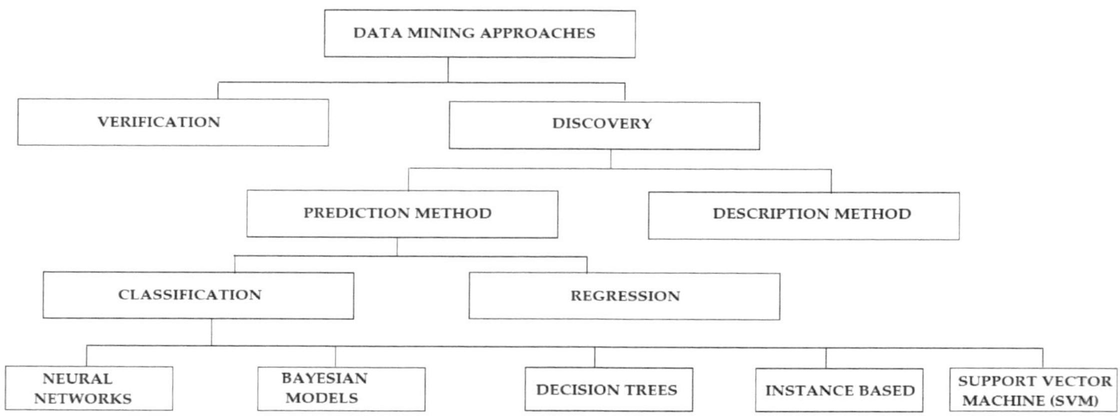

Figure 1. Taxonomy of Data Mining.

The research conducted in [10] revealed that the authors substantiated and built methodology for an ensemble classification of individual students' performance and collective performance quantification. According to [11], educational data mining involves four development phases which are filtering process of the students' data: selection of attributes or variables relating to their performance; extraction of knowledge for the filtered students' data; interpretation and evaluation. The research study by the authors in [12] was conducted by predicting successfully binary academic performance on school students who had number of passed test as 40–60% in both mathematics and computer science with the aim of obtaining correlation between the scores to investigate the student' cognitive abilities. The J48 algorithm is one of the best machine learning algorithms which can examine educational data categorically and continuously; it has been used by most researchers for classification of students' dataset and it usually obtains accurate results [13]. According to research study conducted in [14], the J48 algorithm was utilized for classification on students' dataset also comparing their performances with evaluation principles such as accuracy and implementation time. It revealed that the performance of classification techniques differs with datasets. The study also showed that factors such students' datasets, number of instances, attributes and the type of attributes enhanced the classifier's performance. J48 came out with better results on most educational dataset [13,14]. Researchers have applied decision tree utilizing the J48 classification algorithm to predict academic performances of students in the tertiary institution by simply testing this algorithm on unseen dataset to calculate accuracy. They intend to use this algorithm build model that can be used by the university to predict student performance, evaluate the teaching skills adopted by the lecturers and improve the learning potentials of the students in the other academic specializations [15].

2. Dataset Description

The data of the students' academic record analyzed in WEKA utilized the J48 classification algorithm method to test and predict from the students' future learning outcome using final year students' dataset record from five countries. The analysis was conducted on the students' academic results in C# programming language examinations with a total grade of 100%. The departments considered include Computer Science in Lagos State University Nigeria; Computer Science in University of Kirkuk Iraq; School of computers and systems science in Jawaharlal Nehru University New Delhi India; College of Computer Science and Information Technology in Sudan University of Science and Technology, Khartoum Sudan; and Computer Sciencein University of Cape Town South Africa. The students' dataset obtained consist of five attributes which are "scores obtained in the C-SHARP (C#) examinations", "grade remarks", "grades", "gender" and "department". For the purpose of the J4.8 algorithm analysis in WEKA, only "grades" columns to produce a detailed accuracy class reading. The grades were classified into A (70–100) marks, B (60–69) marks, C (50–59) marks, D (40–49) marks and F (0–39) marks which depicts excellent, very good, average, poor and failed, respectively. The functional requirements for the analysis of the students' data conducted in WEKA can be illustrated pictorial with the aid of program flowchart. Program Flow charts (Figure 2) are data flow that describes the sequence of data operations and decisions for a particular program or algorithm [16].

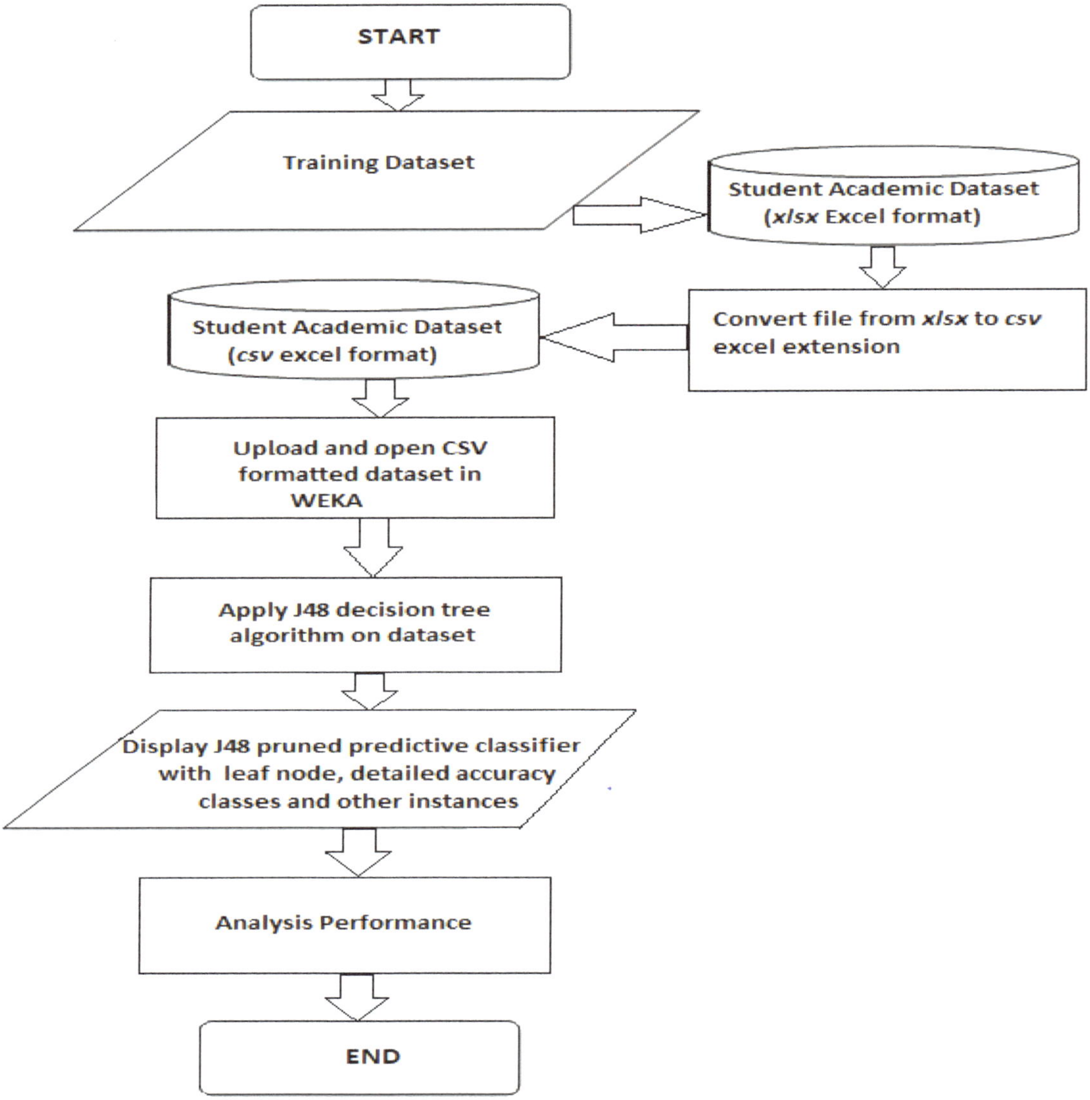

Figure 2. Flow Chart for Students' Dataset analysis in WEKA [16].

3. Methods

The J48 Decision Algorithm is a predictive machine learning model that the dependent variables also known as target value of a new sample based on various attribute values of the data available [17]. The node of a J48 decision tree denotes the different utilized attributes [18]. With the aid of tree classification algorithm, the essential distribution of data become easier to understand and flexible to implement. J48 is an extension of ID3 and it develops a decision node utilizing the expected estimations of the class. J48 algorithm deals with decision trees pruning, lost or missing attribute estimations of the data and varying attribute costs [19]. The J48 algorithm can be generated via the following three stages [20]:

- Stage 1: If an instance belongs to similar class, the leaves are labeled with a similar class;
- Stage 2: For each attribute, the potential data will be figured and the gain in this data will be attained from the test conducted on attribute;
- Stage 3: Finally, the best attribute will be selected in regard to the current selection parameter.

3.1. Students' Dataset Analysis in WEKA

The J48 tree generated in WEKA for the students' academic dataset across the 5 countries utilized 50% percentage split with training set: 25% for the test data and the remaining 25% for validate to obtain the classifier model. The J48 decision tree classifier output algorithm obtained from the students' result for the five universities analyzed is displayed in the Appendix A section of this work.

3.2. Calculations of the Evaluation Measures of the Detailed Accuracy Class Table

In the data analysis conducted, the three standard measures used in the evaluation of the classification qualities include the Recall, Precision and F-Measure. Precision is the ratio of the correctly classified cases of total number of misclassified cases and correctly classified cases [21]. The recall is the ratio of correctly classified samples to the total number of unclassified instances and correctly classified cases. The F-measure is the aggregate of the values of recall and precision [21,22]. Other measures used in the obtaining and evaluation of results include the execution time, TP rate, FP rate, ROC area, PRC area and confusion matrix [23].

The calculations of the precision, F-measure, recall values can be obtained using the Equations (1)–(3), respectively:

$$\text{Precision} = \frac{\text{TP}}{\text{TP} + \text{FP}} \tag{1}$$

$$\text{F} - \text{Measure} = 2 \times \frac{\text{Recall} \times \text{Precision}}{\text{Recall} + \text{Precision}} \tag{2}$$

$$\text{Recall} = \frac{\text{TP}}{\text{TP} + \text{FN}} \tag{3}$$

The TP represents the values of the true positive rate; the FP represents false positive rate value, and the FN represents the false negative rate. The precision, F-measure and the Recall values are some of the evaluation parameters generated in WEKA in the detailed accuracy by class table.

3.3. Outcomes of J48 Decision Tree Generated from Students' Dataset Analysis

This section shows the J48 decision trees generated from the students' academic result imported in WEKA environment platform for the analysis. See Figures 3–7. The Grade_Remarks Attribute Platform for Students' dataset is shown in Appendix A of this research.

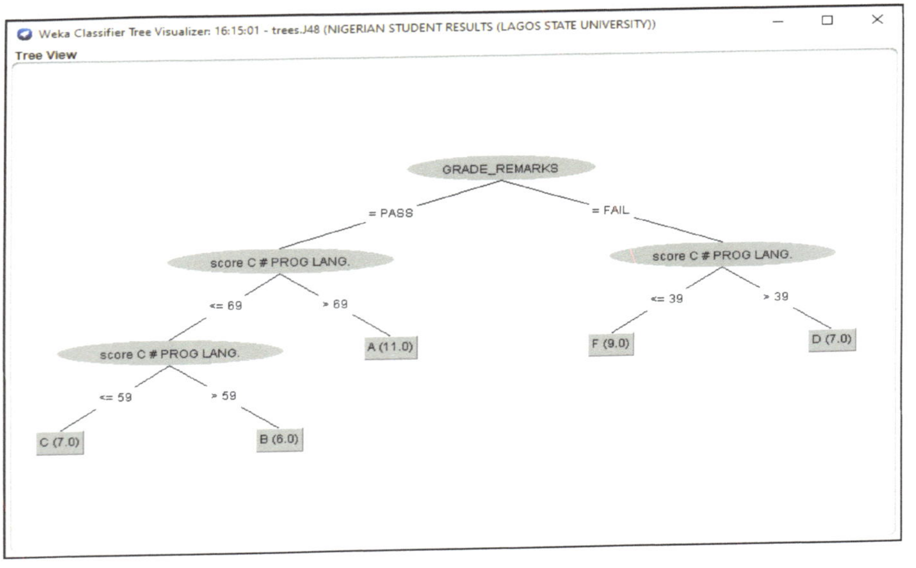

Figure 3. J48 Decision Classifier for Nigerian Students' dataset.

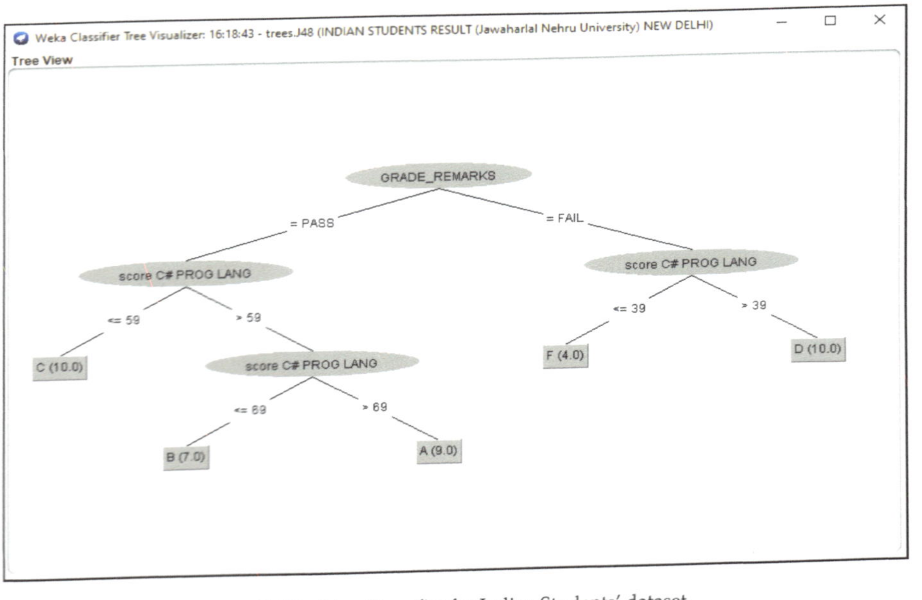

Figure 4. J48 Decision Classifier for Indian Students' dataset.

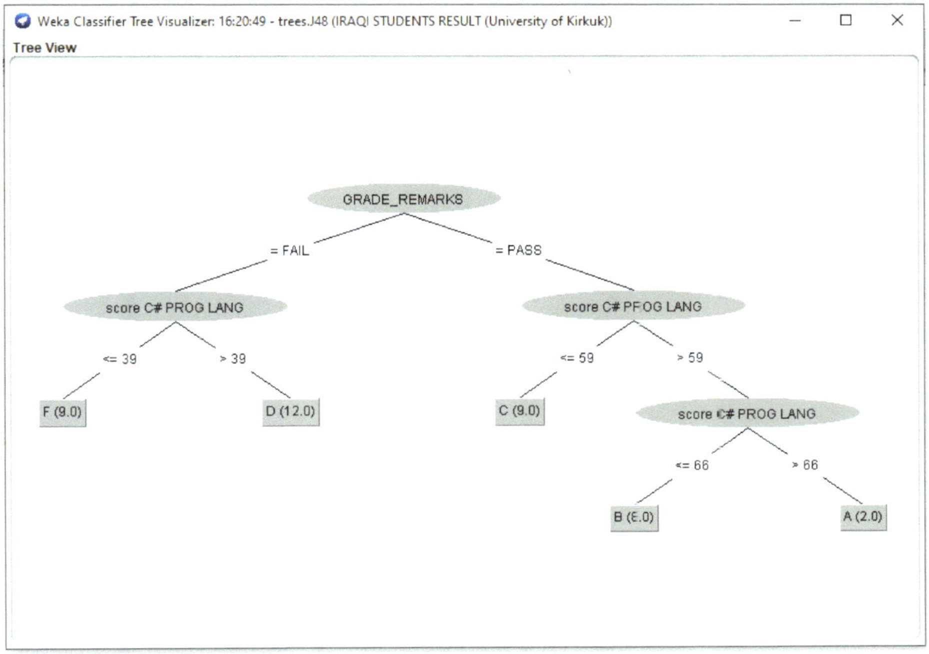

Figure 5. J48 Decision Classifier for Iraqi Students' dataset.

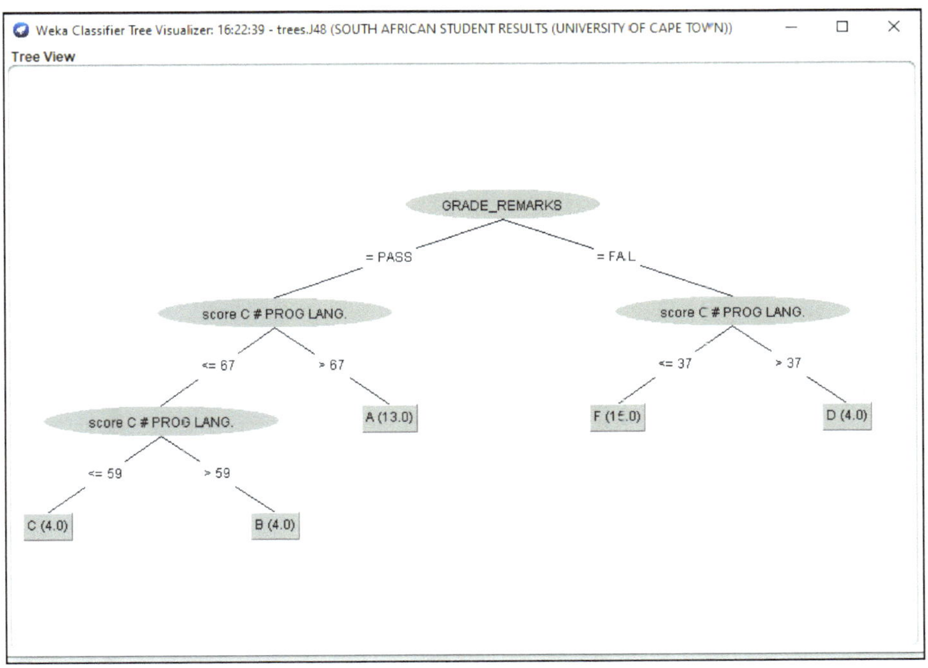

Figure 6. J48 Decision Classifier for South African Students' dataset.

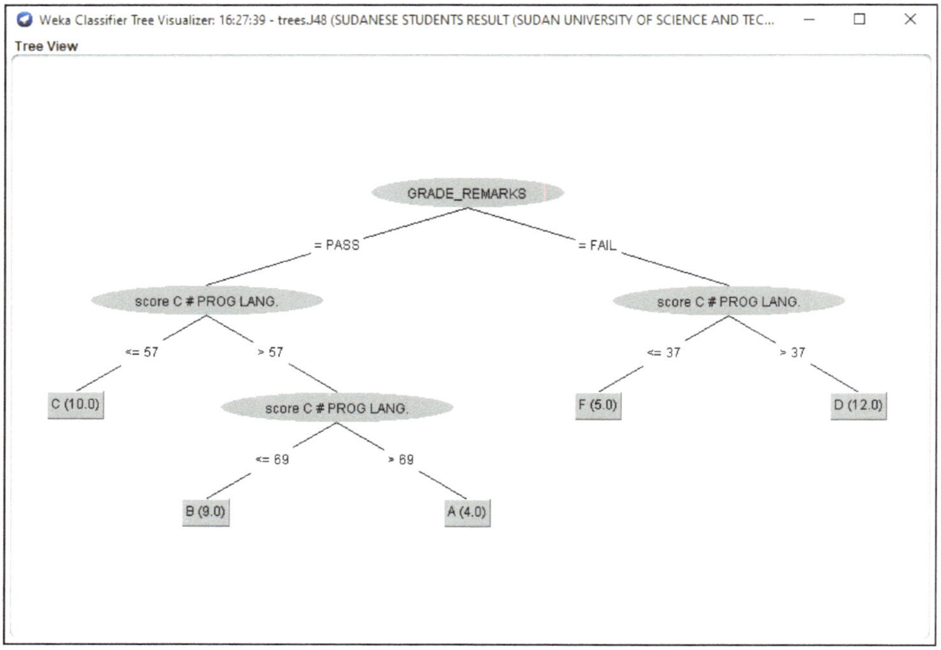

Figure 7. J48 Decision Classifier for Sudanese Students' dataset.

4. Results and Discussion

The results for the analysis, based on the Kappa statistical values, mean that absolute error, recall, Precision, and F-Measure obtained from the five universities can be computed in tabular form. Table 1 shows the values obtained from the student's dataset analysis. The Kappa interpretation obtained revealed a range of 0.9070–0.9582 which perfectly agrees with the general values for most analysis.

Table 1. Values obtained for the Students' dataset across the five universities.

Countries	Mean Absolute Error	Kappa	Recall	Precision	F-Measure
India	0.02	0.9313	0.700	0.850	0.7677
South Africa	0.1489	0.9070	0.950	1.000	0.9744
Sudan	0.04	0.9582	0.950	1.000	0.9744
Iraq	0.02	0.9308	0.650	0.700	0.6741
Nigeria	0.1114	0.9296	0.900	0.967	0.9323

4.1. Plots of Evaluation Parameters from the Analysis Conducted on the Students' Dataset

The parameters (TP Rate, FP Rate, Precision, Recall, F-Measure, MCC, ROC-Area and PRC-Area) obtained in this research work based on detail accuracy class analysis revealed from WEKA, we plotted flow lines that illustrate these parameters for the purpose of obtaining knowledgeable patterns to be displayed in a statistical perspective. These flow lines were illustrated based on values of the evaluation parameter derived from the WEKA analysis conducted on the five universities considered as case study in this work. Figures 8–12 illustrates the plots of the parameters for the five universities.

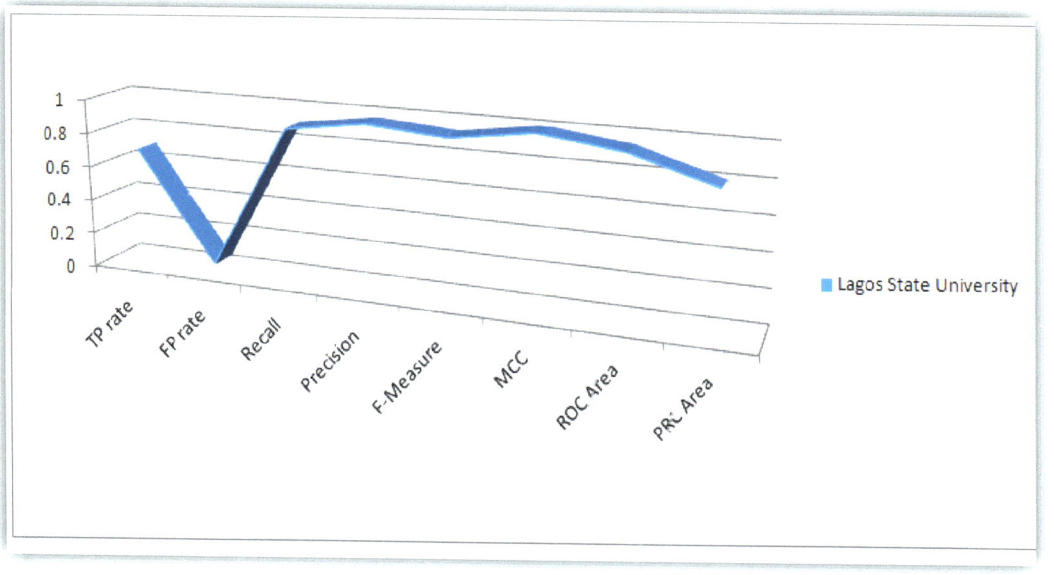

Figure 8. Plot of Evaluation Parameters based on Nigerian Students' dataset analysis.

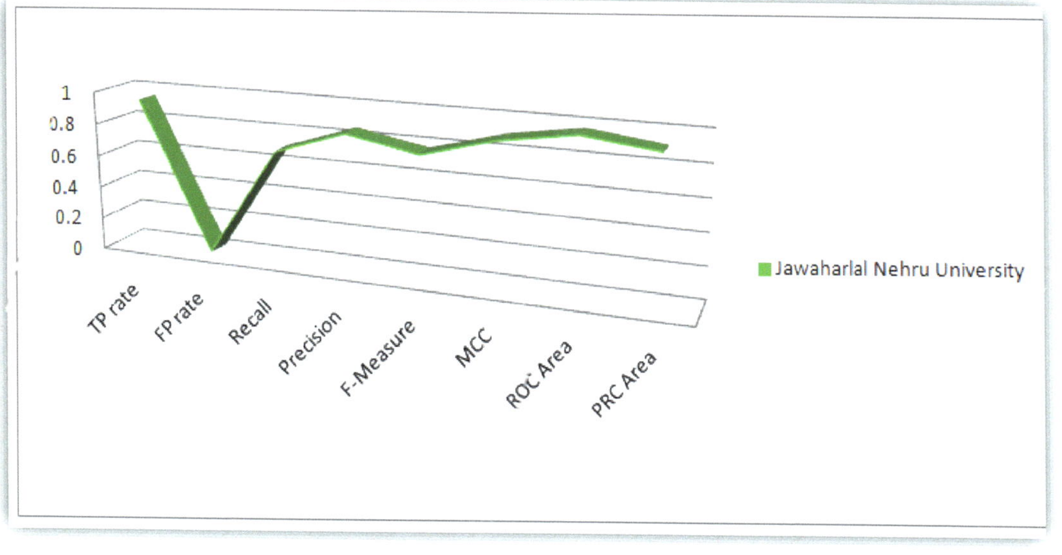

Figure 9. Plot of Evaluation Parameters based on Indian Students' dataset analysis.

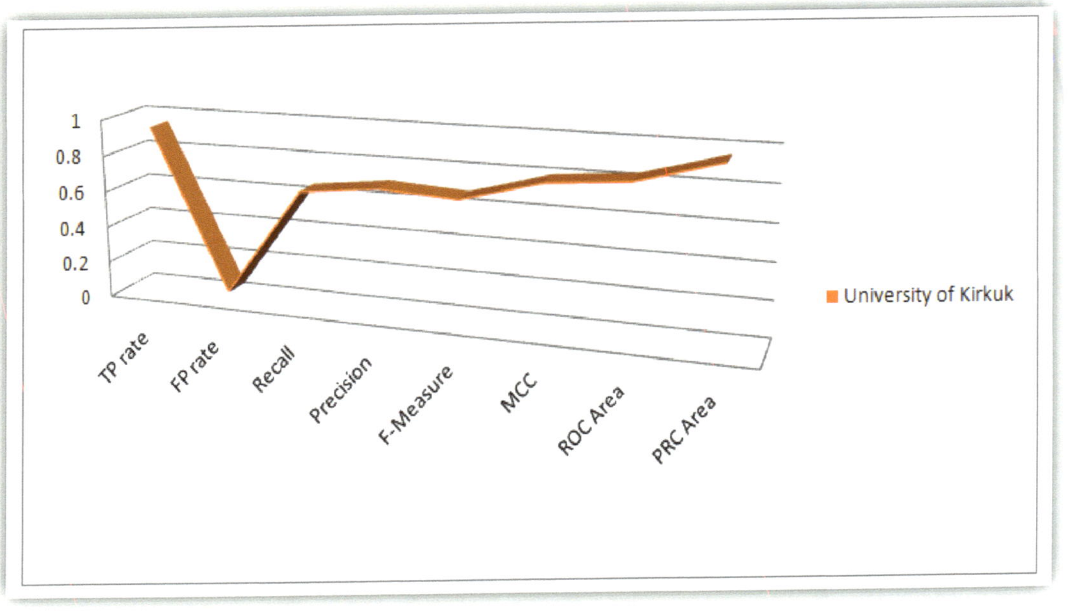

Figure 10. Plot of Evaluation Parameters based on Iraqi Students' dataset analysis.

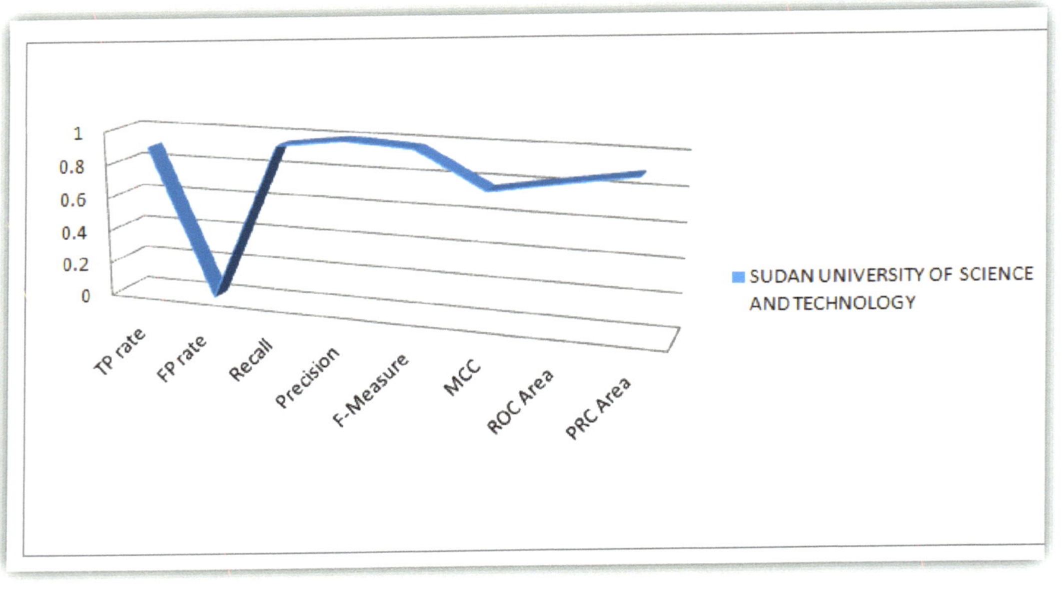

Figure 11. Plot of Evaluation Parameters based on Sudanese Students' dataset analysis.

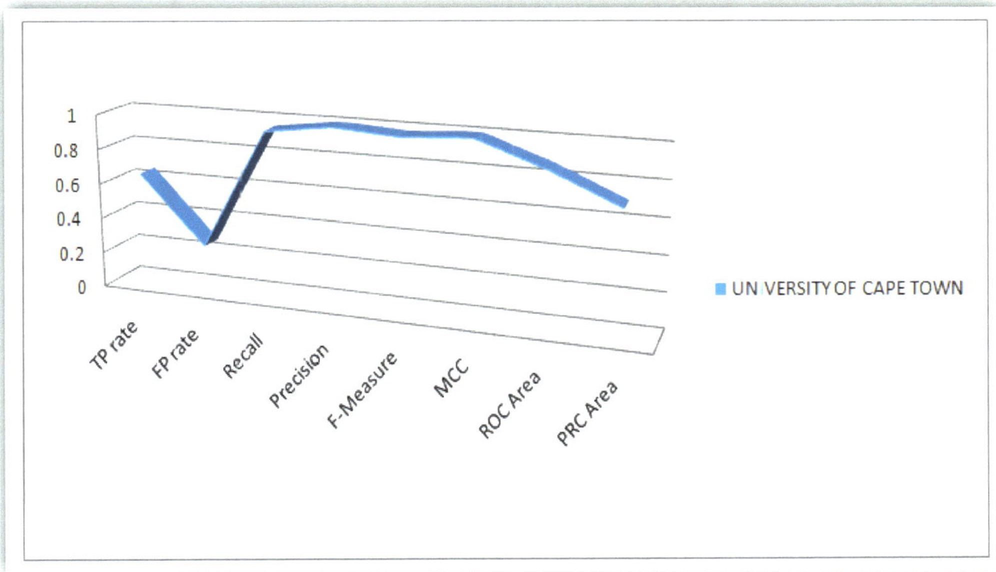

Figure 12. Plot of Evaluation Parameters based on South African Students' dataset analysis.

4.2. Analysis of J48 Decision Trees Generated in WEKA for the Five Universities

The Figures 3–7 shown in this research study illustrates the J48 decision trees generated in WEKA for the five universities. In this section, we provided a detailed explanation of the J48 tree generated in the Section 3.3 of this work. The J48 decision tree classifier shown in Figure 3 illustrates that 11 students had grade A and passed with scores greater than 69 marks; 6 students had grade B, passed with scores greater than 59 marks and less than equal to 69 marks; 7 students had grade C and passed with scores less than or equal to 59 marks; 7 students had grade D and failed with scores greater than 39 marks; and 9 students had grade F with scores less than or equal to 39 marks. In general, a total of twenty-four students were in the category of those who passed while total of sixteen students were in the category of those failed. The J48 decision tree classifier shown in Figure 4 illustrates that 9 students had grade A and passed with scores greater than 69 marks; 7 students had grade B, passed with scores greater than 59 marks and less than equal to 69 marks; 10 students had grade Cand passed with scores less than or equal to 59 marks; ten students had grade D and failed with scores greater than 39 marks; and 4 students had grade F with scores less than or equal to 39 marks. In general, a total of twenty-six students were in the category of those who passed while total of fourteen students were in the category of those who failed. The J48 decision tree classifier shown in Figure 5 illustrates that 2 students had grade A and passed with scores greater than 66 marks; 8 students had grade B, passed with scores greater than 59 marks and less than equal to 66 marks; 9 students had grade Cand passed with scores less than or equal to 59 marks; 12 students had grade D and failed with scores greater than 39 marks; and nine students had grade F with scores less than or equal to 39 marks. In general, a total of nineteen students were in the category of those who passed while total of 21 students were in the category of those who failed. The J48 decision tree classifier shown in Figure 6 illustrates that 13 students had grade A and passed with scores greater than 67 marks; 4 students had grade B, passed with scores greater than 59 marks and less than equal to 67 marks; 4 students had grade C and passed with scores less than or equal to 59 marks; 4 students had grade D and failed with scores greater than 37 marks; and 15 students had grade F with scores less than or equal to 37 marks. In general, a total of twenty-one students were in the category of those who passed while total of 19 students were in the category of those who failed. The J48 decision tree classifier shown in Figure 7 illustrates that 4 students had grade A and passed with scores greater than 69 marks; 9 students had grade B, passed with scores greater than 57 marks and less than equal to 69 marks; 10 students had grade Cand passed with scores less than or equal to 57 marks; 12students had grade D and failed with scores greater than 37 marks; and 5 students had grade F with scores less than or equal to 37 marks. In general, a total of twenty-three students were in the category of those who passed while total of seventeen students were in the category of those who failed.

5. Conclusions and Future Scope

As a result of the rapid increase in extraction of useful knowledge from data, data mining has significantly contributed to most educational institutions in many countries today. The test and prediction conducted on students' academic performance has really helped both learners and educators to improve their learning and teaching skills, respectively. This research work uses the WEKA data analytics platform to perform J48 classification algorithm on the students' result across five universities in five countries on the basis of the Execution time, TP rate, FP rate, Precision, Recall, ROC Area, PRC Area, MCC and the F-measure. WEKA took different attributes based on the stratified cross validation via the J 48 tree algorithm to obtain the correctly classified instances, the incorrectly classified instances and others (which includes the mean absolute, root mean squared, relative absolute and root relative squared) error values. Confusion matrixes were generated for the students' dataset with A, B, C, D and F representing the class labels. The Kappa values obtained from the analysis revealed a range of 0.907–0.9582, which is the perfect reading for most analytical values. Plots such as flow lines and Bar charts were generated on both the evaluation parameters and the attributes, respectively. We discovered that the J48 algorithm provided better results and, in future, we intend to extend our research using different parameters in a different analytic environment.

Author Contributions: Conceptualization, W.I.; methodology, W.I.; software, W.I., S.A., A.A.S. and O.A.A.; validation, W.I. and S.A.; formal analysis, O.A.A.; investigation, W.I. and H.A.; resources, W.I. and O.A.A.; data curation, W.I. and O.A.A.; writing—original draft preparation, W.I.; writing—review and editing, W.I., A.A.S. and S.A.; visualization, H.A., O.A.A. and W.I.; supervision, S.A.; project administration, S.A. All authors have read and agreed to the published version of the manuscript.

Funding: This research received no external funding.

Institutional Review Board Statement: Not Applicable.

Informed Consent Statement: Not Applicable.

Data Availability Statement: Data available upon request to the corresponding authors.

Acknowledgments: The authors are indeed grateful to universities used as case study for providing their students' academic data for the success of this research.

Conflicts of Interest: The authors declare no conflict of interest.

Nomenclatures

TP-rate	True Positive Rate
FP-rate	False Positive Rate
FN-rate	False Negative Rate
ROC Area	Receiver Operating Characteristics Area
PRC Area	Precision Recall Curve Area
MCC	Matthews Correlation Coefficient
PPV	Positive Predictive Value
KDD	Knowledge Discovery in Database

WEKA	Waikato Environment for Knowledge Analysis
EDM/LA	Educational Data Mining and Learning Analytics
ID3	Iterative Dichotomiser 3
J48	Java 48

Appendix A

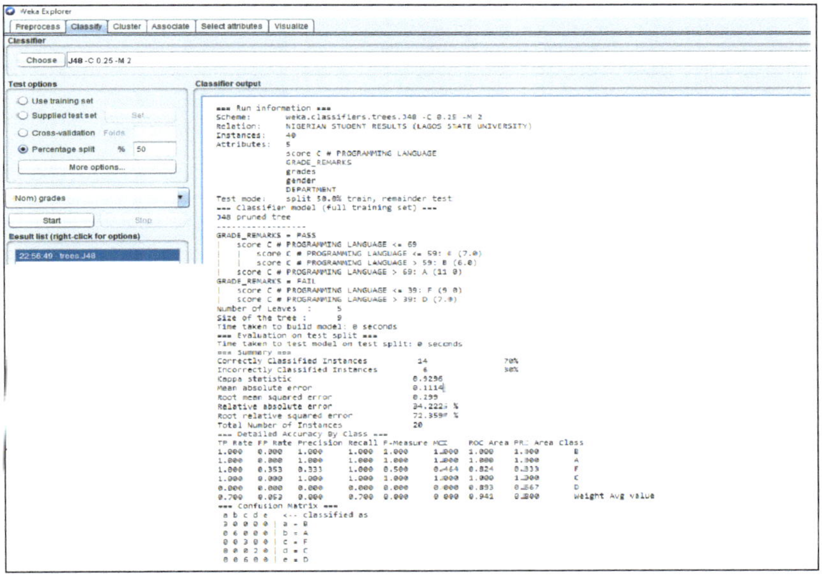

Figure A1. WEKA J48 Decision Tree Classifier Algorithm obtained from the Nigerian Students' Result.

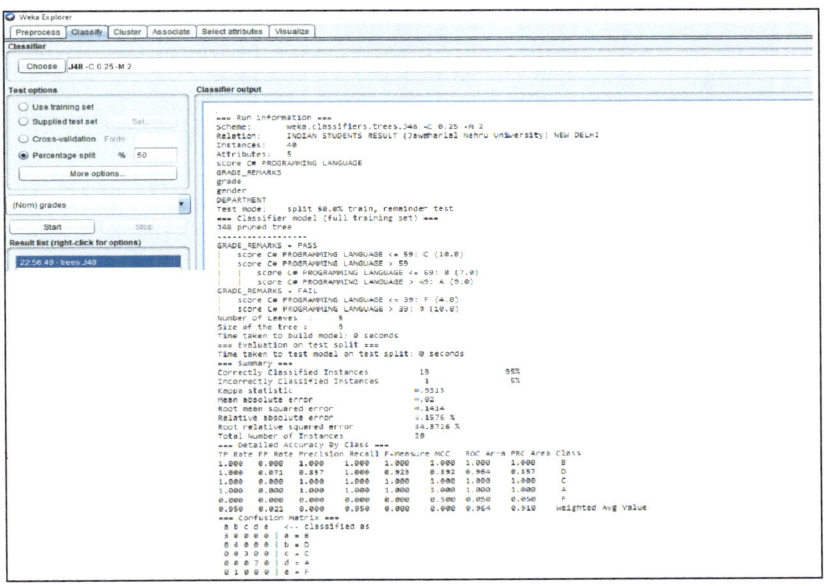

Figure A2. WEKA J48 Decision Tree Classifier Algorithm obtained from the Indian Students' Result.

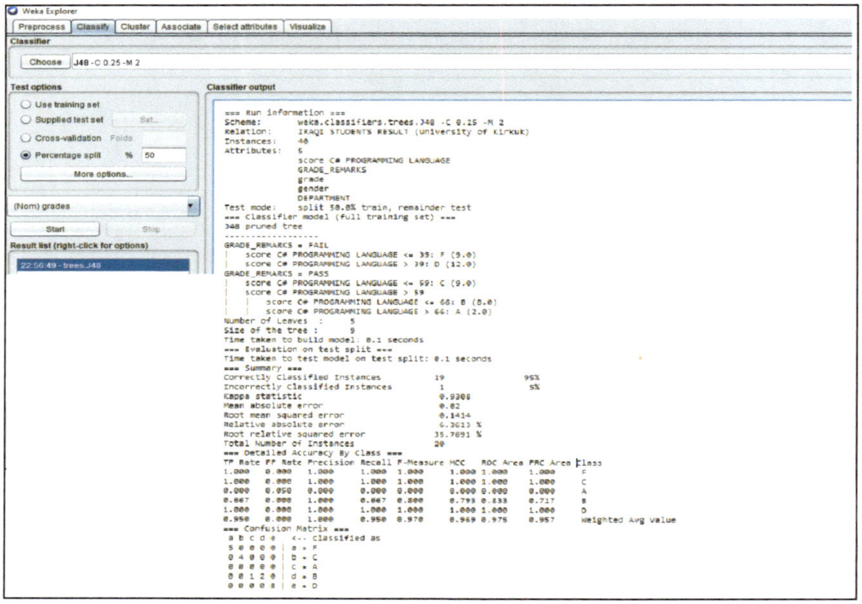

Figure A3. WEKA J48 Decision Tree Classifier Algorithm obtained from the Iraqi Students' Result

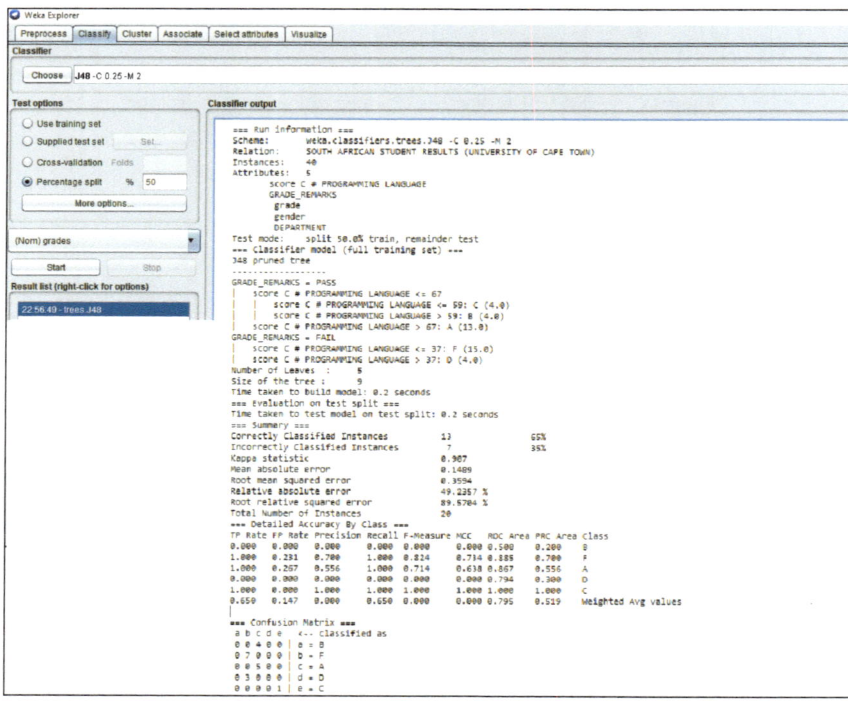

Figure A4. WEKA J48 Decision Tree Classifier Algorithm obtained from the South African Students' Result.

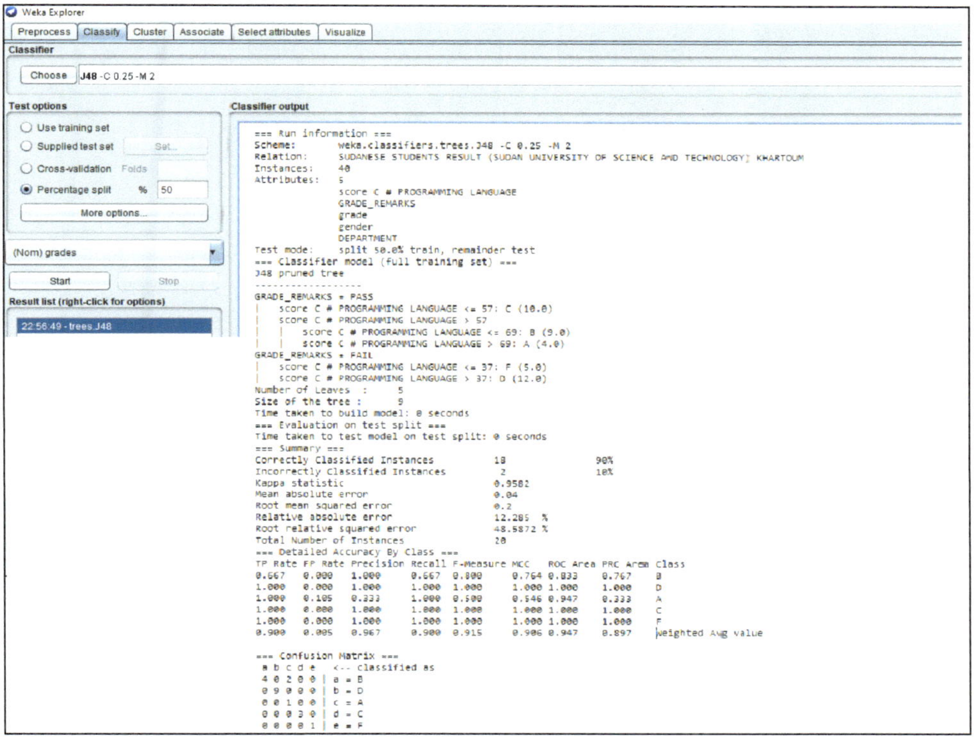

Figure A5. WEKA J48 Decision Tree Classifier Algorithm obtained from the Sudanese Students' Result.

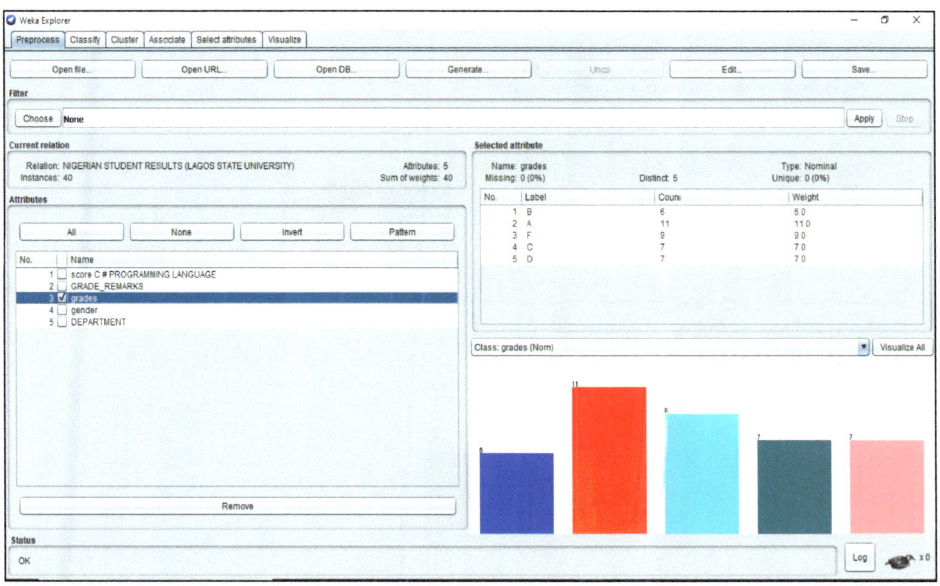

Figure A6. Grade_Remarks Attribute Platform for Nigerian Students' dataset.

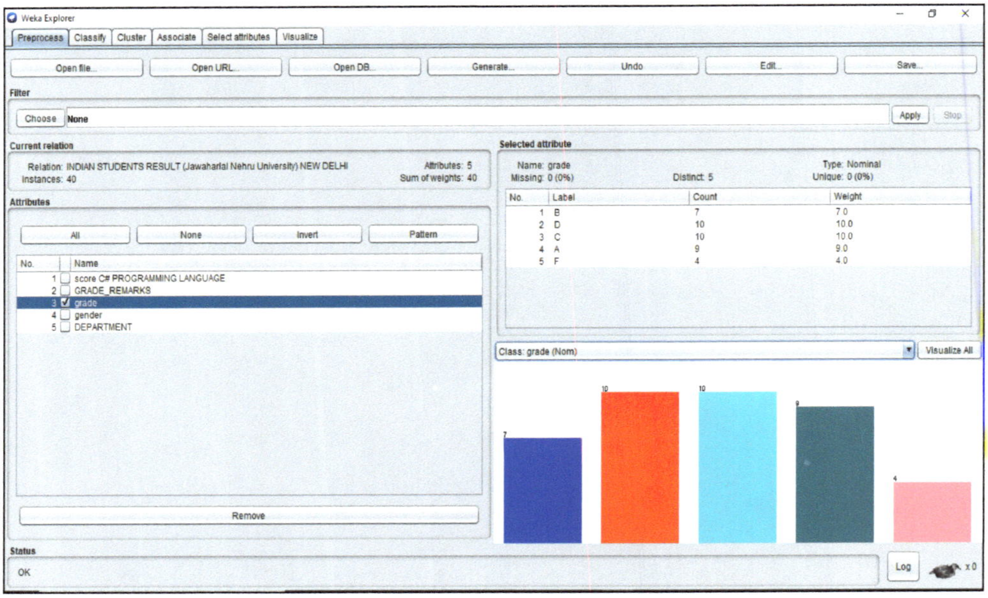

Figure A7. Grade_Remarks Attribute Platform for Indian Students' dataset.

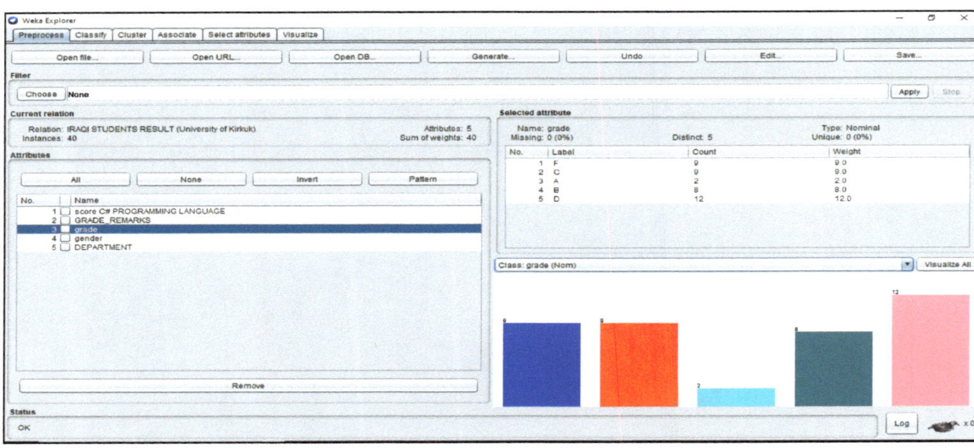

Figure A8. Grade_Remarks Attribute Platform for Iraqi Students' dataset.

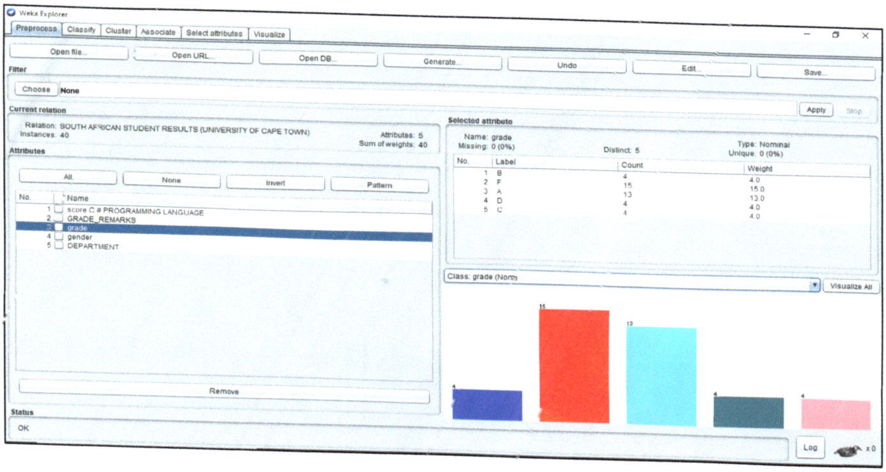

Figure A9. Grade_Remarks Attribute Platform for South African Students' dataset.

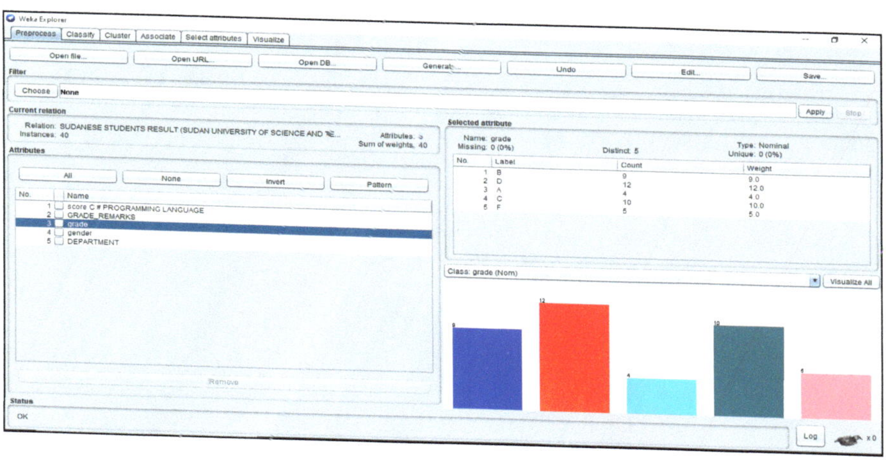

Figure A10. Grade_Remarks Attribute Platform for Sudanese Students' dataset.

References

1. Tsinidou, M.; Gerogiannis, V.; Fitsilis, P. Evaluation of the factors that determine quality in higher education: An empirical study. *Qual. Assur. Educ.* **2010**, *18*, 227–244. [CrossRef]
2. Romero, C.; Ventura, S. Educational data mining: A review of the state of the art. *IEEE Trans. Syst. Man Cybern. Part C* **2010**, *40*, 601–618. [CrossRef]
3. Han, J.; Pei, J.; Kamber, M. *Data Mining: Concepts and Techniques*; Elsevier: Amsterdam, The Netherlands, 2012.
4. Jiawei, H.; Kamber, M. Data mining: Concepts and Techniques. University of Illinois at Urbana-Champaign, 2001. Available online: http://hanj.cs.illinois.edu/bk2/toc.pdf (accessed on 4 April 2022).
5. Aziz, A.A.; Ismail, N.H.; Ahmad, F. Mining Students' Academic Performance. *J. Theor. Appl. Inf. Technol.* **2013**, *53*, 485–495.
6. Romero, C.; Ventura, S.; Espejo, P.G.; Hervas, C. Data mining algorithms to classify students. In *Educational Data Mining*; Computer Science Department, Corbora University: Andalusia, Spain, 2008.
7. Salal, Y.K.; Abdullaev, S.M.; Kumar, M. Educational Data Mining: Student Performance Prediction in Academic. *Int. J. Eng. Adv. Technol. IJEAT* **2019**, *8*, 54–59.
8. Garner, S.R. WEKA: The Waikato Environment for Knowledge Analysis. In Proceedings of the New Zealand Computer Science Research Students Conference, Hamilton, New Zealand, 18–21 April 1995; pp. 57–64.
9. Maimon, O.; Rokach, L. *Data Mining and Knowledge Discovery Handbook*; Springer: Berlin/Heidelberg, Germany, 2005. [CrossRef]

10. Abdullaev, S.M.; Salal, Y.K. An economic deterministic ensemble classifiers with probabilistic output using for robust quantification: Study of unbalanced educational datasets. In Proceedings of the 1st International Scientific and Practical Conference on Digital Economy (ISCDE 2019), Advances in Economics, Business and Management Research, Chelyabinsk, Russia, 7–8 November 2019; Volume 105, pp. 658–665. [CrossRef]
11. Vranić, M.; Pintar, D.; Skočir, Z. The use of data mining in education environment. In Proceedings of the 9th International Conference on IEEE, Winchester, UK, 8–13 July 2007; pp. 243–250.
12. Abdullaev, S.M.; Lenskaya, O.Y.; Salal, Y.K. Computer Systems of Individual Instruction: Background and Perspectives. *Educ. Sci.* **2018**, *10*, 64–71. [CrossRef]
13. Sharma, G.M.; Bhargava, N.; Bhargava, R. Decision Tree analysis on J48 algorithm on Educational Data Mining. *Proc. Int. J. Adv. Res. Comput. Sci. Softw. Eng.* **2013**, *3*, 1114–1119.
14. Anjali, B.; Raut, A.A. Students Performance Prediction Using Decision Tree Technique with J48 algorithm. *Int. J. Comput. Intell. Res.* **2017**, *13*, 1735–1741.
15. Mehta, S.H.; Ashish, A. Predicting Students' Performance using J48 Decision Tree. *Int. J. Sci. Res. Comput. Sci. Eng. Inf. Technol.* **2019**, *4*, 122–129. [CrossRef]
16. Algorithm and Flow Chart. Available online: https://faradars.org/wp-content/uploads/2015/07/Algorithm-and-Flow-Chart.pdf (accessed on 4 April 2022).
17. Farhad, A.; Sanjay, P. Comparative Study of J48, Naive Bayes and One-R Classification Technique for Credit Card Fraud Detection using WEKA. In *Advances in Computational Sciences and Technology*; Research India Publications: Rohini, India, 2017; Volume 10, pp. 1731–1743. ISSN 0973-6107.
18. Ihya, R.; Namir, A.; Sanaa, E.F.; Mohammed, A.D.; Fatima, Z.G. J48 Algorithms of Machine Learning for Predicting User's the Acceptance of an E-Orientation Systems. In Proceedings of the 4th International Conference on Smart City Applications, Casablanca, Morocco, 2–4 October 2019.
19. Kaur, G.; Chhabra, A. Improved J48 Classification Algorithm for the Prediction of Diabetes. *Int. J. Comput. Appl.* **2014**, *98*, 13–17. [CrossRef]
20. Adhatrao, K.; Gaykar, A.; Dhawan, A.; Jha, R.; Honrao, V. Predicting Students performance using ID3 extension and C4.5 classification algorithms. *Int. J. Data Min. Knowl. Manag. Process IJDKP* **2013**, *3*, 39–52. [CrossRef]
21. Chen, T.Y.; Kuo, F.C.; Merkel, R. On the Statistical Properties of the F-Measure. In Proceedings of the Fourth International Conference on Quality Software, 2004. QSIC 2004, Braunschweig, Germany, 8–9 September 2004; pp. 46–153.
22. Srivastava, S.K.; Singh, S.K. Multi-Parameter Based Performance Evaluation of Classification Algorithms. *Int. J. Comput. Sci. Inf. Technol. IJCSIT* **2015**, *7*, 115–125. [CrossRef]
23. Hussain, S.; Dahan, N.A.; Ba-Alwib, F.M.; Ribata, N. Educational Data Mining and Analysis of Students' Academic Performance Using WEKA. *Indones. J. Electr. Eng. Comput. Sci.* **2018**, *9*, 447–459. [CrossRef]

MDPI AG
Grosspeteranlage 5
4052 Basel
Switzerland
Tel.: +41 61 683 77 34

Data Editorial Office
E-mail: data@mdpi.com
www.mdpi.com/journal/data